III-Nitride Materials Devices and Nano-Structures

III-Nitride Materials Devices and Nano-Structures

Editor

Zhe Chuan Feng

Guangxi University, P. R. China

World Scientific

NEW JERSEY • LONDON • SINGAPORE • BEIJING • SHANGHAI • HONG KONG • TAIPEI • CHENNAI • TOKYO

Published by

World Scientific Publishing Europe Ltd.
57 Shelton Street, Covent Garden, London WC2H 9HE
Head office: 5 Toh Tuck Link, Singapore 596224
USA office: 27 Warren Street, Suite 401-402, Hackensack, NJ 07601

British Library Cataloguing-in-Publication Data
A catalogue record for this book is available from the British Library.

III-NITRIDE MATERIALS, DEVICES AND NANO-STRUCTURES

ISBN 978-1-78634-318-5

Desk Editor: Rhaimie Wahap

Typeset by Stallion Press
Email: enquiries@stallionpress.com

Printed in Singapore

PREFACE

Group III-Nitrides semiconductor materials, including GaN, InN, AlN, InGaN, AlGaN and AlInGaN, i.e. (Al,In,Ga)N, are excellent semiconductors, covering the spectral range from deep ultraviolet (DUV) to UV, visible and infrared, with unique properties very suitable for modern electronic and optoelectronic applications. Remarkable breakthroughs have been achieved in recent years for research and development (R&D) in these materials and devices, such as high-power and high brightness UV-blue-green-white light emitting diodes (LEDs), UV-blue-green laser diodes (LDs), photo-detectors and various optoelectronics and electronics devices and applications.

The Nobel Prize in Physics 2014 was awarded jointly to Isamu Akasaki, Hiroshi Amano and Shuji Nakamura "*for the invention of efficient blue light-emitting diodes which has enabled bright and energy-saving white light sources*". Red and green diodes had been invented since 1960s–70s but without blue LED. Despite considerable efforts, the blue LED had remained a challenge for a long time. The success and inventions on GaN-based LEDs were revolutionary and benefiting of the mankind. III-Nitrides-based industry has formed and acquired rapid developments over the world. Incandescent light bulbs lit the 20th century and the 21st century will be lit by LED lamps.

Recently, I have edited two books, *III-Nitride Semiconductor Materials* (2006) and *III-Nitride Devices and Nanoengineering* (2008), both published by Imperial College Press, in III-Nitride fields. The developments of these materials and devices are moving so rapid that many data or knowledge, even those just published quite recently, need to be modified and upgraded. This new book, *III-Nitride Materials, Devices and Nano-Structures* is the third book after the two, will cover the rapidly new developments and achievements in the field, in particular those made since 2009. It will lean more towards the engineering and materials side than pure science.

The book is organized for a wide range of audiences and covers each of the basic and critical aspects of III-Nitrides science and technology. Each chapter, written by experts in the field, reviews the important topics and achievements in recent years, discusses progresses made by different groups, and suggests further works that need to be done. This book provides useful information on the device and nano-scale process, fabrication of LEDs, LDs, photodetectors and nano-devices, characterization, application and development on the III-Nitrides semiconductor devices and nano-engineering.

This book consists of 11 well-written review chapters in three parts:

Part I — General:

(1) Comprehensive Theoretical and Experimental Studies on III-nitrides, Doping, Nano-structures and LEDs
(2) Waste Energy Harvesting using III-Nitrides Materials
(3) III-Nitrides for Infrared Optoelectronics
(4) GaN-based photodetectors

Part II - III-Nitride Materials:

(5) Single Crystal AlN: Growth by Modified Physical Vapor Transport and Properties
(6) Indium-Rich III-Nitride Alloys Growth by Super-Atmospheric and Migration-Enhanced, Plasma-assisted MOCVD and Their Properties analysis
(7) Combined Synchrotron X-ray absorption and optical spectroscopic studies on III-nitrides

Part III - III-Nitride Devices and Nano-Structures:

(8) III-Nitride Nano-structures and Improving the Luminescence Efficiency from Quantum Well LEDs,
(9) Fabrication and Characterization of Green Resonant-Cavity Light-Emitting Diodes Prepared by Wafer Transfer Technologies

(10) Nanotexturing Effects in GaN/InGaN Multi-Quantum-Wells LED Planar Structures
(11) Group III-Nitride Nanostructures for Light Emitting Devices and Beyond.

The current book presents the key properties of III-Nitrides materials, devices and nano-structures, describes key technologies and demonstrates the remaining challenging issues in material preparation and device fabrication for future R&D in the 21st century. This book can serve well to material growers and evaluators, device designers and processing engineers as well as potential users of III-Nitrides technologies, including new comers, postgraduate students, engineers and scientists in the III-Nitrides field.

Zhe Chuan Feng
Distinguished Professor in Optoelectronic Materials
and Detection Technology
College of Physics Science & Technology, Guangxi University
Guangxi Key Laboratory for Relativistic Astrophysics

CONTENTS

Part I

General

CHAPTER 1

COMPREHENSIVE THEORETICAL AND EXPERIMENTAL STUDIES ON III-NITRIDES, DOPING, NANO-STRUCTURES AND LEDS

Jinmin Li*, Zhiqiang Liu†, Xiaoyan Yi and Junxi Wang

Research and Development Center for Solid State Lighting,
Institute of Semiconductors, Chinese Academy of Science, Beijing

China State Key Laboratory of Solid State Lighting
A35, Qinghua East Road, Haidian District, Beijing, China
and
College of Materials Science and Opto-Electronic Technology,
University of Chinese Academy of Sciences, China
**jmli@semi.ac.cn*
†lzq@semi.ac.cn

Even though III-Nitrides have demonstrated their ample application in solid-state light, display, laser, detector and high-power RF/switching device, many other new breakthroughs on fundamental physics and fabrication techniques are still highly desired to lead to their next expansion of the research and application field in the near future. In this chapter, the general properties of III-nitrides as well as their ternary and quaternary were briefly reviewed. Specifically, we mainly focused on the underlying physics of p doping challenge, nitrides nano-structures and their application on LEDs. We presented an overview of some main results and progress in highly effective p doping for nitrides. Some promising strategies to address the issue of p doping challenge were also introduced. Furthermore, we describe the prevailing GaN nanowire fabrication methods including wet etching and selective area growth and their application in novel LED structures.

1. Introduction

III-Nitrides (InN, GaN, AlN) as well as their ternary and quaternary are considered one of the most important groups of semiconductors after Si and arsenide. Owing to their excellent optical and electrical properties III-Nitrides have demonstrated their ample application in solid-state light, display, laser, detector and high-power RF/switching device. A series of fundamental breakthroughs in nitrides semiconductor material led to the successes of high efficiency blue LEDs in the last two decades. In 2014, the Royal Swedish Academy of Sciences awarded the Nobel Prize for Physics to Isamu Akasaki, Hiroshi Amano, and Shuji Nakamura "for the invention of efficient blue light emitting diodes which has enabled bright and energy-saving white light sources". Despite these tremendous achievements, the potential of nitrides definitely should not be limited to blue LEDs, but should also expand to other fields. Many other new breakthroughs are also highly desired to lead to the next expansion of the research and application field of III-nitrides material systems in the near future.

This chapter will give a brief review of general properties of nitrides, fundamental physics in p-doping and new development of the growth and characterization methods of nitrides nano-materials.

2. General Properties of Nitrides

2.1. *Crystal structure of nitrides*

III-Nitrides can be of wurtzite (Wz), zincblende (ZB), and rocksalt structure. For bulk AlN, GaN, and InN, wurtzite is the thermodynamically stable structure at under ambient conditions, which has a hexagonal unit cell and thus two lattice constants c and a. The space grouping for the wurtzite structure is P63mc in the Hermann–Mauguin notation. The Wz structure consists of two interpenetrating hexagonal close-packed sub-lattices with different type of atom, offset along the c-axis by 5/8 of the cell height (5c/8). Three surfaces are of special importance in nitrides, which are (0001) c-planes, (11–20) a-planes, and (1–100) m-planes. Other surfaces could also be observed in nitrides nano-structures, i.e. (1–102) r-planes, (11–23) n-planes. For epitaxy GaN thin film or single crystals, determining the orientation is not straight-forward and will depend on growth conditions, substrates or doping.

2.2. *Electronic band structure*

In a crystal with Wz symmetry, the conduction band wavefunctions are mainly composed by the s-orbitals of cation and transform the Γ point congruent with the Γ7 representation of the space group C46V. While, the valence band states are constructed out of appropriate linear combinations of products of p-like (px, py, and pz-like) orbitals of anion. Under the influence of the crystal field and spin–orbit interactions, the hallmark of the wurtzite structure, the six-fold degenerate Γ15 level, splits into Γv9, upper Γv7, and lower Γv7 levels. Calculations indicate that the group-III Nitrides in the wurtzite and cubic phase have all direct energy band gaps. At room temperature the value of bandgap is 6.13 eV for AlN.[1] 3.44 eV for GaN[2] and 0.65–0.7 eV for InN.[3]

2.3. *Polarization effects*

Plasma-induced molecular-beam epitaxy (PIMBE) on c-plane sapphire GaN/sapphire substrate,5,6 and observed N-face GaN.

Polarization is an important property of the III–V nitride semiconductors resulting in huge electric fields built-in in the structures grown in the (0001) direction. Thus, it is very important for the design of nitride heterostructures to achieve device with different function. III-Nitrides with Wz structures exhibit two different sequences of atomic layering in the two opposing directions parallel to c axes. Consequently, spontaneous crystallographic polarity along these axes can be observed. Meanwhile, III-Nitrides lack center of inversion symmetry and exhibit large piezoelectric effects when strained. In the case of hetero-epitaxial growth of thin films of nitrides, the polarity of the material cannot be predicted in a straightforward way, and must be determined by experiments. For example, both types of polarity were reported to be found in GaN (0001) layers grown by metal-organic chemical vapor deposition (MOCVD) on c-plane sapphire if the layers exhibited rough morphology, while for smooth films Ga-face polarity was exclusively concluded from the experimental results.[4] While, Smith *et al.* reported on investigations of surface reconstructions of GaN grown by plasma-induced molecular-beam epitaxy (PIMBE) on c-plane sapphire GaN/sapphire substrate,[5,6] and observed N-face GaN.

Polarization is an important property of the III–V nitride semiconductors resulting in huge electric fields built-in in the structures grown in the (0001) direction. Thus, it is very important for the design of nitride heterostructures to achieve device with different function.

3. p-type Doping

Application of semiconductors as electric and/or optoelectronic devices depends on their dopalility with shallow donors and acceptors. Unfortunately, most wide band gap (WBG) semiconductors, experience a serious doping asymmetry problem, i.e. they can easily be doped n- or p-type, however not both. Doping asymmetry in WBG arises from the fact that wide-band-gap semiconductors have either a low valence band maximum or a high conduction band minimum. Some semiconductors such as ZnTe, and diamond in which the valence band is relatively close to the vacuum level have preferable p-type conductivity.[7–9] Consequently, achieving n-type conductivity in these semiconductors is very difficult. In contrast, GaN, ZnO, ZnS, with their valence bands relatively far from the vacuum level, have preferable n-type conductivity.[10–12] For materials with very larger band gap such as AlN, both p- and n-type doping are difficult.[13] The doping difficulty also exists in nanostructure semiconductors where the band gaps increase due to the quantum confinement. These doping problems have hindered the potential applications of many WBG and nanostructure materials.

3.1. *Origin of p doping difficulties*

Extensive research has been done to understand the origin of the p doping difficulties in nitrides semiconductors and to find possible solutions to overcome the doping difficulty. Generally speaking, there are three main factors that could cause the doping limit in a semiconductor material[14–17]:

(a) The desirable dopants have limited solubility, and difficult to incorporate into the lattice of host materials. The solubility of Mg in the GaN lattice is limited to about 1020/cm^3, otherwise Mg^3N^2 will prefer to form.
(b) The desirable dopants have sufficient solubility, but they produce deep levels, which are not ionized at working temperatures. In another word, the ionization energy of acceptor is too high.
(c) There is spontaneous formation of compensating defects. First principle calculation by C. G. Van de Walle[18] shows self-interstitial and anti-site defects have very high formation energies and are thus unlikely to occur

in GaN during growth. A di-vacancy (VGa-VN) has relatively high formation energy in GaN and as such is also unlikely to form in large concentrations. While, it should be point out that the formation energy of nitrogen vacancies is high in n-type GaN, however low in p-type GaN. And the formation energy of nitrogen vacancies decreases as Fermi level move close to valence band of GaN, making it a likely compensating center during acceptor doping.

(d) In p-type GaN, hydrogen is known to passivate the dominant acceptor (MgGa). In Mg-doped GaN grown by MOCDD, the electrically neutral Mg–H complex has a binding energy of 0.7 eV. During the post-growth annealing, the Mg–H complex dissociates, and H diffuses either to the surface or to extended defects.

3.2. *Present solution for p doping difficulties*

Extensive research has been done to understand the origin of the asymmetrical doping difficulties in WBG to find possible solutions to overcome the issue in nitrides.[19–23] Recent advances in crystal growth technology have shown that the issues of low solubility and hydrogen passivation can, at least to some extent, be overcome by using non-equilibrium growth techniques and high-temperature annealing. However, alleviating the more fundamental problem of high activation energies has, to date, not yet been satisfactorily achieved. The underlying physical mechanism in this problem is attributed to the electronic structure of the host material. Nitrogen is strongly electronegative and has a deep 2p atomic orbital. Thus, the valance band maximum (VBM) of nitrides, which contain mostly N 2p orbitals, is at relatively low energies. This leads to a relatively deep acceptor energy level which makes it very inefficient for thermal activation. To date, the most promising acceptor for III-Nitrides continues to be Mg. Unfortunately, even with Mg dopant ions, the activation energy Ea of the Mg dopant in GaN is still in the range of 160 and 200 meV. For AlN, the activation energy can be as high as 630 meV. Consequently, only a small fraction of Mg can be activated at room temperature. It is, therefore, reasonable to believe that modulating the energy position of acceptor states related to the VBM of host materials should be a promising solution for high effective p-type doping.

Based on this consideration, various approaches have been sought to lower the acceptor levels and reduce the acceptor ionization energy in nitrides. Recently, B. Gunning *et al.* proposed a strategy for lowering the acceptor impurity states by extremely high doping4. They argue that, as the electrically active acceptor concentration increases, the isolated deep acceptor levels begin to interact and split into an impurity band, which is closer to the valence band thus lowering the effective activation energy. Peter and Schubert[24,25] demonstrated another strategy and found that by polarization induced modulation of the valence band edge in a superlattice, the low doping efficiency could be partially overcome. Simon and Jena[26] also suggested that a 3D hole gas could be produced using the built-in electronic polarization in nitrides. B. Gunning *et al.* proposed another strategy to lower the acceptor impurity states by extremely high doping.[27] Recently, co-doping was regarded as another important strategy to address this issue,[19,28–30] for example by Si-Mg co-doping and other mutually passivated defect pair co-doping. However, intensive theoretical analyses show that this type of energy level coupling is too small to significantly reduce the acceptor ionization energy due to different symmetries and wave-function characteristics.[23] Therefore, although partial successes have been achieved, the mechanisms of those methods are still controversial and poorly understood. Better approaches or alternative strategies to create more stable and shallower acceptors in nitrides are highly desired.

3.3. *New strategy based on band-engineering*

As mentioned above, modulating the energy position of acceptor states related to the VBM of host materials should be a promising solution for high effective p-type doping. In the past, we have proposed various approaches to overcome the bipolar doping difficulty in WBG semiconductors based on band-engineering. These approaches have been tested by both experiment and systematical calculation of defect formation energies and transition energy levels of intrinsic and extrinsic defects using first-principles density-functional theory.

(1) Impurity Resonant States p-type Doping in Wide-Band-Gap Nitrides

The behavior of Mg as an acceptor in nitride semiconductors is strongly linked to the position of the Mg impurity states related to the VBM of the host materials. Besides co-doping, a periodic oscillation of the valance band

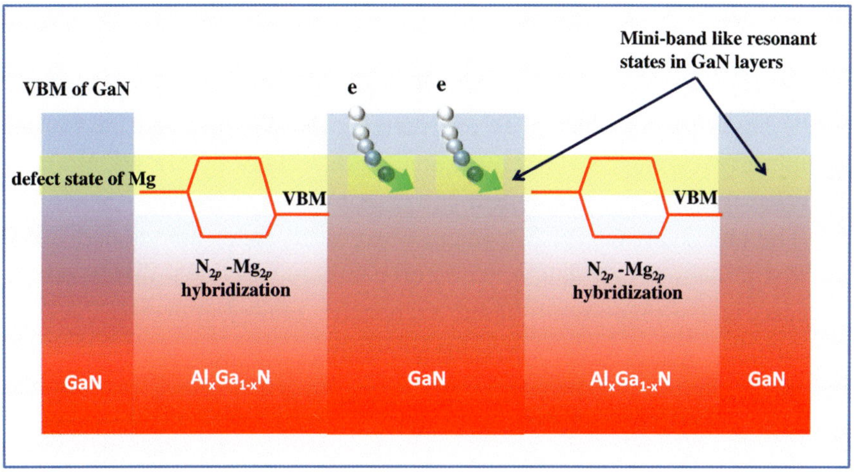

Fig. 1. Schematic model showing the mechanism of impurity resonant states p-type doping.[30] Reprinted from Macmillan Publishers Ltd: [Scientific Reports] (Ref. [30]), copyright (2016).

edge produced by a superlattice structure, such as AlxGa1-xN/GaN, can also modify the characteristics and energy position of the VBM.[31,32] Based on this consideration, a novel strategy for efficient p-type doping is proposed to overcome the fundamental problem of high activation energy by inducing impurity resonant states in an Mg doped AlxGa1-xN/GaN superlattice structure. As schematically shown in Fig. 1, in the structure developed using our proposed strategy, the discrete wave-functions of Mg impurity states are able to overlap to form continuous miniband-like impurity states.[33,34] Therefore, the initially localized impurity states in AlxGa1-xN barrier layers form resonant states in the GaN layer (i.e. with energy levels below or close to the GaN VBM, it strongly depends on the Al percentage in AlxGa1-xN). To see the exact energy position of Mg impurity state, one would need to use a pretty large cell. Alternatively, in this work we offer the above qualitative band-diagram to explain the idea of resonant state p-type doping. In the case of considerable acceptor density, these impurity states are broadened,[35–38] which can further enhance the coupling between them. In this new scenario, electrons are able to drop from the VBM of GaN into the impurity states or band without any energy barrier, which means the acceptors are self-ionized. Hence, high concentration of the acceptors can be expected. In addition, as proposed by previous reports, the polarization effect also enhances the ionization of the deep acceptors and leads to the accumulation of carriers as a hole sheet, which further increase the effective hole concentration in the host materials.[31,32,39] To test these proposed concepts, the impact of impurity resonant states on the ionization energy of Mg acceptors is analyzed through both theoretical and experimental methods.

The characteristics of Mg impurity resonant states are studied using the first principles calculation, based on a density functional theory (DFT) encoded in the plane-wave based Vienna Ab initio Simulation Package (VASP).[40] In these calculations, the generalized gradient approximations (GGA) of Perdew-Burke-Ernzerhof (PBE) functionals are used for the exchange correlation potential.[41] The cutoff energy is chosen to be 800 eV. For relaxed structures, the atomic forces are less than 0.03 eV/A. For simplicity, an AlN/GaN superlattice structure was examined using $2 \times 2 \times 10$ supercell models rather than an AlxGa1-xN/GaN structure.

To understand the characteristics and distribution of Mg impurity states, the charge density of the Mg impurity states at the Γ point are plotted in Fig. 2. As can be expected, most of the charge density is accumulated around the Mg atoms. However, it cannot be ignored that a significant amount of Mg impurity states become delocalized and distributed in both barrier and

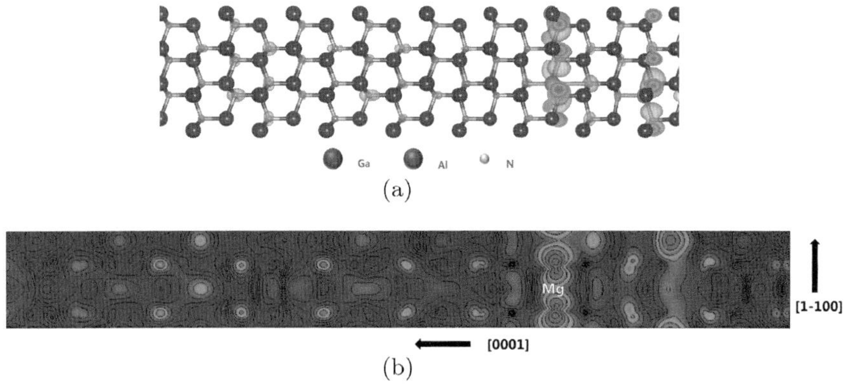

Fig. 2. Isosurface charge density plots of Mg impurity states at Γ point in AlxGa1-xN/GaN. (a) atomic configuration and isosurface charge density of Mg impurity states, (b) isosurface charge density of Mg impurity states in m plane.[30] Reprinted from Macmillan Publishers Ltd: [Scientific Reports] (Ref. [30]), copyright (2016).

well. In term of the well, such Mg impurity orbitals lie inside the valance band and act as resonant states. Now, the only question left to consider is whether the discrete impurity states in different barrier layers can couple with each other.

In a previous report, E. F. Schubert assumed that the acceptor-effective Bohr radius is much smaller than the period of the superlattice, and argued that the accepter levels in the barriers are not influenced by the adjacent barriers.[32] In fact, the acceptor Bohr radius is not directly relevant to the Mg impurity level and its coupling with the host.[42] In the nitride matrix environment, the Mg impurity states and host N 2p states can couple strongly with each other since they each share the same t2p symmetry, and hybrid orbitals are formed. As a result, now the Mg impurity states will also contain the characteristics of N 2p orbitals and become delocalized to some extent. The distribution of Mg impurity states in both barrier and well is direct evidence to support our theory proposed above. To understand the mechanism of orbital hybridization between Mg and N, projected densities of states (DOSs) were analyzed and are shown in Fig. 3. As can be seen, several peaks of Mg 2p states, especially near the VBM overlap with that of N 2p are observed, which indicates the coupling between them.[43,44] Therefore, we suggest that hybridization of the Mg and N 2p

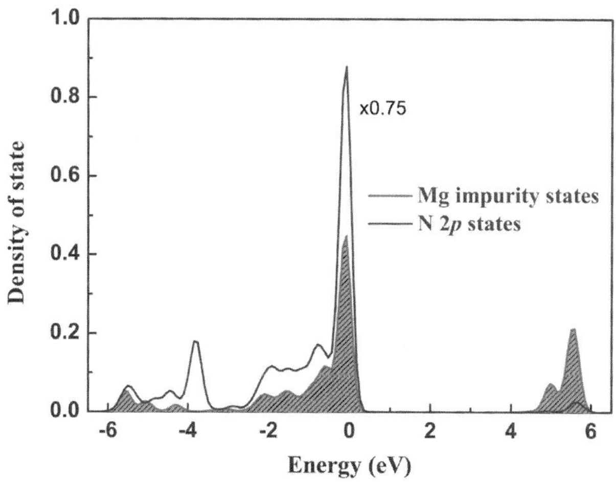

Fig. 3. Calculated projected density of states of Mg 2p impurity states and N 2p states.[30] Reprinted from Macmillan Publishers Ltd: [Scientific Reports] (Ref. [30]), copyright (2016).

states should be the reason for the occurrence of impurity states lying in the well.

To further test the concept of impurity resonant state p-type doping, AlxGa1-xN/GaN superlattice structures were grown by metal-organic chemical vapor deposition (MOCVD) on a c-plane sapphire substrate. After depositing a low-temperature GaN nucleation layer on the sapphire substrate, a 3 μm undoped GaN layer was grown. Then, the AlxGa1-xN/GaN superlattice was deposited upon the undoped GaN layer. The barrier and well thickness were both 10 nm with total 10 periods. To avoid the conventional thermal ionization mechanism of Mg dopant in GaN layers, only AlxGa1-xN/GaN was Mg doped. The aluminum percentage in the barrier was fixed at 30%, a value typically used in GaN LED structures. The sample structures were characterized by TEM and asymmetrical (105) X-ray reciprocal space mapping (RSM). As shown in Fig. 4(a), the main GaN peak and the zero-order diffraction satellite peak of the AlxGa1-xN/GaN MQWs are aligned in a vertical line parallel to the Qy axis, indicating the 30% AlGaN films is almost completely strained without relaxation along the plane direction. The high crystalline quality of our sample can also be confirmed by TEM image shown in Fig. 4(b).

Furthermore, as shown in Fig. 4(c), secondary ion mass spectrometry measurements were performed to verify the incorporation and distribution of Mg atoms. As can be seen, Mg is mostly distributed in AlxGa1-xN as intended.

In many previous reports[31,32,35,39] the standard Hall model was used to analyze the carrier concentration and mobility in superlattice-like structures. However, it should be noted that a simple Hall measurement gives no thickness information, therefore can only determine sheet hall concentration. As a result, it is difficult to delineate the contribution from bulk carriers and two-dimensional carrier gases.

To address this issue, we quantitatively determine both the bulk and two-dimensional carrier properties by firstly applying a two-carrier-species (2D and bulk carriers) Hall-effect model. The measured hole concentrations are shown in Fig. 5, which exhibits a very weak dependence on temperature. However, if one closely looks at the hole concentration as a function of temperature, a more complicated behavior can be revealed. At relative high temperatures, 300 K to 200 K, a slight freeze-out effect is

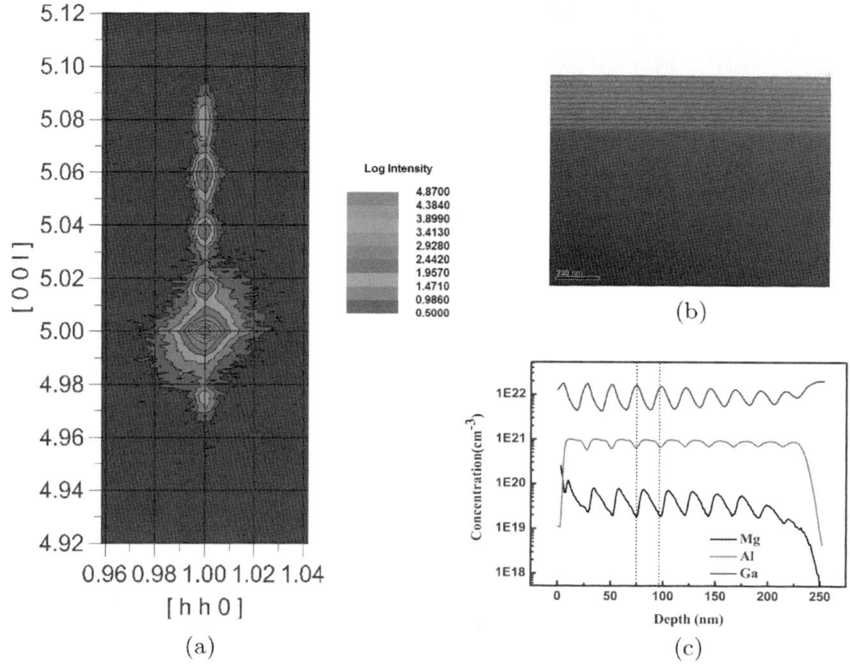

Fig. 4. Structure and crystalline quality of AlxGa1-xN/GaN sample. (a) asymmetrical (105) X-ray reciprocal space mapping, (b) TEM image of our AlxGa1-xN/GaN sample, (c) SIMS depth profiles of Mg for AlxGa1-xN/GaN sample.[30] Reprinted from Macmillan Publishers Ltd: [Scientific Reports] (Ref. [30]), copyright (2016).

observed. However, on further decrease of the temperature, an abnormal increase of hole concentration (usually known as the freeze-in effect) is observed. Similar hole freeze-in behavior at low temperatures has also been observed in many previous reports.[35,39,45] Unfortunately, most earlier observations of this effect are not discussed in detail or are simply attributed to donor compensation. They argue that, as the thermally activated acceptors freeze out with decreasing temperature, compensating donors begin to have more effect on the conduction. However, based on previous reports even in n-type GaN without obvious compensating effects, such abnormal freeze-in behaviors can also be observed.[45] This phenomenon therefore deserves further attention. Furthermore, in our superlattice-like structures, besides bulk holes, parallel sheets of 2D hole gases can also be created at the interface of heterojunctions. It does not

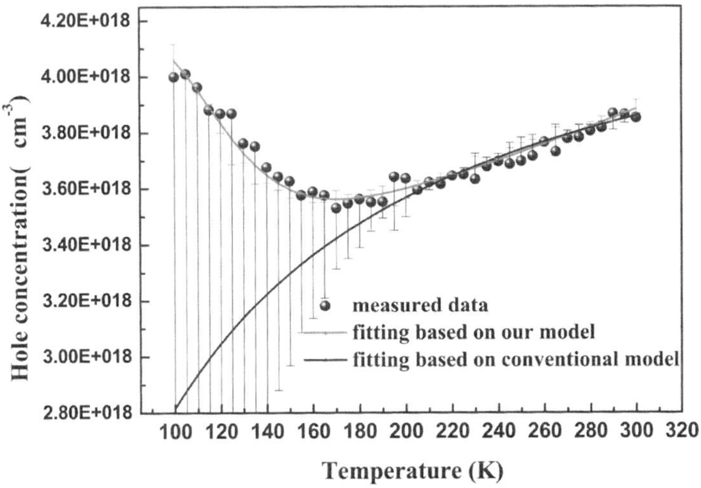

Fig. 5. Hole concentration as a function of temperature. The fitting curves are shown as solid lines using conventional hall-effect model and two-carrier-species Hall-effect model.[30] Reprinted from Macmillan Publishers Ltd: [Scientific Reports] (Ref. [30]), copyright (2016).

make sense to ignore the obvious differences in the electrical properties between them.

Here, a two-carrier-species Hall-effect model is proposed to analyze the electrical behaviors of our AlxGa1-xN/GaN sample. The relevant relationships for a two-carrier-species Hall-effect analysis can be expressed as:[46]

$$\sigma_{\square} = \sum_i \sigma_{\square i} = \sum_i e\mu_i p_{\square i} = \sum_i e\mu_{Hi} p_{H\square i}, \tag{1}$$

$$R\sigma_{\square}^2 = \sum_i R_i \sigma_{\square i}^2 = \sum_i e\mu_{Hi}^2 p_{H\square i}, \tag{2}$$

where: $\sigma_{\square i}$, $p_{\square i}$, $p_{H\square i}$ respectively represent the true sheet conductivity, the sheet concentration and the sheet Hall concentration of carrier i; and μ_i, and $\mu_{Hi}(= R_i\sigma_{\square i})$ are respectively the mobility and Hall mobility of carrier i. In our model, $i = 1$ represents bulk holes, and $i = 2$ represents 2D holes. The Hall factor is close to unity as many other previous studies have used. We divide Eqs. (1) and (2) by d, the thickness of AlxGa1-xN/GaN structure to obtain the normally measured quantity, Hall concentration p_H as:

$$p_H = \frac{p_{H\square}}{d} = \frac{\sigma_{\square}^2/d^2}{eR_{\square}\sigma_{\square}^2/d} = \frac{(\mu_{H1}p_{H1} + \mu_{H2}p_{H\square 2}/d)^2}{(\mu_{H1}^2 p_{H1} + \mu_{H2}^2 p_{H\square 2}/d)}$$

$$= \frac{(Qp_{H1} + p_{H\square 2}/d)^2}{(Q^2 p_{H1} + p_{H\square 2}/d)}, \tag{3}$$

Here, Q is equal to μ_{H1}/μ_{H2}, and is the ratio of bulk and 2D carrier Hall mobility.

The bulk hole concentration p_{H1} in a semiconductor with acceptor concentration NA and acceptor ionization energy E_a can be expressed as[47,48]:

$$p_{H1} = \frac{(N_A - N_D)}{N_D}\frac{N_V}{4}\exp\left(-\frac{E_a}{kT}\right), \tag{4}$$

where $N_V = 2\frac{(2\pi m_p^* k)^{3/2}}{h^3}$ is the effective density of states at the valance band edge of GaN; mp* is the effective mass for holes; g is the acceptor degeneracy (g = 4); T is the temperature; and h and kB are Planck's and Boltzmann's constants respectively. Here, we assume that the concentration of 2D hole gases is temperature independent. Consequently, substituting Eq. (4) into Eq. (3) we obtain the numerical relationships between the Hall concentration p_H and T. From this expression, the value of E_a and $p_{H\square 2}$ can be obtained iteratively

As shown in Fig. 5, the value of (acceptor ionization energy of bulk holes) and (sheet Hall concentration of 2D holes) can be obtained iteratively. It is observed that the conventional Hall-effect model is in agreement with experimental data at high temperatures (above 200 K), but decreasing temperature leads to a significant departure of the calculated concentration from that observed experimentally. Meanwhile, our model does agree with the measured experimental data very well at both low and high temperatures. The fitting parameters acceptor ionization energy and hole concentration are about 60 meV and 8.36 × 1013 cm^{-2}, respectively. Based on the single acceptor model, the bulk hole concentration is calculated to be about 1.14 × 1018 cm^{-3} at 300 K, which is about one order of magnitude higher than that of the normal p-type sample prepared by the same tools.

Similar 2D carrier gases have been widely reported in many previous works in both p-type and n-type materials,[31,45] which can be attributed to polarization doping. In this work, we are more concerned with the abnormal high bulk hole concentration observed. As discussed above, we propose that this is the result of impurity resonant state p-type doping, which increases the overall bulk hole concentration by transforming the localized impurity states in barriers into resonant states in wells through orbital hybridization between Mg 2p and host N 2p orbitals. The underlying physical mechanism of this effect can also be understood in another way: in this new scenario, the deep acceptors in the barrier layers ionize into the valence band of the neighboring narrow band-gap material, rather than into its own, deeper, valance band. The high bulk hole concentration is strong evidence to support our theory that high efficiency p-type doping can be achieved by impurity resonant states in superlattice structures. Furthermore, we would like to point out that our approach can be considered as one special form of modulation doping. However, the purpose here is to generate high concentration carriers in wide band gap nitrides, which is otherwise difficult by directly doping the material itself, whereas the modulation doping is typically used to separate the dopant ions from the carriers in order to achieve high carrier mobility in the well, as wildly studied in arsenide or Ge/Si.[49,50]

In summary, a novel strategy for efficient p-type doping was proposed to overcome the fundamental problem of high activation energy in high bandgap III-nitrides by introducing impurity resonant states in an Mg doped AlxGa1-xN/GaN superlattice structure. This structure can be used to achieve efficient nitride based optoelectronic devices, especially in the deep ultraviolet wavelength range. The concept of impurity resonant state p-type doping presented here could also be applied to the production of highly p-type conductors in other wide-band-gap materials. Furthermore, the two-carrier-species Hall-effect model was used to extract the electrical parameters of bulk holes and 2D hole gases in superlattice-like structures, respectively. The model reported here can also be used to explain the abnormal and seldom analyzed freeze-in effect observed in many previous reports.

(2) Ga-face Polarization induced 3D hole gas

P-type conductivity is always a technique difficulty for wide-band-gap III-Nitride semiconductor materials. In 2010, a p-type doping method named polarization-doping has been proposed by Simon and Jena[26] in N-face graded AlGaN layer. However, N-face GaN-based structures tend to have poor surface morphology and high impurity concentration, no matter grown by MOCVD or molecular beam epitaxy.

Three-dimensional hole gas (3DHG) in (0001)-oriented graded AlGaN by using polarization-doping has been achieved by our group. The hole concentration in graded AlGaN layer is as high as $\sim$1018/cm^3 at RT and shows weak temperature dependence.

The metal-face graded AlGaN is grown by single-wafer home-made MOCVD. Before growing the graded AlGaN layer, a 1 um thick un-doped AlN layer is deposited on the c-plane sapphire at 1200°C. Then a 100-nm-thick Mg-doped graded AlxGa1$-$xN layer with Al composition from x $=$ 0.3 to 0 is grown on the AlN layer.

As shown in Fig. 6, for the metal-face graded AlGaN layer grown on the AlN layer, it can be deemed as N unit cells with a Al composition linearly decreasing from x $=$ x0 to 0. The unbalance bound sheet charge density is Al composition dependent, which decreases as the Al composition decreases. The net bound sheet charge density at the interface of the ith and i $+$ 1th unit cell is then summed as [(Pisp$+$Pipe) $-$ (Pi$+$1sp$+$Pi$+$1pe)]/q. Pisp $-$ Pi$+$ 1sp $<$ 0, and Pipe $-$ Pi $+$ 1pe $<$ 0, mean that the net bound sheet charges are negative at the interface of each unit cell. Consequently, equivalent holes at the interface of each unit cell are induced from the acceptors (Mg) by the polarization field to neutralize the negative bound sheet charges. These holes spread over the graded AlGaN layer continuously, thus form a mobile 3DHG.

As shown in Fig. 7, the RT hole concentration of the graded AlGaN layer is 2.6 $\times$ 1018 cm^{-3}, one order of magnitude higher than that of the GaN layer (2.5 $\times$ 1017 cm^{-3}). As the temperature decreases, the hole concentration of the GaN layer drops exponentially caused by carrier freezeout due to the high activation energy of Mg in GaN. While for the graded AlGaN layer, the hole concentration shows weak temperature dependence, it is as high as 4.5 $\times$ 1017 cm^{-3} when the temperature down to 100 K. This reveals that the holes in the graded AlGaN layer are induced by polarization, because the polarization-doping does not need thermal ionization energy to activate. The RT resistivity of the graded AlGaN layer is 0.6 Ω cm, lower than that of the GaN layer (1.8 Ω cm), indicating an enhanced p-type conductivity in the graded AlGaN layer. Besides, as the temperature decreases, the

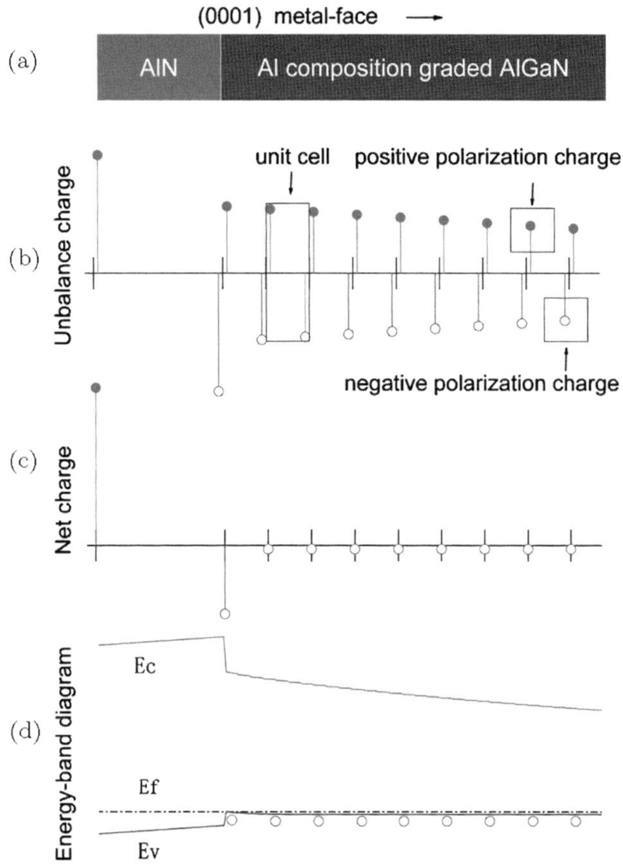

Fig. 6. Schematic illustration of polarization-induced 3DHG in (0001)-oriented metal-face III-nitride structure.[51] Reproduced from Ref. [51], with the permission of AIP Publishing.

resistivity of the graded AlGaN layer increases slower than that of the GaN layer.

This p-type doping method overcomes the problem of high Mg activation energy in III-nitrides and paves a way for achieving high-efficiency in wide-band-gap semiconductor light-emitting devices. The polarization-doped (0001)-oriented GaN-based LEDs was studied by using APSYS software. As shown in Fig. 8, the calculated results indicated that, the EL intensity and the IQE in the polarization-doped LEDs are enhanced significantly compared with the conventional LED due to the polarization-doped hole as well as the smooth valence band in the graded AlGaN

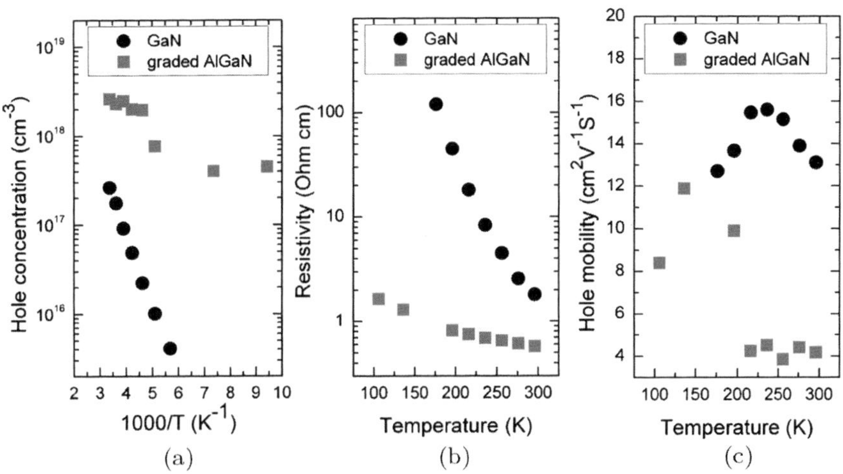

Fig. 7. Temperature-dependent Hall-effect measurements for the GaN layer and the graded AlGaN layer.[51] Reproduced from Ref. [51], with the permission of AIP Publishing.

layer, and consequently the enhanced hole injection. On the other hand, the efficiency droop of the polarization-doped LEDs is also improved as compared with the conventional LED.

The influence of the degree of AlGaN gradation on polarization-doped LEDs is also studied and shown in Fig. 9. Because of the enhanced hole concentration caused by the increased degree of AlGaN gradation, the EL intensity and EQE of the polarization-doped LED are improved when x0 is increased from 0.15 to 0.2. However, when x0 is further increased to 0.25 and 0.3, the EL intensity and EQE are decreased due to the increased alloy scattering in the graded AlGaN and the larger strain in the MQWs, which reduce the hole injection and the radiative recombination rate.

(3) Co-doping Mg with d-orbital elements

As shown in Fig. 10, the behavior of Mg as an acceptor in nitride semiconductors is linked to the character of the valance band maximum (VBM) of host, which primarily originates from anion p orbitals. A deep, flat VBM promotes hole localization and increases the ionization energy of acceptors. Therefore, by introducing other additional energy levels, which have the same symmetry and can couple to the N 2p orbitals, the relative

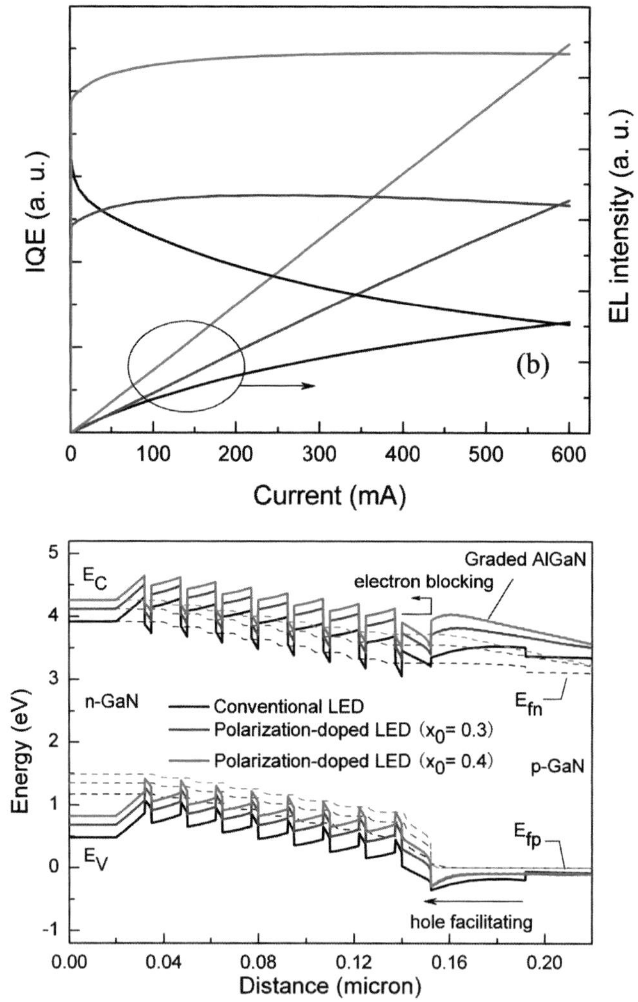

Fig. 8. Optical performances and band diagrams of the polarization-doped and the conventional LEDs.[52] Reproduced from Ref. [52], with the permission of AIP Publishing.

position and character of the GaN VBM can be changed. As a result, lower ionization energy of acceptors may be expected. In the calculation presented here, the additional energy level is introduced by substitutional In, which has d states that are fully occupied and merged deeply in the valance band. We anticipate that the characters of effective acceptor levels in GaN can be

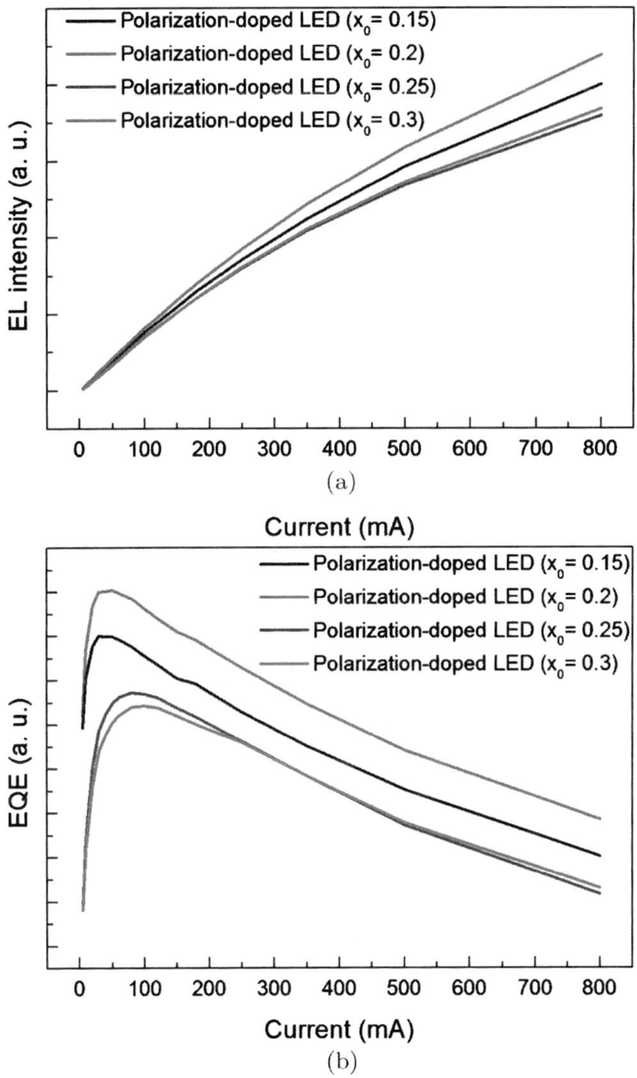

Fig. 9. The EL intensity as a function of the injection current for all the polarization-doped LEDs.[53] Reproduced from Ref. [53], with the permission of AIP Publishing.

modified by the coupling between the defect states of the substitutional In and Mg, which have the same t2 symmetry.

To understand the underlying physical mechanism of In-Mg doping, projected densities of states (DOSs) were analyzed to examine the orbital

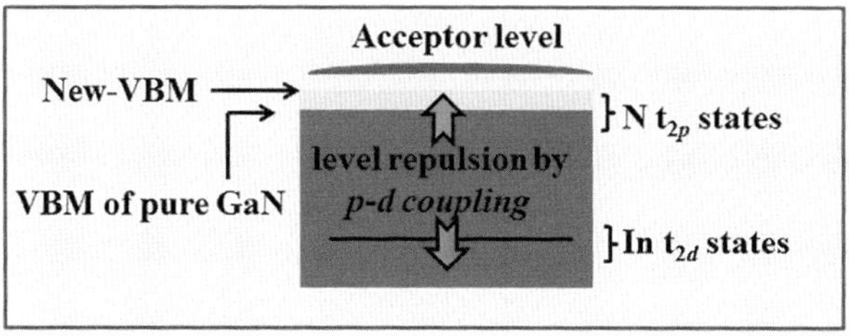

Fig. 10. Schematic model showing the position and the hybridization and level repulsions between In t2d and N t2p states, and the hybridization and level repulsions between p-like defect level of the substitutional In and Mg.[29] Reproduced from Ref. [29] with permission of The Royal Society of Chemistry.

hybridization between doping atoms and the host matrix using the projector-augmented-wave method with generalized gradient approximations, as implemented in the Vienna *ab initio* simulation package.[54]

Wurtzite GaN structures with In-Mg co-doping were examined using $2 \times 2 \times 2$ supercell models. Two Ga atoms are substitute by In and Mg randomly. Actually, we have tried several different configurations. The calculation results (transition energy and densities of states) did not change too much. Here, the generalized gradient approximations (GGA) of Perdew-Burke-Ernzerhof (PBE) functionals are used for the exchange correlation potential.[54] A reciprocal space k-point mesh of $5 \times 5 \times 5$ is employed. The cutoff energy is chosen to be 800 eV. All the atoms are fully relaxed until the Hellmann-Feynman force is less than 0.03 eV/ang.

The projected densities of states (DOSs) of In t2d, N t2p states and Mg t2p were shown in Fig. 11.

As can be seen, several peaks of the In t2d states overlap with that of N t2p, which indicated the strong coupling between them.[44,55] As know, In 4d states will split into two-fold e.g. states and three-fold t2d states in the environment of the GaN matrix. Due to sharing the same t2 symmetry, the In t2d and N t2p states can couple with each other. Based on the band coupling model developed by Dalpian and Wei,[56] such kind of states (t2p, t2d,) coupling will lead to a level repulsion. As results, the VBM (mostly composed by N 2p obitals) will be pushed up and the In t2d states will be pushed down. The calculated total density of states for pure GaN and In doped GaN could also be used to further clarify the band structure modulation effect of orbital hybridization between doping atoms and the

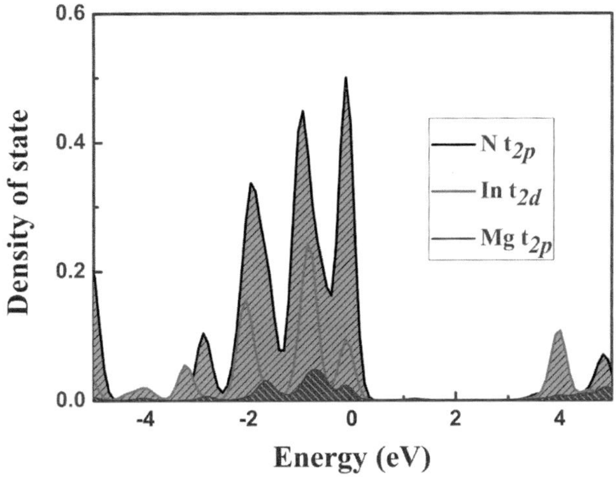

Fig. 11. Calculated projected density of state of In 4d, N 2p and Mg 2p obtitals.[29] Reproduced from Ref. [29] with permission of The Royal Society of Chemistry.

host matrix.[29] It is clearly visible that state distribution of In doped GaN (blue line) shift to higher energies due to the coupling and level repulsion between In t2d and N t2p. As a result, a "new" higher VBM, which is composed of the hybrid p-d orbitals, is formed. As discussed above, the upward shift of the VBM will facilitate acceptor activation and thus better p-type dopability in GaN. Furthermore, several peaks of Mg t2p state also overlap with that of In t2d. Such kind of overlapping is coming from the fact that Mg t2p and In t2d. share the same coupling with N t2p states. Considering the limited atom concentration and large average distance, the direct coupling between In t2d. and Mg t2p state should be rather weak.

The underlying physic of the above observation can also be understood in another way. We chose the planes containing In-N bond (a) and Ga-N bond (b) and plotted calculated charge density of the VBM in Fig. 12. It could be noticed that the VBM states tend to be more localized on the In-N site and shows strong p-d hybridization orbital character. As know, the VBM of InN is 1.11 eV higher than that of GaN due to larger In-N bond length and strong coupling between N 2p orbital and In 4d orbitals.[57] In this new scenario, the VBM of In doping system turns to be more InN-like. Therefore, the MgGa defect level is closer to the VBM. As a result, the acceptor ionization energy is smaller.

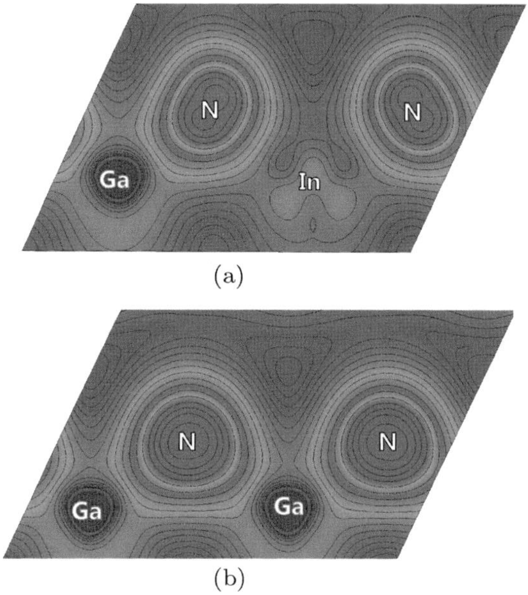

Fig. 12. Isosurface charge density plots of VBM of In-Mg co-doping GaN at Γ point in the planes of (a) N-In-N bonds and (b) N-Ga-N bonds.[29] Reproduced from Ref. [29] with permission of The Royal Society of Chemistry.

To determine the defect transition energy levels, the total energy for a supercell containing In-Mg complex in charge state q (E(α, q)) and the same supercell of pure GaN (E(host)) were calculated. The Brillouin zone integration using special k-points[18] is consistent and can give good results on the total energy and converged charge density. However, this method can only give a poor description on the symmetry, energy levels of the defect state and VBM of system. Furthermore, the obtained results are sensitive to the selection of the k-points sampling. In this work, the hybrid scheme is used to combine the advantages of both the special k-point and gamma point-only approaches.[18] The transition energy level of Mg acceptor with respect to the VBM is obtained by:

$$\begin{aligned}\varepsilon(q/0) &= [\varepsilon_D^\Gamma(0) - \varepsilon_{VBM}^\Gamma(host)] \\ &\quad + [E(\alpha, q) - (E(\alpha, 0) - \varepsilon_D^k(0))]/(-q) \qquad (1)\end{aligned}$$

which acts as the control sample. After the growth, the samples were annealed at 650°C for 30 min in nitrogen to activate the Mg dopant.

Secondary ion mass spectrometry measurements were performed with a Physical Electronics with a 6 kV 100 nA Cs+ beam, at a detection angle of 450 to verify the incorporation and distribution of Mg and In atoms. Temperature-dependent Hall-effect measurements performed at a magnetic field of 0.5 T using the van der Pauw method. NiAu acts as the ohmic contact layer. High resolution x-ray diffraction (HRXRD) was performed to characterize the crystalline quality of the samples using a Philips X'Pert system.

Secondary ion mass spectrometry measurements were performed to verify the incorporation and distribution of Mg and In atoms. Furthermore, the hole concentration confirmed by Hall measurement is as high as $1.4 \times 1018\,\text{cm}^{-3}$ at room temperature, which is nearly one order of magnitude higher than the typical value by direct doping GaN in our lab. The acceptors activation can be determined by fitting the concentration data. It decreased to about 160 meV and can be act as a direct evidence to support our calculation results. The hole mobility of In-Mg codoping sample is about $9.6\, cm^2/V{\cdot}s$ at room temperature, which is slightly decrease compared with the control sample (about $10.2\, cm^2/V \cdot s$). We argue that the decrease of hole where $\varepsilon_D^{\Gamma}(0)$, $\varepsilon_D^{k}(0)$ are the defect levels at the gamma point and at special k point (averaged); $E(\alpha, q)$, $E(\alpha, 0)$ are total energies for supercells with defect α in the q charge state and in the neutral charge state, respectively; $\varepsilon_{VBM}^{\Gamma}(host)$ means the VBM energy of the supercell at the gamma point. The $\varepsilon_D^{\Gamma}(0)$ and $\varepsilon_{VBM}^{\Gamma}(host)$ are eigenvalues in two different systems. Therefore, to get accurate descriptions of the VBM and defect transition energy levels it is very important to align the potential of defect cell with pure host, especially for shallow levels. Previous calculations have exhibited that the transition energy levels have been severely underestimated without the potential alignment in wide band gap material systems.[14]

Based on equation (1), we calculated the ionization energy of the isolated Mg acceptor in bulk wurtzite GaN, which is about 208 meV. It agrees well with previous both calculations and experimental observations.[18] For comparison, the ionization energy $\varepsilon(-/0)$ of In-Mg acceptor is only 135 meV, which is about 70 meV shallower.

Samples in this study were grown on 2 inch (0001) sapphire substrates by a Vecco 3×2 inch metal organic chemical vapor deposition (MOCVD). Trimethylindium (TMIn), trimethylgallium (TMGa), bis(cyclopentadienyl) magnesium (Cp2Mg) and ammonia (NH3) were used as precursors for In, Ga, Mg and N, respectively. The TMIn/Cp2Mg molar flow ratio is remaining at about 5. For the whole growth process, the carrier gas was hydrogen. Firstly, 30 nm GaN was deposited at the surface of sapphire substrate as the buffer layer. Then, a 1 μm undoped GaN layers were deposited followed by 0.5 μm In-Mg codoping GaN layer at 1050°C. To suppress the formation of compensation center (nitrogen vacancies), a high V/III molar flow ratio of 10,000 was used. The TMIn/Cp2Mg molar flow ratio is remaining at about 5. For comparison, conventional Mg-doped GaN was also prepared using the same grow parameters, mobility should be mainly ascribed to the increase of impurity scattering and/or dislocation scattering.

In summary, the underlying physics and the acceptor ionization process in co-doping Mg with d-orbital elements, specifically indium, was analyzed experimentally and theoretically. It showed that the effective acceptors ionization for In-Mg co-doped systems is mainly coming from the band structure tuning effect of In atom, practically for VBM of the nitrides. Using first-principle band structure calculation, we investigated the orbital hybridization between doping atoms and the host matrix. A "new" higher VBM, which is composed of the hybrid p-d orbitals, is formed due to the strong coupling and level repulsion between In t2d and N t2p states. The elevating of the VBM will facilitate acceptor activation and thus better p-type dopability in GaN.

4. Nitrides Nano-structures

Recently, the nanowire LED,[58–62] mainly core/shell LED, has attracted large attention because of their potential advantages. In comparison to the conventional thin film LEDs, core/shell nanorod structure LEDs have higher aspect ratio and larger active regions. Besides, the InGaN/GaN multi-quantum wells (MQWs) grown on nonpolar or semipolar GaN facets are suggested to be able to effectively increase the light extraction efficiency and reduce the quantum confined stark effect (QCSE), which increases the internal quantum efficiency of LEDs through enhancing the radiative recombination rate in MQWs.[63–65] What more, nitride nanowires could improve the material quality with respect to thin films, a better handling of the strain induced by thermal expansion mismatch.

For the structure of core/shell LED, the growth of high density ordered GaN nanowires (core) is necessary for the follow-up metal contacting and homogeneous radial growth. The top-down etching and bottom-up selective area MOCVD growth[61,62] are two prevailing GaN nanowire growth methods. We have widely studied these two methods.[66,71] GaN dry plasma etching is widely used in the III nitrides industry. However, reactive ion etching (RIE) suffers from relatively low etch rates and plasma damage in devices. Compared with RIE, wet-etching has the advantages of reducing surface damage and low cost, and providing a simple method for device fabrication. Figure 13 shows the nanostructure mainly developed by wet etching (a) and selective area growth (b). We found the suppression of the etch rate in the vertical direction could help to form he uniform nanowire. A thin AlGaN which was stable in the KOH solution was used to suppress the vertical direction. The wetting nanowires were surrounded by six $\{1\bar{1}00\}$ m plane. On the other hand, the existed large vertical etching could develop the pyramid structure with six $\{11\bar{2}1\}$ facets. The top-down may be an effective method to obtain the array nanowires. However, the nanowire crystal quality was poor because of the film growth and wetting. The selective area growth (SAG) was the most effective GaN nanowire growth methods because the high crystal quality and no introduction of foreign metals. Our team also systematic studied the growth technology. We found the hydrogen content has a large impact on the morphology of the nanostructure. Interesting, the results were similar to the wet etching. The high hydrogen content benefited to the development of long one-dimensional nanorods while lack or no hydrogen carrier gas lead to form the pyramid structure. The morphology change of GaN nanostructures with H2 flow rate could be explained by the two functions of hydrogen, that is, the passivation and etching effects, during the growth process. The semipolar $\{1\bar{1}01\}$ r-plane is N-terminated surface, which is easily passivated by the hydrogen atoms from NH3 or H2. Thus, the low hydrogen content could benefit to the r-plane. However, with increasing H2 flow rate, the etching effect of H2 starts to dominate. Consequently, some pyramids are etched and the m-plane is exposed, which leads to the presence of nanorods.

The high-current driven, high brightness InGaN/GaN core–shell LED is the final aim for this study. Figure 14 shows the pyramid array InGaN/GaN core–shell light-emitting diodes (LEDs) which were fabricated by using a

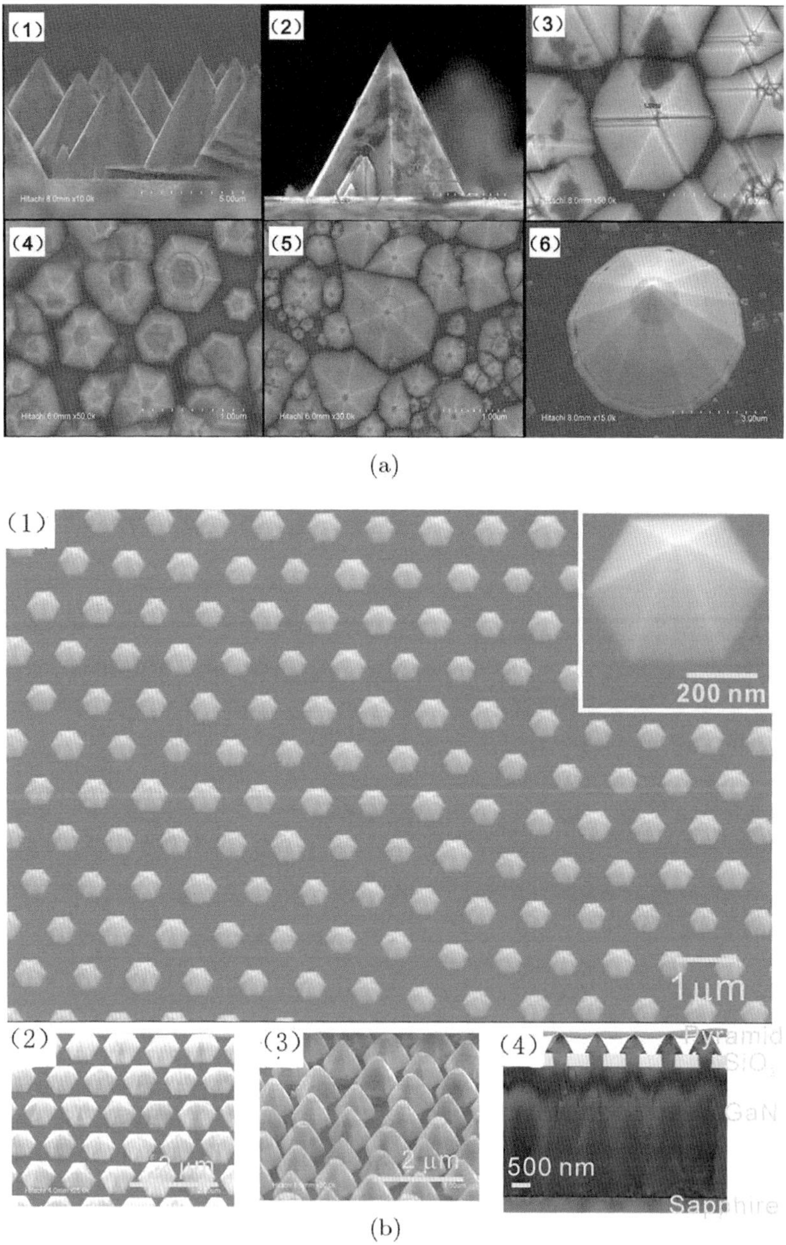

Fig. 13. The development of GaN nanostructure by wet etching (a).[67] Reproduced from Ref. [67], with the permission of AIP Publishing. and selective area growth (b).[71] Reproduced from Ref. [71], with the permission of AIP Publishing.

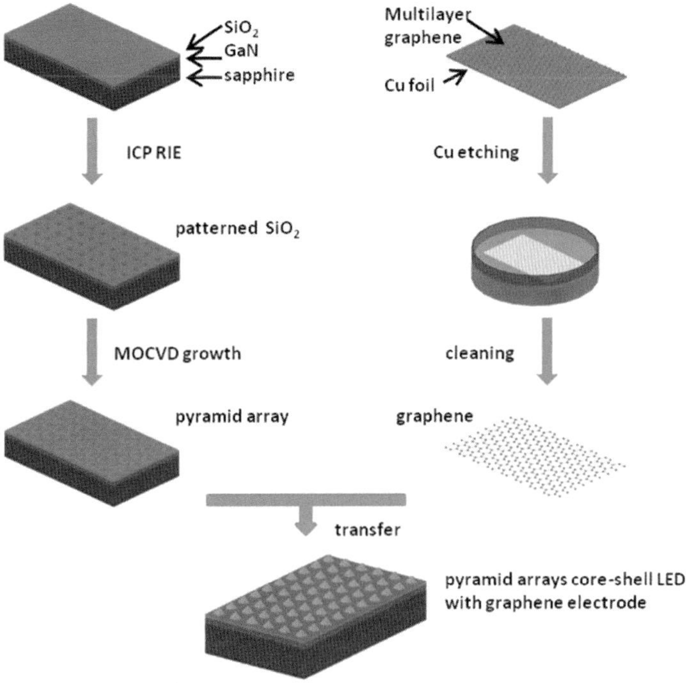

(a)

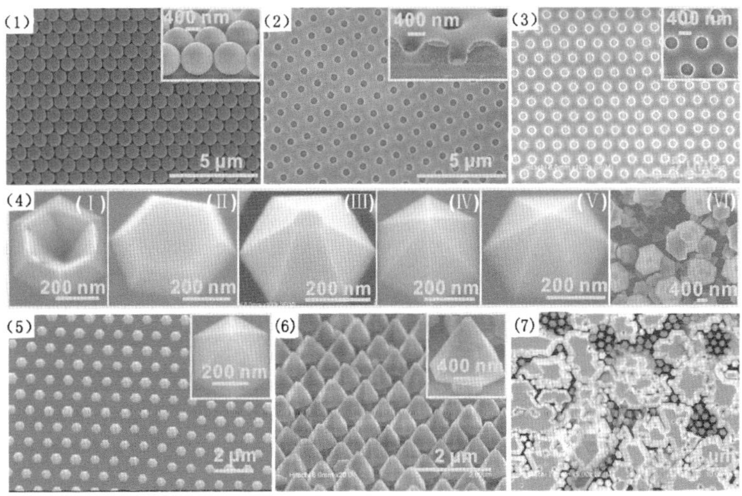

(b)

Fig. 14. Pyramid array core–shell LED with graphene electrodes: (a) Schematic illustration of the key fabrication procedures[70,71] DOI:10.7567, with the permission of IOPScience. (b) SEM image of large-area pyramid core–shell LED.[73] Reproduced from Ref. [72], with the permission of AIP Publishing.

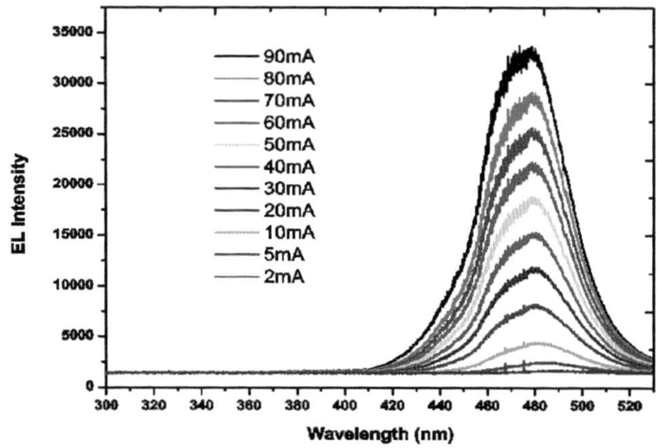

(a)

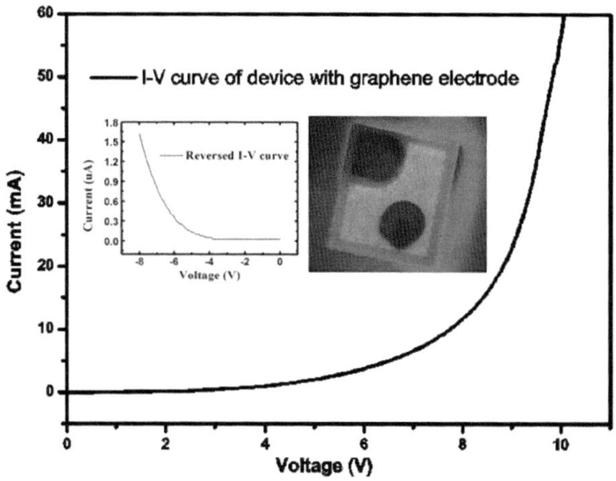

(b)

Fig. 14. (*Continued*) Pyramid array core–shell LED with graphene electrodes: (c) Room-temperature EL spectra of pyramid array core–shell LED at various current injections[70,71] DOI:10.7567, with the permission of IOP Science. (d) I–V curve of LEDs with multilayer graphene electrode. The inset is a photograph of the light emission from a single LED chip and reversed bias voltage I–V curve[70,71] DOI:10.7567, with the permission of IOP Science.

highly homogeneous multilayer graphene transparent conducting electrode. The key fabrication procedures are the SAG growth of the pyramid n-doped GaN, low temperature P-GaN growth, translating graphene acted as p-type electrode and depositing the p- and n-GaN contact pads. The

EL spectra [Fig. 14(c)] of graphene interconnect pyramid array core–shell LEDs exhibits a dominant emission peak at 478 nm with no noticeable spectral shift as the injection current is increased from 2 to 90 mA. This result is attributed to the fact that the built-in strain in the MQWs is very small in these pyramid array InGaN/ GaN core–shell LEDs and the piezoelectric field also decreases effectively by growing MQWs on a semipolar facet $(1\bar{1}01)$ GaN pyramid sidewall. The current–voltage (I-V) curve shows rectifying characteristics with a dynamic resistance at various current. The dynamic resistance originates from the contact resistance of p-GaN and graphene. The reversed leakage current of pyramid array core–shell LEDs is about 1.6 uA@8V [inset of Fig. 14(d)], which indicates that the device has good reliability. The illumination photo [inset of Fig. 14(d)] of the LED chip at 20 mA shows uniform luminosity, which indicates the good current spreading by using graphene electrodes.

5. Summary

In this chapter, we mainly focused on the underlying physics of p doping challenge, nitrides nano-structures and their application on LEDs. We presented an overview of some main results and progress in highly effective p doping for nitrides. Some promising strategies to address the issue of p doping challenge were also introduced, including: (1) impurity Resonant States p-type Doping in Wide-Band-Gap Nitrides, (2) Ga-face Polarization induced 3D hole gas, (3) Co-doping Mg with d-orbital elements. Though these new p doping strategy, the hole concentration in GaN systems have been improve by one order of magnitude higher than what typically achievable by direct Mg-doping. The results will be helpful to understand the underlying physics of p-doping challenge and beneficial for the researchers to further address the issue of p-doping in wide-band-gap materials. Parts of these works have been applied in GaN based LED structures and demonstrated improved performance. Furthermore, series of nitrides nano-structrues have been fabricated by wet etching and selective area growth. The pyramid array InGaN/GaN core–shell light-emitting diodes (LEDs) was achieved by using a highly homogeneous multilayer graphene transparent conducting electrode.

References

1. H. Angerer, D. Brunner, F. Freudenberg, O. Ambacher, M. Stutzmann, R. Hopler, T. Metzger, E. Born, G. Dollinger, A. Bergmaier, S. Karsch, and H. J. Korner, *Appl. Phys. Lett.*, **71**(11), 1504–1506 (1997).
2. B. Monemar, *Physical Review B*, **10**(2), 676–681 (1974).
3. B. Maleyre, S. Ruffenach, O. Briot, and A. van der Lee, *Superlattices and Microstructures*, **36**(4–6), 527–535 (2004).
4. B. Daudin, J. L. Rouviere, E. Ligeon, M. Arlery, R. Niebuhr, and K. H. Bachem, *Blue Laser and Light Emitting Diodes*, CRC Press, 496–499 (1996).
5. A. R. Smith, R. M. Feenstra, D. W. Greve, M. S. Shin, M. Skowronski, J. Neugebauer, and J. E. Northrup, *Appl. Phys. Lett.*, **72**(17), 2114–2116 (1998).
6. A. R. Smith, R. M. Feenstra, D. W. Greve, J. Neugebauer, and J. E. Northrup, *Phys. Rev. Lett.*, **79**(20), 3934–3937 (1997).
7. S. Koizumi, K. Watanabe, M. Hasegawa, and H. Kanda, *Science*, **292**(5523), 1899–1901 (2001).
8. J. Isberg, J. Hammersberg, E. Johansson, T. Wikstrom, D. J. Twitchen, A. J. Whitehead, S. E. Coe, and G. A. Scarsbrook, *Science*, **297**(5587), 1670–1672 (2002).
9. Z. Teukam, J. Chevallier, C. Saguy, R. Kalish, D. Ballutaud, M. Barbe, F. Jomard, A. Tromson-Carli, C. Cytermann, J. E. Butler, M. Bernard, C. Baron, and A. Deneuville, *Nature Materials*, **2**(7), 482–486 (2003).
10. A. Tsukazaki, M. Kubota, A. Ohtomo, T. Onuma, K. Ohtani, H. Ohno, S. F. Chichibu, and M. Kawasaki, *Japanese Journal of Applied Physics Part 2-Letters & Express Letters*, **44**(20–23), L643–L645 (2005).
11. M. Joseph, H. Tabata, and T. Kawai, *Japanese Journal of Applied Physics Part 2-Letters*, **38**(11a), L1205–L1207 (1999).
12. D. C. Look, B. Claflin, Y. I. Alivov, and S. J. Park, *Physica Status Solidi a — Applications and Materials Science*, **201**(10), 2203–2212 (2004).
13. G. F. Neumark, *Materials Science & Engineering R-Reports*, **21**(1), 1–46 (1997).
14. S. B. Zhang, S. H. Wei, and A. Zunger, *J. Appl. Phys.*, **83**(6), 3192–3196 (1998).
15. Y. F. Yan, C. S. Jiang, X. Z. Wu, R. Noufi, S. H. Wei, and M. M. Al-Jassim, *PVSC: 2008 33rd Ieee Photovoltaic Specialists Conference*, Vols 1–4, 1513–1516 (2008).
16. Y. F. Yan, J. B. Li, S. H. Wei, and M. M. Al-Jassim, *Phys. Rev. Lett.*, **98**(13), 135506 (2007).
17. S. H. Wei and S. B. Zhang, *Physical Review B*, **66**(15), 155211 (2002).
18. C. G. Van de Walle and J. Neugebauer, *J. Appl. Phys.*, **95**(8), 3851–3879 (2004).
19. Y. Aoyagi, M. Takeuchi, S. Iwai, and H. Hirayama, *Appl. Phys. Lett.*, **99**(11), 112110 (2011).
20. K. S. Kim, M. S. Han, G. M. Yang, C. J. Youn, H. J. Lee, H. K. Cho, and J. Y. Lee, *Appl. Phys. Lett.*, **77**(8), 1123–1125 (2000).
21. T. Tanaka, A. Watanabe, H. Amano, Y. Kobayashi, I. Akasaki, S. Yamazaki, and M. Koike, *Appl. Phys. Lett.*, **65**(5), 593–594 (1994).
22. S. Nakamura, M. Senoh, N. Iwasa, S. Nagahama, T. Yamada, and T. Mukai, *Japanese Journal of Applied Physics Part 2-Letters*, **34**(10b), L1332-L1335 (1995).
23. Y. F. Yan, J. B. Li, S. H. Wei, and M. M. Al-Jassim, *Phys. Rev. Lett.*, **98**(13), 135506 (2007).

24. E. F. Schubert, W. Grieshaber, and I. D. Goepfert, *Appl. Phys. Lett.*, **69**(24), 3737–3739 (1996).
25. P. Kozodoy, Y. P. Smorchkova, M. Hansen, H. L. Xing, S. P. DenBaars, U. K. Mishra, A. W. Saxler, R. Perrin, and W. C. Mitchel, *Appl. Phys. Lett.*, **75**(16), 2444–2446 (1999).
26. J. Simon, V. Protasenko, C. X. Lian, H. L. Xing, and D. Jena, *Science*, **327** (5961), 60–64 (2010).
27. B. Gunning, J. Lowder, M. Moseley, and W. A. Doolittle, *Appl. Phys. Lett.*, **101**(8), 082106 (2012).
28. Y. Q. Gai, J. B. Li, S. S. Li, J. B. Xia, Y. F. Yan, and S. H. Wei, *Physical Review B*, **80**(15), 153201 (2009).
29. Z. Q. Liu, B. L. Fu, X. Y. Yi, G. D. Yuan, J. X. Wang, J. M. Li, Luna, and I. Ferguson, *RSC Advances*, **6**(6), 5111–5115 (2016).
30. Z. Q. Liu, X. Y. Yi, Z. G. Yu, G. D. Yuan, Y. Liu, J. X. Wang, J. M. Li, N. Lu, I. Ferguson, and Y. Zhang, *Scientific Reports*, **6**, 23950 (2016).
31. P. Kozodoy, Y. P. Smorchkova, M. Hansen, H. L. Xing, S. P. DenBaars, U. K. Mishra, A. W. Saxler, R. Perrin, and W. C. Mitchel, *Appl Phys Lett*, **75**(16), 2444–2446 (1999).
32. E. F. Schubert, W. Grieshaber, and I. D. Goepfert, *Appl Phys Lett*, **69**(24), 3737–3739 (1996).
33. S. Fafard, Y. H. Zhang, and J. L. Merz, *Phys Rev B*, **48**(16), 12308–12311 (1993).
34. R. A. Davies, M. J. Kelly, and T. M. Kerr, *Appl Phys Lett*, **53**(26), 2641–2643 (1988).
35. B. Gunning, J. Lowder, M. Moseley, and W. A. Doolittle, *Appl Phys Lett*, **101**(8), 082106 (2012).
36. N. F. Mott and W. D. Twose, *Adv Phys*, **10**(38), 107–163 (1961).
37. D. C. Look, D. C. Reynolds, J. W. Hemsky, J. R. Sizelove, R. L. Jones, and R. J. Molnar, *Phys Rev Lett*, **79**(12), 2273–2276 (1997).
38. N. F. Mott, *Rev Mod Phys*, **40**(4), 677–683 (1968).
39. J. Simon, V. Protasenko, C. X. Lian, H. L. Xing, and D. Jena, *Science*, **327**(5961), 60–64 (2010).
40. G. Kresse and J. Furthmuller, *Comp Mater Sci*, **6**(1), 15–50 (1996).
41. J. P. Perdew, K. Burke, and M. Ernzerhof, *Phys Rev Lett*, **77**(18), 3865–3868 (1996).
42. Y. Zhang and J. W. Wang, *Phys Rev B*, **90**(15), 155201 (2014).
43. L. Liu, P. Y. Yu, Z. X. Ma, and S. S. Mao, *Phys Rev Lett*, **100**(12), **127203** (2008).
44. G. M. Dalpian and S. H. Wei, *Phys Rev B*, **72**(11), 115201 (2005).
45. S. B. Li, M. E. Ware, J. Wu, V. P. Kunets, M. Hawkridge, P. Minor, Z. M. Wang, Z. M. Wu, Y. D. Jiang, and G. J. Salamo, *J Appl Phys*, **112**(5), 053711 (2012).
46. D. C. Look, *33-Electrical Characterization of GaAs Materials and Devices*. (Wiley, Chichester; New York, 1989).
47. C. A. Hurni, J. R. Lang, P. G. Burke, and J. S. Speck, *Appl Phys Lett*, **101**(10), 102106 (2012).
48. Y. D. Chen, H. L. Wu, G. L. Yue, Z. M. Chen, Z. Y. Zheng, Z. S. Wu, G. Wang, and H. Jiang, *Appl. Phys. Express*, **6**(4), 4 (2013).

49. R. Dingle, H. L. Störmer, A. C. Gossard, and W. Wiegmann, *Appl Phys Lett*, **33**(7), 665–667 (1978).
50. D. C. Dillen, K. Kim, E. S. Liu, and E. Tutuc, *Nat Nanotechnol*, **9**(2), 116–120 (2014).
51. L. Zhang, K. Ding, J. C. Yan, J. X. Wang, Y. P. Zeng, T. B. Wei, Y. Y. Li, B. J. Sun, R. F. Duan, and J. M. Li, *Appl. Phys. Lett.*, **97**(6), 062103 (2010).
52. L. Zhang, K. Ding, N. X. Liu, T. B. Wei, X. L. Ji, P. Ma, J. C. Yan, J. X. Wang, Y. P. Zeng, and J. M. Li, *Appl. Phys. Lett.*, **98**(10), 101110 (2011).
53. L. Zhang, X. C. Wei, N. X. Liu, H. X. Lu, J. P. Zeng, J. X. Wang, Y. P. Zeng, and J. M. Li, *Appl. Phys. Lett.*, **98**(24), 241111 (2011).
54. G. Kresse and J. Furthmuller, *Physical Review B*, **54**(16), 11169–11186 (1996).
55. L. Liu, P. Y. Yu, Z. X. Ma, and S. S. Mao, *Phys. Rev. Lett.*, **100**(12), 127203 (2008).
56. G. M. Dalpian, S. H. Wei, X. G. Gong, A. J. R. da Silva, and A. Fazzio, *Solid State Commun.*, **138**(7), 353–358 (2006).
57. F. Wang, J. B. Li, S. S. Li, J. B. Xia, and S. H. Wei, *Physical Review B*, **77**(11), 113202 (2008).
58. M. Tchernycheva, P. Lavenus, H. Zhang, A. V. Babichev, G. Jacopin, M. Shahmohammadi, F. H. Julien, R. Ciechonski, G. Vescovi, and O. Kryliouk, *Nano letters*, **14**(5), 2456–2465 (2014).
59. J. R. Riley, S. Padalkar, Q. Li, P. Lu, D. D. Koleske, J. J. Wierer, G. T. Wang, and L. J. Lauhon, *Nano letters*, **13**(9), 4317–4325 (2013).
60. H. P. Nguyen, S. Zhang, A. T. Connie, M. G. Kibria, Q. Wang, I. Shih, and Z. Mi, *Nano letters*, **13**(11), 5437–5442 (2013).
61. R. Koester, J. S. Hwang, D. Salomon, X. Chen, C. Bougerol, J. P. Barnes, S. Dang Dle, L. Rigutti, A. de Luna Bugallo, G. Jacopin, M. Tchernycheva, C. Durand, and J. Eymery, *Nano letters*, **11**(11), 4839–4845 (2011).
62. F. Qian, S. Gradecak, Y. Li, C. Y. Wen, and C. M. Lieber, *Nano Letters*, **5**(11), 2287–2291 (2005).
63. P. Waltereit, O. Brandt, A. Trampert, H. T. Grahn, J. Menniger, M. Ramsteiner, M. Reiche, and K. H. Ploog, *Nature*, **406**(6798), 865–868 (2000).
64. S. P. Chang, Y. C. Chen, J. K. Huang, Y. J. Cheng, J. R. Chang, K. P. Sou, Y. T. Kang, H. C. Yang, T. C. Hsu, H. C. Kuo, and C. Y. Chang, *Appl Phys Lett*, **100**(6), 061106 (2012).
65. S. F. Li and A. Waag, *J Appl Phys*, **111**(7), 071101 (2012).
66. Liancheng Wang Jun Ma, Zhiqiang Liu, Guodong Yuan, Xiaoli Ji, Ping Ma, and Xiaoyan Yi Junxi Wang, Guohong Wang, and Jinmin Li, *Opt Express*, **21**(3), 3547 (2013).
67. Liancheng Wang, Jun Ma, Zhiqiang Liu, Xiaoyan Yi, Guodong Yuan, and Guohong Wang, *J Appl Phys*, **114**(13), 133101 (2013).
68. Junjie Kang, Zhi Li, Zhiqiang Liu, Hongjian Li, Yongbing Zhao, Yingdong Tian, Ping Ma, Xiaoyan Yi, and Guohong Wang, *Journal of Crystal Growth*, **386**, 175–178 (2014).
69. Peng Ren, Gang Han, Bing-Lei Fu, Bin Xue, Ning Zhang, Zhe Liu, Li-Xia Zhao, Jun-Xi Wan, and Jin-Min Li., *Chin Phys Lett*, **33**(6) (2016).
70. Kui Wu, Tongbo Wei, Ding Lan, Xuecheng Wei, Haiyang Zheng, Yu Chen, Hongxi Lu, Kai Huang, Junxi Wang, Yi Luo, and Jinmin Li, *Appl. Phys. Lett.*, **103**(24), 241107 (2013).

71. Junjie Kang, Zhi Li, Hongjian Li, Zhiqiang Liu, Xiao Li, Xiaoyan Yi, Ping Ma, Hongwei Zhu, and Guohong Wang, *Appl Phys Express*, **6**(7), 072102 (2013).
72. Kui Wu, Tongbo Wei, Haiyang Zheng, Ding Lan, Xuecheng Wei, Qiang Hu, Hongxi Lu, Junxi Wang, Yi Luo, and Jinmin Li, *J. Appl. Phys.*, **115**(12), 123101 (2014).

CHAPTER 2

WASTE ENERGY HARVESTING USING III-NITRIDE MATERIALS

E. Ghafari*, E. Witkoske†, Y. Liu*,‡, C. Zhang*, X. Jiang*, A. Bukowski§, B. Kucukgok*, M. Lundstrom†, I. T. Ferguson¶,‖ and N. Lu*,‖

**SMART Laboratory,*
Lyles School of Civil Engineering School of Materials Engineering,
Birck Nanotechnology Center, Purdue University
luna@purdue.edu

†School of Electrical and Computer Engineering,
Purdue University, West Lafayette, Indiana 47907, USA

‡College of Physics Science & Technology,
Guangxi University, 530004 Nanning, China

§Department of Physics and Astronomy,
Purdue University, West Lafayette, Indiana 47907, USA

¶College of Engineering and Computing,
Missouri University of Science and Technology, 305 McNutt Hall,
1400 N. Bishop, Rolla, MO 65409, USA
ianf@mst.edu

The field of thermoelectrics has grown significantly in the last few decades by a combination of developing new materials and advances in engineering novel devices. III-Nitride materials and their related alloys have come upon the scene relatively recently, garnering commercial success in a wide range of fields and applications. The III-Nitride materials exhibit excellent temperature stability due to their wide-bandgaps, making them highly desirable alternatives for high temperature thermoelectric electricity generation from waste energy harvesting.

III-Nitride materials provide an opportunity for not only achieving high power factors, but also maintaining excellent temperature stability. These unique features of the III-nitrides, make them highly desirable alternatives for high temperature thermoelectric power generation from a variety of waste energy sources.

1. Introduction

The world currently now relies heavily on non-renewable sources such as oil, coal and natural gas for its energy. Non-renewable fossil fuels draw upon assets that will inevitably be depleted, harming the environment as well as being almost impossible to recover. In contrast, renewable energy resources such as wind and solar are constantly being restored by the sun's interaction with our environment. The conversion of waste heat into electrical power has a significant potential to impact renewable energy technologies. This will help us reduce our dependence on fossil fuels and to solve the environmental issues associated with their consumption. According to a report by Lawrence Livermore National Laboratory's nearly 60% of energy produced in US is lost as a form of heat each year.[1] Of the loss, 20–50% of waste heat is generated from industrial processes, such as power plants, metal furnaces, and chemical production. Thermoelectric devices can harvest waste heat via the Seebeck effect: the conversion of a temperature differential into electricity at the junction of two materials. Thermoelectric devices have drawn attention due to increased efficiency of the devices for renewable power sources. Bismuth Telluride (Bi_2Te_3) is one of the most common thermoelectric materials that is widely used in thermoelectric cooling applications. The first report of Bi_2Te_3 as a Peltier cooler was in 1954 by H. J. Goldsmid.[2] However, Bi_2Te_3 is toxic and rare, which makes them expensive and not environmental friendly. In addition, this material is an efficient thermoelectric material only around room temperature which limits its commercial applications at higher temperatures. The thermoelectric field has grown significantly by developing new materials and novel device structures. III-Nitride materials and their alloys are relatively new materials that have seen recent commercial success in optoelectronic devices as well as in power electronics. Advances in III-Nitride materials technology have resulted in many applications for solid state lighting,[3] solar cells,[4] light-emitting diodes (LEDs)[5] and most recently, thermoelectricity.[6,7] This book chapter will review the recent advancement in using III-Nitrides for thermoelectric applications and discuss the future prospects. The binary

III-Nitride and their ternary alloys will be reviewed theoretically and experimentally for room temperature to high temperature thermoelectric applications.

2. III- Nitride Thermoelectric Materials

This section will review the thermoelectric properties of the most common types of III-Nitride materials and their alloys.

2.1. *GaN*

Semiconducting materials have been widely used in electronic applications since the 1960s. Early semiconducting materials like silicon (Si) and germanium (Ge) are those that have been most widely applied. Subsequently, alloy semiconductors, including gallium arsenide (GaAs), indium antimonide (InSb), were introduced. With a bandgap of 3.4 eV,[8] Gallium nitride (GaN) is one of those newest generation semiconductors which have wide bandgaps. GaN most commonly has a wurtzite crystal structure or sometimes zinc blende structure. The lattice constants of GaN are 3.186 Å and 5.186 Å for a and c, respectively[9] for the wurtzite structure. Thermoelectric properties of several different GaN materials reported by different research groups are presented. Figure 1 and Fig. 2 are the plots of Seebeck coefficient S, electrical conductivity σ, and electronic thermal conductivity K_E against carrier concentration. The black squares represent the data reported from an InN/GaN core/shell nanowire whose main heat carriers are electrons. In Fig. 1, it can be seen that all materials attain their highest absolute value of Seebeck Coefficient at low carrier concentrations. The highest Seebeck value for both thin film GaN and InN/GaN nanowire is around 500 μV/K, while bulk GaN is slightly lower, which may be attributed to the lower density of states of bulk materials. In Fig. 1, the nanowire structure has a higher electrical conductivity than the other two at the same carrier concentration, as the core InN has a smaller band gap. In Fig. 2, only thc InN/GaN nanowire's electronic thermal conductivity was reported, but the relationship between electrical and thermal conductivity of other materials should remain the same. The highest ZT for the InN/GaN nanowire was reported as 0.8598 with a carrier concentration of $7 \times 10^{17}\,\text{cm}^{-3}$ at 300 K.[10] The study also predicted that when the doping concentration is

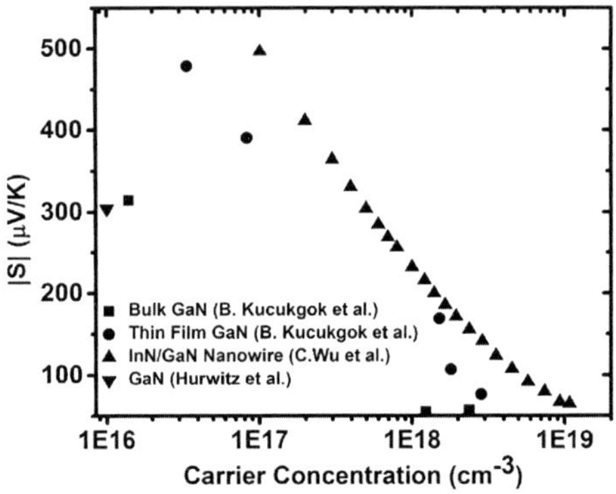

(a)

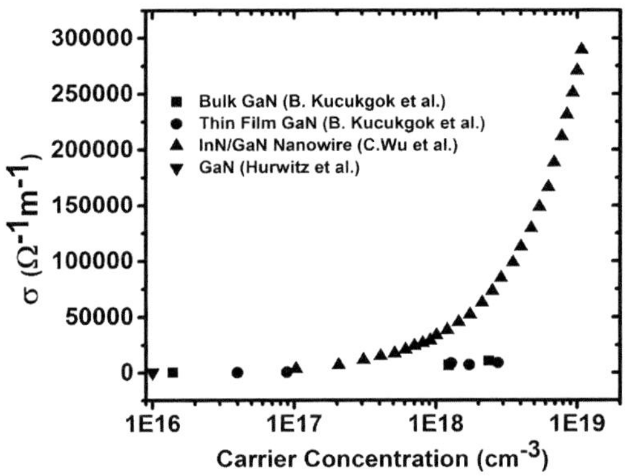

(b)

Fig. 1. (a) Seebeck coefficient (b) electrical conductivity vs carrier concentration.

$1 \times 10^{19}\,\mathrm{cm}^{-3}$, ZT can theoretically reach 1.713 at 1000 K, indicating it can potentially be a good thermoelectric material at high temperatures.

2.2. *InGaN*

Indium gallium nitride (InGaN) has the same crystal structure as GaN, but with slightly different lattice constant since indium atom is larger than gallium atom. The utility of this material arises from the ability to vary its bandgap. InN has a bandgap of only 0.7 eV, by replacing some Ga atoms with indium atoms, one can have a tunable ban gap between 0.7 eV and 3.4 eV. This change in band structures can significantly affect

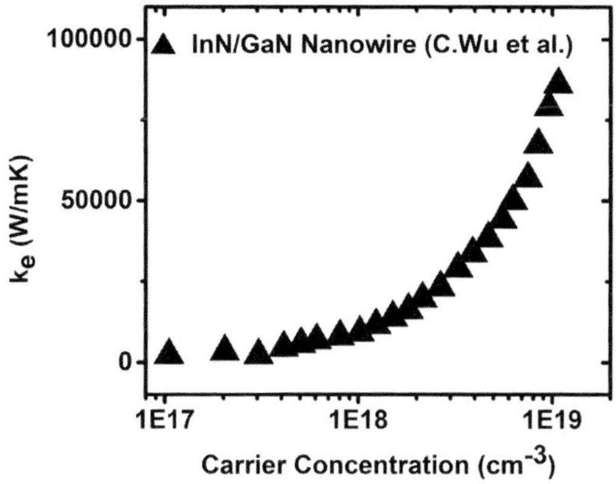

Fig. 2. Electronic thermal conductivity vs carrier concentration.

all the thermoelectric properties of a material. Figure 3 shows how the InN fraction changes the thermoelectric properties, all the data presented in black square are with $1.1 \times 10^{19}\,cm^{-3}$ carrier concentration, while that for the red circles varies. Carrier concentrations are $7.68 \times 10^{17}\,cm^{-2}$, $8.97 \times 10^{17}\,cm^{-2}$, $1.27 \times 10^{18}\,cm^{-2}$, $1.62 \times 10^{18}\,cm^{-2}$ and $3.4 \times 10^{18}\,cm^{-3}$ with respect to InN fraction in an increasing order. All the samples are tested at room temperature.[13,14] For Fig. 3, an obvious negative correlation can be seen between Seebeck coefficient and InN fraction. However, although the same trend can be observed according to Sztein's work, it only holds for data from Kucukgok *et al.* if the InN fraction is below 0.2. ZT, on the other hand, shows an inversed pattern. Also, one thing needs to be mentioned is that although the electrical conductivity from Kucukgok *et al.*'s is almost half of that from Sztein's, due to a significant improvement in Seebeck coefficient, the overall ZT is still much higher under the same InN fraction. The optimal In content is 0.17 and 0.2, resulting in $ZT = 0.04$ and 0.072 at room temperature for Sztein *et al.*'s and Kucukgok *et al.*'s respectively.

2.3. *AlGaN*

AlGaN has not been extensively studies as a high temperature thermoelectric material, with only a few groups referencing a thorough study of its thermoelectric properties. Most studies on $Al_xGa_{1-x}N$ investigate the effect

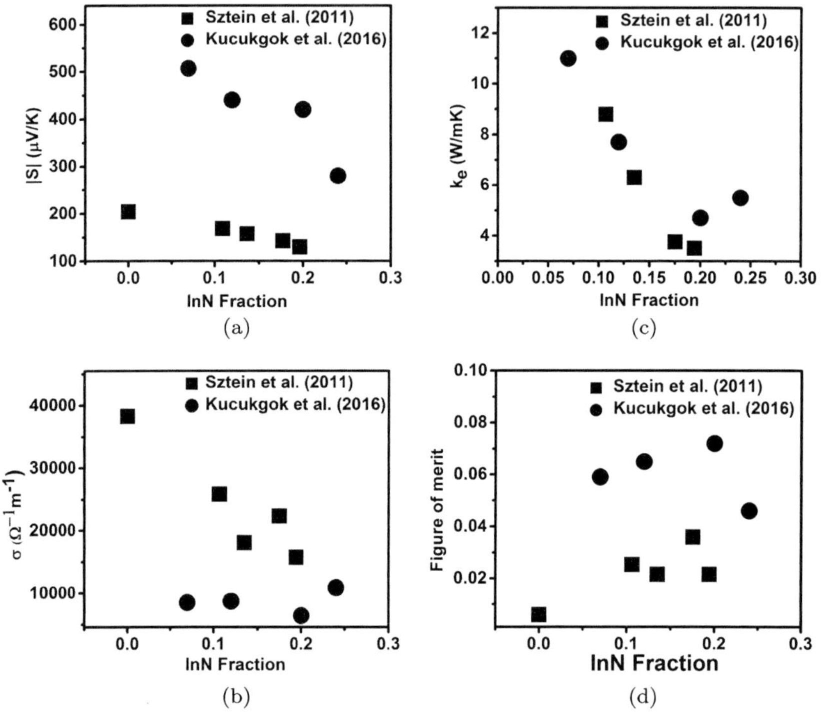

Fig. 3. (a) Seebeck coefficient (b) electrical conductivity (c) thermal conductivity (d) figure of merit vs. InN fraction.

of the mass fraction of Aluminum, x, or the carrier density, n, on the thermoelectric properties and ZT (figure of merit) value. From Fig. 4 shows that as the temperature increases, the thermal conductivity, κ, decreases. These data reported from different groups show a slight discrepancy, however, because Sztein *et al.*[15] and Liu *et al.*[16] used slightly different methods to calculate the thermal conductivity. Liu used a method known as Callaway's approximation,[17] and Sztein used Callaway's approximation along with a virtual crystal extension developed by B. Abeles.[18] This discrepancy is especially evident at lower temperatures, where we see the dramatic increase in thermal conductivity from the work that did not use the virtual crystal expansion.

While the low temperature discrepancy does exist, $Al_xGa_{1-x}N$ is a thermoelectric material primarily used in higher temperature environments.

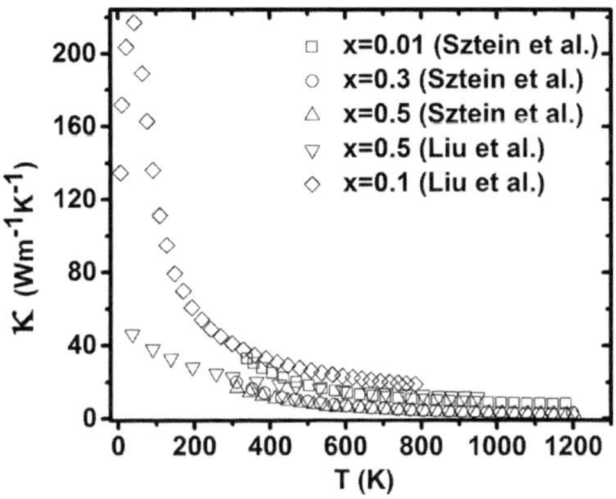

Fig. 4. The thermal conductivity of $Al_xGa_{1-x}N$ vs temperature at several different mass fractions of Al.

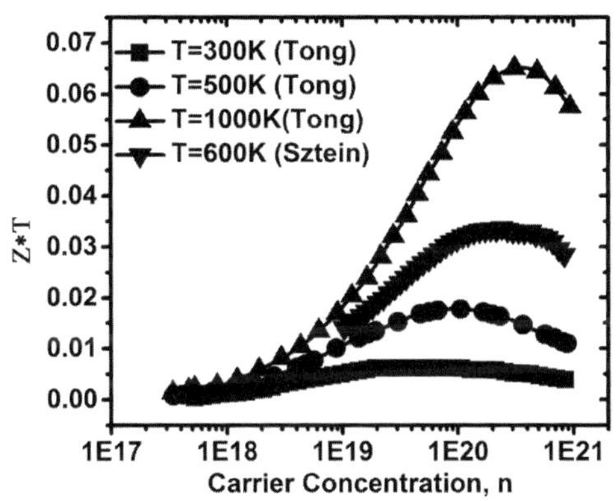

Fig. 5. $Al_xGa_{1-x}N$ ZT as a function on carrier concentration.

Both works in Fig. 4 show that the higher mass fractions of Al will result in lower thermal conductivities, and thermal conductivity will decrease as the temperature is increased due to increased phonon scattering at higher temperatures. Figure 5 shows the change of ZT of AlGaN with a function of the carrier concentration. It is evident that for different temperatures, there are different peaks for ZT value. For higher temperatures, a higher carrier concentration would yield a more effective thermoelectric material, and the lower temperatures have peaks at lower carrier concentrations. This shows optimum carrier concentration varies from $3 \times 10^{19}\,cm^{-3}$ at 300 K to

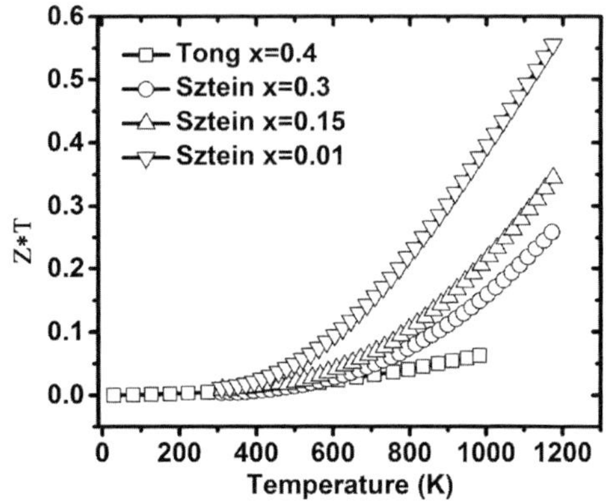

Fig. 6. $Al_xGa_{1-x}N$ ZT vs temperature from 0 to 1200 K.

3×10^{20} cm^{-3} at 1000 K. While it cannot be used as a direct comparison due to the different carrier densities, Fig. 6 compares the effect of different mass fractions of Al, x, to ZT, from temperatures ranging from 0 K to 1200 K. Interestingly, as the value of x is decreased, ZT will increase dramatically. This plot is most useful in the analysis of the thermoelectric properties of $Al_xGa_{1-x}N$ due to the ability to evaluate the thermoelectric effectiveness of the material at various temperatures. A recent study has explored a way to modify the ZT value for $Al_{0.2}Ga_{0.8}N$ by achieving high electrical conductivity and low thermal conductivity. This is done by polarizing the electrical field of the GaN/AlN/ $Al_{0.2}Ga_{0.8}N$ superlattices, leading to a much higher electron mobility. This increases the ZT value to 0.08 at room temperature.[20]

2.4. *AlN*

AlN has the widest band gap of all III-Nitrides (~6.1 eV) and also has high temperature stability, thermal conductivity and deep-UV transparency, making AlN a great candidate for high temperature/power applications or UV emitters or detectors down to 200 nm. While bulk AlN is available commercially it is still typically grown as an epitaxial on sapphire or SiC wafers.[21] The thermal conductivity of AlN has been thoroughly studied theoretically. Figure 7 shows just how high the thermal conductivity in AlN is at lower temperatures. It is this reason that AlN is ideal for higher temperature applications, where the thermal conductivity is lower,

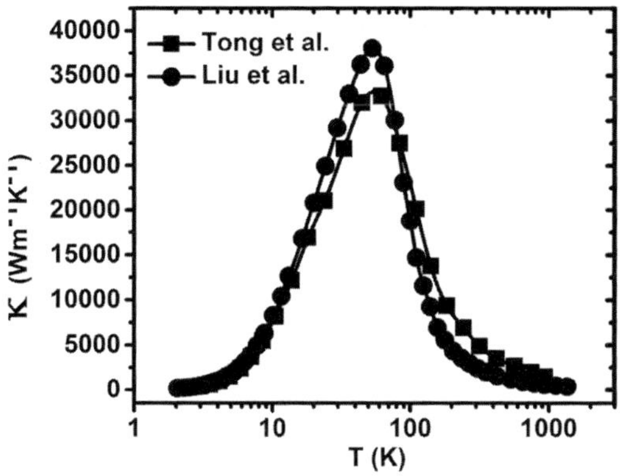

Fig. 7. Thermal Conductivity of AlN as a function of temperature.

and ZT will be higher. Due to the low thermal expansion coefficient, high electrical sensitivity, and low toxicity, AlN is ideal for many high power applications. Along with many other III-Nitrides, AlN can be grown hexagonally (wurtzite structure) or cubically (zinc-blende structure), each with their own benefits and drawbacks. AlShaikhi *et al.* compared the phonon properties between these two different structures and reported that the average mean free phonon path in the zinc-blende phase is roughly four times that in the wurtzite phase. This means that AlN will have a significantly higher thermal conductivity in the zinc-blende phase, which implies that the wurtzite phase structure is better suited in high temperature applications.[22]

2.5. *InN*

Unlike GaN which has a wide band gap of 3.4 eV, InN only has a ~0.7 eV band gap.[9] This means that it is more like a traditional narrow band gap material, with good electrical conductivity at undoped conditions. Similar to GaN, InN has a wurtzite crystal structure with lattice constant 3.574 Å and 5.704 Å for a and c.[23] Figure 8 to Fig. 10 show the thermoelectric properties for InN with respect to temperature or carrier concentration.

From Fig. 8, one can see that as carrier concentration goes up, the absolute value of the Seebeck coefficient drops up to the feasible range of carrier concentration. Also data shows at higher temperatures InN has a lower Seebeck coefficient than the one at lower temperatures. This phenomenon is caused by so-called "bipolar effect" due to the minority

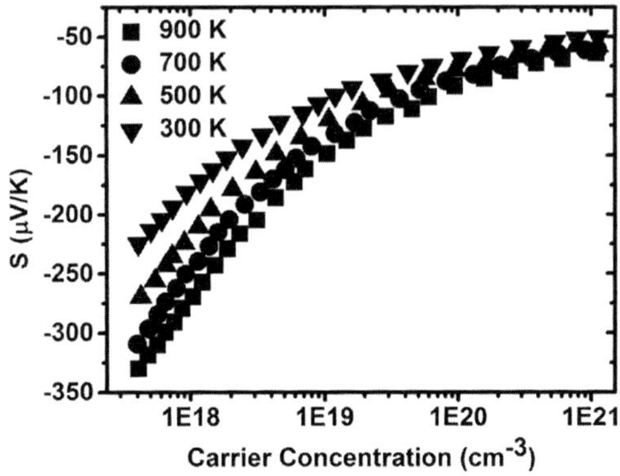

Fig. 8. Seebeck Coefficient vs Carrier Concentration at different temperature.

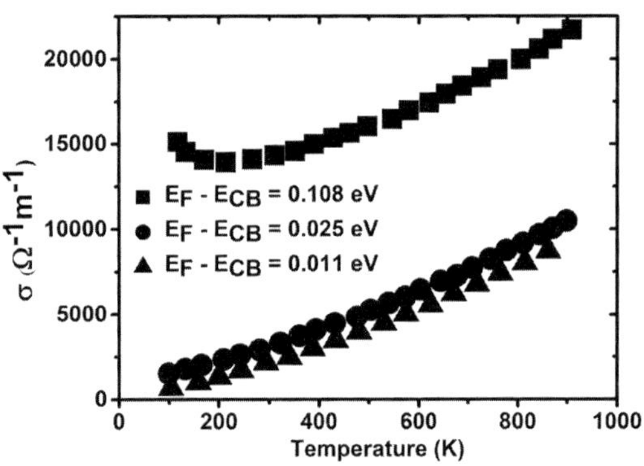

Fig. 9. Electrical conductivity vs temperature for different Fermi levels.

carriers hopping to the conduction band at high temperature, which cancels out the n- or p- characteristic of materials. Figure 9 shows electrical conductivity at different temperatures for different Fermi level values due to different free carrier concentrations. The three values are 0.108 eV, 0.025 eV and 0.011 eV corresponding to $6 \times 10^{18}\,cm^{-3}$, $9 \times 10^{17}\,cm^{-3}$ and $4 \times 10^{17}\,cm^{-3}$.

The larger the difference between Fermi level and the lowest conduction band energy in a degenerate semiconductor, the higher the electrical conductivity. The overall trend for electrical conductivity with respect to temperature is similar to that of the Seebeck coefficient, excluding the initial

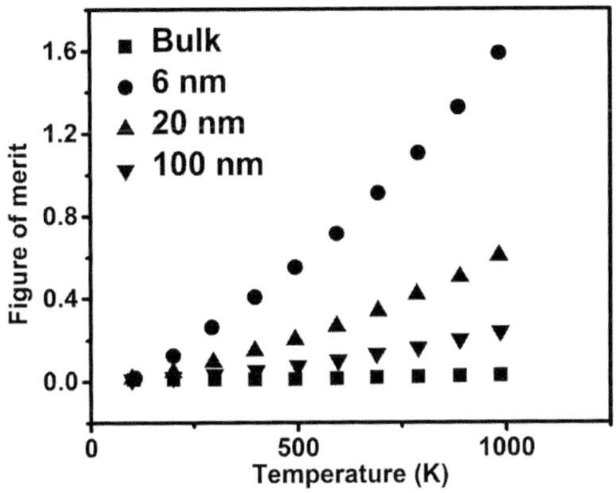

Fig. 10. Figure of Merit ZT vs Temperature.

drop in conductivity corresponding to the freeze out level. In Fig. 10, the thermoelectric figure of merit ZT is summarized for bulk InN and some InN nanowires with different sizes. As shown in the figure, the smaller nanowires have higher ZT values. The highest ZT is about 1.6 for a 6 nm InN nanowire at 1000 K, which is actually 100 times larger than that at room temperature.[25]

Also, there is still no indication that ZT will decrease for >1000 K, meaning that ZT could be even higher than 1.6 when the temperature is above 1000 K, which makes the InN nanowire an excellent thermoelectric material for high temperature usage.

To summarize, the thermoelectric properties of III-Nitride materials are discussed above, including GaN, InGaN, AlGaN, AlN and InN. Although these materials do not have thermoelectric properties, they all show a huge increase in thermoelectric figure of merit ZT at high temperature, which means that III-Nitride materials may potentially become the best for high temperature applications.

2.6. *Transition metal nitrides thermoelectric materials*

Transition-metal nitrides are commonly used for wear-resistant coatings and electronic contact materials because of their thermal and mechanical stability, high electrical conductivity, and chemical inertness. In recent years, transition-metal nitrides are gaining attention as good materials for thermoelectric applications because of their wide range of electrical properties, varying from metallic to semiconducting depending on the types of transition metal elements and their stoichiometry. This opens up a great opportunity to tailor materials' electronic properties thus enhance the efficiency of thermoelectric materials and devices.

Scandium nitride (ScN) possesses excellent electrical and mechanical properties suitable for mid-temperature thermoelectric application (400–700°C).[26–28] Apart from its high hardness and chemical inertness, it also has very high temperature stability with a melting point of 2550°C. ScN is stable up to 550°C in air and can be oxidized to Sc_2O_3 above 600°C.[28–30] It has a rocksalt NaCl (B1)-type crystal structure with a lattice parameter of 4.50–4.52 Å.[26,31] For electrical properties, theoretical studies reported that ScN is an indirect semiconductor with energy band gap from 0.9 to 1.6 eV.[26,32–35] Measurements on as-deposited ScN show n-type behavior,[36,37] and the carrier concentration of ScN has been reported to vary from 10^{18} to 10^{22} cm^3 with electron mobility of 100–180 $cm^2\,V^{-1}\,s^{-1}$[35,38–40] This carrier concentration range corresponds to the typical ideal range for thermoelectrics,[41] while also retaining a high carrier mobility.[39] Furthermore, as shown in Figs. 11(a) and (b) Kerdsongpanya *et al.*[42] measured the electrical resistivity, Seebeck coefficient, and a power factor values as of 2.94 $\mu\Omega$m, $-86\,\mu$V/K at 800 K, and 2.5×10^{-3} W/mK^2, respectively using ScN thin films grown by reactive magneton sputtering. By using a literature value for the thermal conductivity of ScN,[27] the ZT value can be estimated to

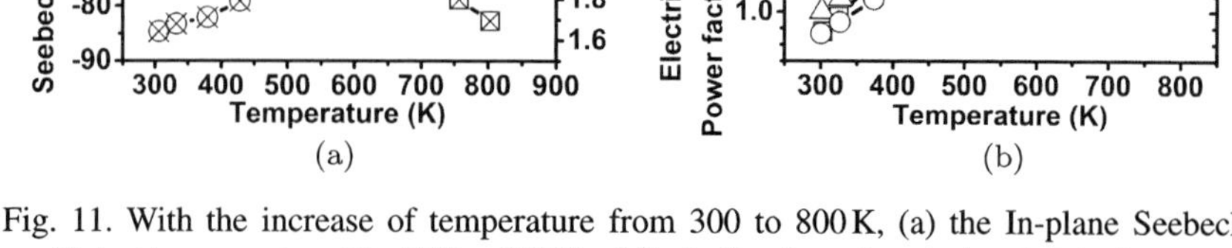

(a) (b)

Fig. 11. With the increase of temperature from 300 to 800 K, (a) the In-plane Seebeck coefficient increases to $-86\,\mu$V/K at 800 K, while the In-plane electrical resistivity increases to 2.94 $\mu\Omega$m at 800 K, and (b) power factor increases linearly up to 2.5×10^{-3} W/mK^2 at 800 K.

be 0.2 at 800 K, which is five times larger than other transition-metals (like CrN).[43] Recently, Burmistrova *et al.* has reported the thermoelectricity performance of ScN films grown on MgO (001) including power factor of 3.3×10^{-3} W/mK^2 and relatively large Seebeck coefficient of $-156\,\mu$V/K at 840 K.[44] Additionally, they investigated the temperature-dependent power factor, Seebeck coefficient, and electrical conductivity from 300 to 840 K, as shown in Figs. 12(a) and (b) and Fig. 13.

The measured power factor increases with temperature, reaching a maximum of 3.5×10^{-3} W/mK^2 at 600 K before saturating, as can be seen from Fig. 13. These power factors are comparable to state of the art thermoelectric material such as pure Bi_2Te_3 or PbTe.[41,45,46] Furthermore, based on the experimental results in Ref. 47, combining the in-plane power factor and cross-plane thermal conductivity, a ZT value of about 0.3 was obtained at 800 K, slightly larger than the ZT value of 0.2 reported by Kerdsongpanya *et al.*[42]

Moreover, influence of deposition pressure on thermoelectric properties of the ScN thin films has been studied. For instance, as shown in Fig. 14, Burmistrova *et al.* measured the in-plane Seebeck coefficient and power factor values of ScN (001) thin films grown by dc reactive magnetron sputtering at different pressures and found a maximum power factor 3.3×10^{-3} W/mK^2 at 2 mTorr. Additionally, the trend; that is, power factor decreases while pressure increases was observed.

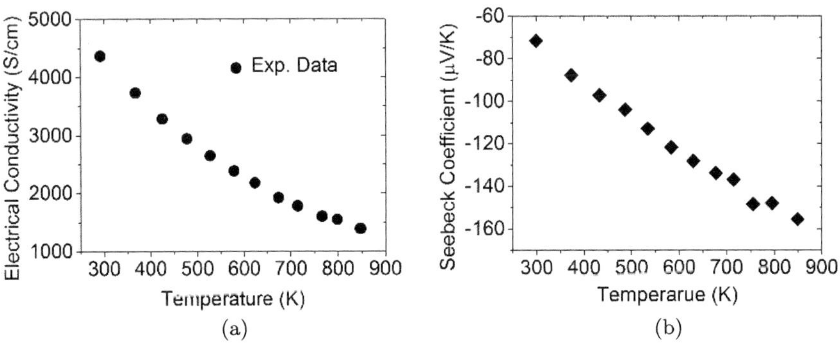

Fig. 12. The temperature-dependent, (a) In-plane electrical conductivity decreased non-linearly to the minimum of 1353 S/cm at 840 K, and (b) in-plane Seebeck coefficient increased to $-156\,\mu$V/K at 840 K, in the temperature range of 300–840 K.

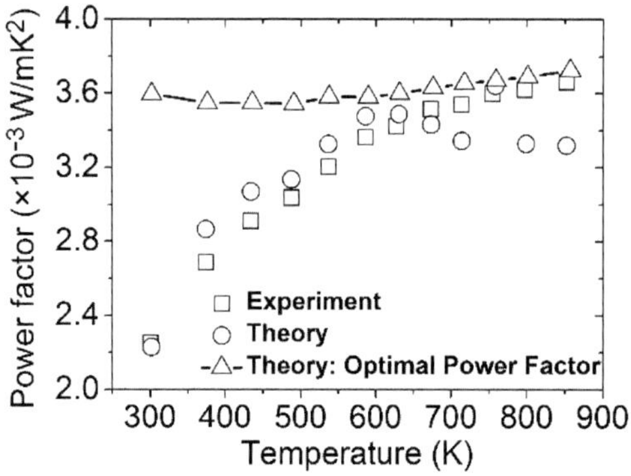

Fig. 13. Power factor of a ScN film deposited on MgO increased with increasing temperature from 300 to 840 K, reaching a maximum value of 3.5×10^{-3} W/mK^2 at 600 K and $\sim 3.3 \times 10^{-3}$ W/mK^2 at 800 K.

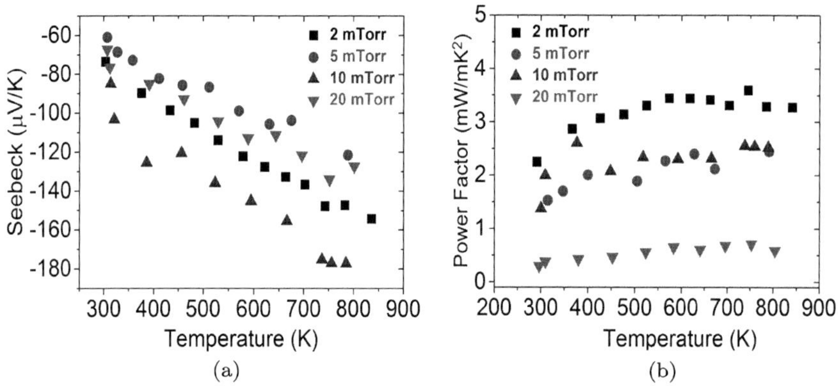

Fig. 14. For ScN thin films deposited at 2, 5, 10, and 20 mTorr, respectively, (a) Seebeck coefficients decreased to $-120\sim -180\,\mu$V/K range, and (b) power factor increased slowly to the maximum value of 3.3×10^{-3} W/mK^2 at 800 K.

Multilayers and superlattices[48–50] grown by alternate deposition of metal and semiconductor materials with nanoscale periods have been both experimentally and theoretically shown as potential promising candidates to obtain ZT values higher than 2. In a previous study,[51] nitride metal/semiconductor multilayers and superlattices were identified as candidate materials for solid-state thermionic generators that operate at temperatures up to 1000°C, along this line, (Zr,W)N/ScN superlattices have been selected as the desirable metal/semiconductor combination because of their crystallographic compatibility and thermal stability.[27] Vijay Rawat *et al.* successfully prepared a set of ZrN (3 nm)/ScN (3 nm) and Zr,WN/ScN multilayers and superlatticcs grown on a MgO (100) substrate by using reactive dc magnetron sputtering to study the room temperature cross-plane thermal conductivities,[27] as shown in Fig. 15. Compared to the thermal conductivities for films ZrN (47 W/mK) and ScN (10.6 W/mK), a distinct low thermal conductivity, 5.25 W/mK, at a period of 6 nm is observed for ZrN/ScN multilayers. Alloying the metal layers with tungsten (Zr,WN/ScN multilayers) also leads to the thermal conductivity values as low as 1.8 W/mK due to the added effects of interface scattering, impurity scattering and/or acoustic impedance mismatch. Apart from the lower thermoelectricity, the simultaneously high Seebeck coefficients and electrical conductivity of superlattices are necessary to obtain a high ZT value. Zebarjadi *et al.* experimentally reported a Seebeck coefficient of 820 μV/K for ZrN (4 nm)/ScN (6 nm) metal/semiconductor superlattices at room temperature.[52] and predicted their figure of merit can increase to 1.5 at high temperatures. They argued the figure of merit for ZrN/ScN superlattices can reach 3 at higher temperatures assuming a nonconserved lateral momentum at the junctions. Bivas Saha *et al.* also calculated the temperature-dependent electrical conductivity and the Seebeck coefficients

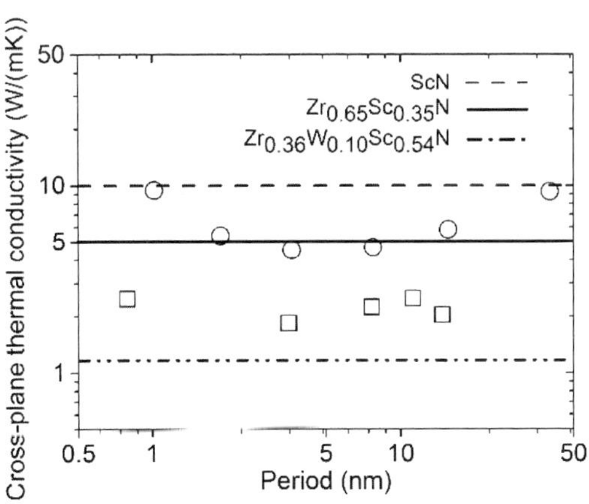

Fig. 15. Dark solid circles and squares indicate the period-dependent cross-plane thermal conductivity of 300-nm-thick ZrN/ScN and $Zr_{0.64}W_{0.36}N$/ScN multilayers superlattices, respectively.

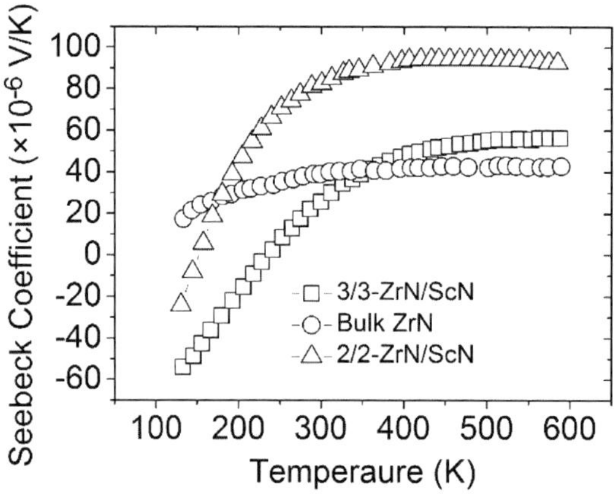

Fig. 16. U The temperature-dependent Seebeck coefficients of bulk ZrN, 2/2 and 3/3 ZrN/ScN superlattices.

of superlattices and the bulk materials (ZrN and ScN).[53] by using the first principles, as shown in Fig. 16. Unlike other conventional thermoelectric materials, the electrical conductivity of the superlattices ZrN/ScN increases linearly with temperature from 120 K to 600 K. The Seebeck coefficient of the superlattices increases with temperature and saturates after 400 K. Base on the above reports, it is proved that a metal/semiconductor multilayer superlattice structure can have an enhanced ZT and can be designed and grown for high-temperature applications. In conclusion, it is found that ScN films grown on MgO (001) substrate tend to possess better thermoelectric properties with a larger Seebeck coefficient of $-156\,\mu V/K$ and power factor up to $3.3 \times 10^{-3}\,W/mK^2$ at 800 K, comparing with ScN films grown on Al_2O_3. It is also observed that low deposition pressures (<10 mTorr) result in enhanced thermoelectric properties for ScN thin films. Furthermore, a metal/semiconductor nanoscale multilayer superlattice structure (e.g., ZrN/ScN) can have a large Seebeck coefficient and lower thermal conductivity at high temperatures, presenting a promising structure worthy of exploration for high-temperature thermoelectric applications.

2.7. *Layered complex metal nitrides*

Layered complex nitrides, AMN_2 (A = Sr, Ba; M = Ti, Zr, Hf),[54–56] with $KCoO_2$-type and α-$NaFeO_2$-type crystal structures, have been considered as novel thermoelectric materials due to superior electrical and physical properties, such as superconductivity,[57,58] ionic conduction,[59] and thermoelectric transport.[60] The layered complex metal AMN_2 compounds mainly consist of two types, $KCoO_2$-type (two-dimensional layered complex nitrides) and α-$NaFeO_2$-type (three-dimensional layered complex nitrides), due to different electronic and crystal structures.[61] In this section, thermoelectric properties of layered complex nitrides will be reviewed and discussed.

2.7.1. *Two-dimensional layered complex metal nitrides with $KCoO_2$-type structure*

Ohkubo *et al.* have studied two-dimensional layered complex nitrides, AMN_2 (A = Sr, Ba; M = Ti) using electronic structure and transport calculations based on density-functional theory (DFT) and Boltzmann theory.[62] Their studies have revealed that two dimensional layered $SrTiN_2$ have anisotropic properties in that their thermoelectric parameters can vary along zz, xx, and yy-directions.[63] Until now, there is no experimental report on ZT for $SrTiN_2$. However, the value of ZT in $SrTiN_2$ is calculated as 0.6 in Ref. 62, far larger than its true value due to the lack of consideration for the lattice thermal conductivity in the calculation process. Compared to $SrTiO_3$ with good thermoelectricity, Isao Ohkubo *et al.* theoretically investigated high-temperature thermoelectric performance and anisotropy for the typical two-dimensional $SrTiN_2$,[62] in the temperature range from 0 to 1300 K, as shown in Fig. 17. With increasing temperature, it is observed that the anisotropic transport properties for $SrTiN_2$ still exist and thermoelectric parameters in the zz-direction are found to be lower than in the xx-direction. Additionally, the power factor of $SrTiN_2$ is lower in the xx-direction compared to $SrTiO_3$, while its figure of merit is higher than $SrTiO_3$ above 200 K in the xx-direction. Apart from the temperature factor, the carrier concentration is another important parameter affecting the thermoelectricity of materials. Thermoelectric transport coefficients for $SrTiO_3$ and AMN_2 as a function of carrier concentration at 300 K and 1000 K were studied by Isao Ohkubo and Takao Mori[63] using density-functional theory and Boltzmann theory, as shown in Fig. 18(a) and (b). It is clearly observed that the electrical conductivities and the electronic part of the thermal conductivities increase with increasing carrier concentrations at

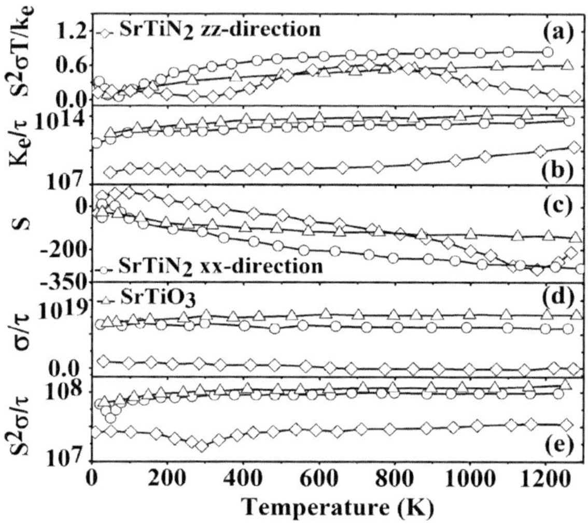

Fig. 17. Calculated temperature-dependent transport properties of $SrTiO_3$ and $SrTiN_2$ with carrier densities corresponding to 6.7×10^{20} and 5.2×10^{19} cm^{-3}, respectively. (a) Figure of merit, (b) electronic part of the thermal conductivities, (c) Seebeck coefficients, (d) electrical conductivities, and (e) power factors. xx and zz labels indicate the in-plane and out-of-plane transport properties along the c-axis, respectively.

the two different temperatures. It can be seen from Fig. 18(a) that figures of merit decrease significantly as the carrier concentration is higher than 10^{19} cm^{-3}; however, at the same time the maximum power factors were achieved owing to the inverse relationship between the thermoelectric parameters.

Figure 18 (b) exhibits that with carrier concentration below 10^{18} cm^{-3}, Seebeck coefficient of $SrTiN_2$ at 1000 K is significantly smaller than those at 300 K, which can be attributed to the effect of thermal expansion on electronic and thermoelectric transport properties.[63] Furthermore, it can be attributed from Figs. 18 (a) and (b) that Seebeck coefficient of two-dimensional layered AMN_2 compounds are much smaller while temperature becomes higher.

Moreover, Rabih Orabi *et al.* investigated the band structures and electronic transport coefficients of two-dimensional layered nitrides AMN_2 to evaluate their potential for thermoelectric applications,[64] Using first principles theoretical methods, they argued $KCoO_2$-type AMN_2 compounds

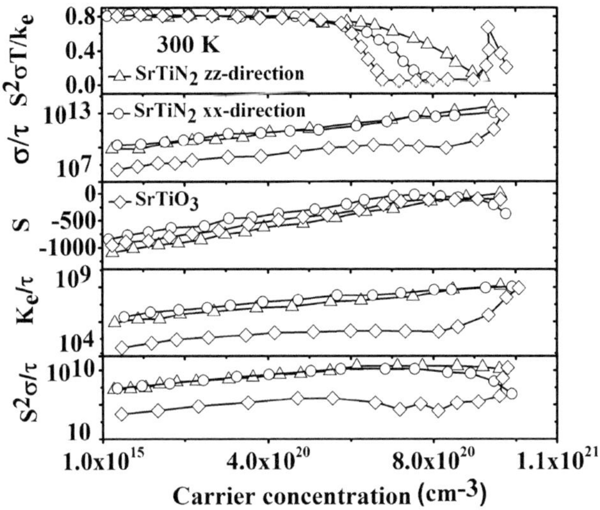

(a)

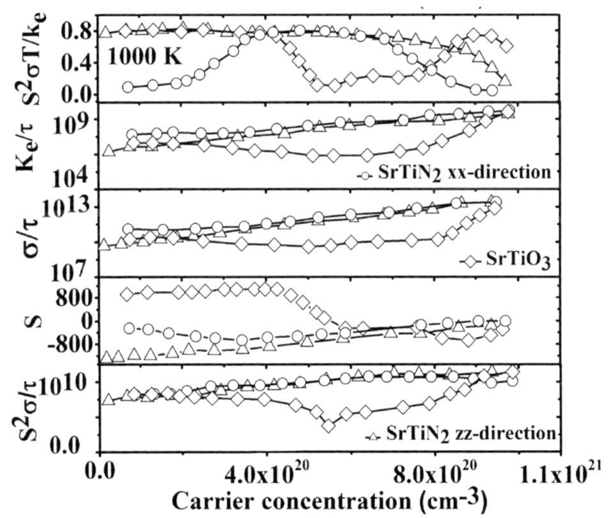

(b)

Fig. 18. (a) Transport coefficients {Seebeck coefficients, electronic conductivities, electronic part of the thermal conductivities, power factors, and electronic figures of merit}as a function of the carrier concentration (n) for $SrTiO_3$ and $KCoO_2$-type $SrTiN_2$ at 300 K and (b) 1000 K, respectively. xx- and zz directions indicate the in-plane and out-of-plane transport properties along the c-axis, respectively.

(A = Sr, Ba; M = Ti) can have high Seebeck coefficients in excess of 150 μV/K and good electrical conductivities.

2.7.2. *Three-dimensional layered complex metal nitrides with $NaFeO_2$-type structure*

Three-dimensional layered complex metal nitrides AMN_2 (A = Sr or Na; M = Zr, Hf, Nb, Ta) with α-$NaFeO_2$ crystal structure are particularly interesting because they are used for energy conversion and storage applications, For example, Na_xCoO_2 exhibits excellent thermoelectric properties[65] and is a base material for superconductors,[66] and $LiCoO_2$ is used as a cathode material in lithium ion batteries.[59] α-$NaFeO_2$ materials exhibit isotropic transport,[65,67,68] which results in a large thermoelectric effect. In terms of the thermoelectricity for three-dimensional AMN_2 compounds, Isao Ohkubo and Takao Mori made a great contribution for better understanding of thermoelectric transport of the layered complex nitrides.[63,69,70] The doping level (carrier concentration) is regard as a critical factor affecting thermoelectricity of materials. Figure 19 shows how the carrier density

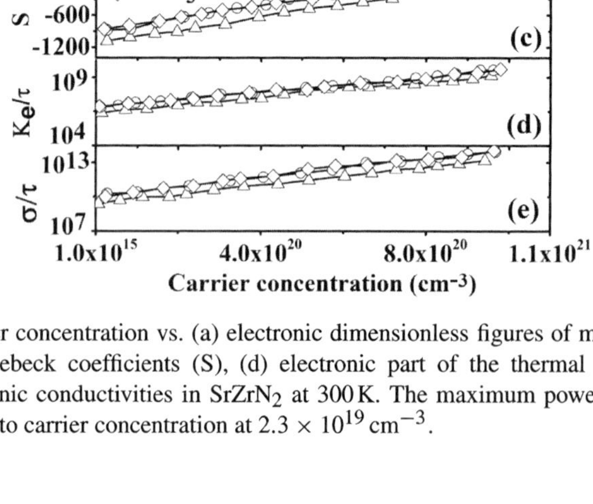

Fig. 19. Carrier concentration vs. (a) electronic dimensionless figures of merit, (b) power factors, (c) Seebeck coefficients (S), (d) electronic part of the thermal conductivities, and (e) electronic conductivities in $SrZrN_2$ at 300 K. The maximum power factor values corresponding to carrier concentration at 2.3×10^{19} cm^{-3}.

affects the electronic conductivity, electronic part of the thermal conductivity, the Seebeck coefficients, and the power factors in addition to the electronic figure of merit, in $SrZrN_2$ at 300 K. The Seebeck coefficients and power factors show a logarithmic dependence on carrier concentrations, which follows Pisarenko behavior. The power factor in $SrZrN_2$ reaches its maximum value at a carrier density of $2.3 \times 10^{19}\,cm^{-3}$, and in $SrHfN_2$ at $8.2 \times 10^{18}\,cm^{-3}$. It can be clearly observed that electrical conductivity, Seebeck, power factor, thermal conductivity in $SrZrN_2$ is not anisotropic because the xx direction transport properties appear to be similar to those for the zz direction with increasing carrier concentration. For the AMN_2 compounds with three-dimensional electronic structure, temperature is still an important factor affecting the values of electrical conductivity, Seebeck, power factor, thermal conductivity and ZT. Figure 20 shows the temperature dependences of several calculated transport coefficients of $SrZrN_2$ and $SrHfN_2$, respectively. The calculated transport coefficients of $SrTiO_3$ are for comparison, because it is a three-dimensional perovskite with good thermoelectric properties. With increasing temperature, the estimated transport

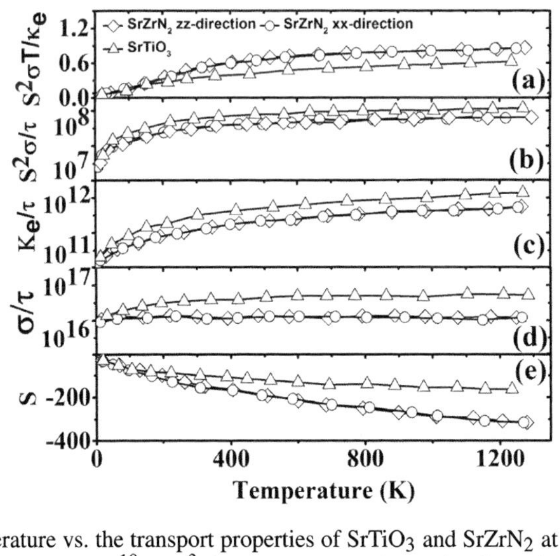

Fig. 20. Temperature vs. the transport properties of $SrTiO_3$ and $SrZrN_2$ at electron carrier concentrations of $2.3 \times 10^{19}\,cm^{-3}$ for $SrZrN_2$. (a) electronic dimensionless figures of merit, (b) power factors, (c) electronic part of the thermal conductivities, (d) electronic conductivities, and (e) Seebeck coefficients.

properties of $SrZrN_2$ and $SrHfN_2$ are not anisotropic,[69] as the in-plane xx and yy direction transports are very similar to that in the zz direction.

Seebeck coefficients in $SrZrN_2$ and $SrHfN_2$ reach $-140\,\mu V/K$ ($SrZrN_2$) and $-145\,\mu V/K$ ($SrHfN_2$) at 300 K, and $-300\,\mu V/K$ ($SrZrN_2$) and $-370\,\mu V/K$ ($SrHfN_2$) at 1200 K, and are larger than those of $SrTiO_3$ ($-100\,\mu V/K$ at 300 K, $-160\,\mu V/K$ at 1200 K) as shown in Fig. 20. The calculated electronic dimensionless figures of merit for $SrTiO_3$ and $SrZrN_2$ are 0.4 and 0.47 at 300 K and 0.6 and 0.83 at 1200 K, respectively. Until now, no experimentally determined and calculated thermal conductivities have been reported for $SrZrN_2$ and $SrHfN_2$. The experimentally measured thermal conductivity values at room temperature of insulating compounds containing Sr and Zr or Hf ions are smaller than the thermal conductivities for $SrTiO_3$,[71,72] so lower thermal conductivities and larger dimensionless figures of merit can be expected in $SrZrN_2$ and $SrHfN_2$.

In summary, the temperature and the carrier concentration are two critical factors affecting the thermoelectric performance of the layered complex metal nitrides AMN_2 (A = Sr, Ba; M = Ti, Zr, Hf) compounds including $KCoO_2$-type and α-$NaFeO_2$ crystal structures. Predicted by density functional theory and Boltzmann transport theory, the calculated electronic dimensionless figures of merit for three-dimensional $SrZrN_2$ with α-$NaFeO_2$ crystal structure are 0.47 at 300 K and 0.83 at 1200 K, respectively. For two-dimensional $SrTiN_2$, the figures of merit reach 0.6 at 300 K and 0.86 at about 1200 K. Comparing with the figures of merit (0.4 at 300 K, 0.6 at 1200 K) for $SrTiO_3$, a good thermoelectric material, layered complex metal AMN_2 (A = Sr, Ba; M = Ti) compounds owning $KCoO_2$-type and α-$NaFeO_2$-type crystal structures show excellent thermoelectric properties with larger Seebeck coefficients and electronic figures of merit. In conclusion, the layered complex nitrides AMN_2 (A = Sr or Na; M = Zr, Hf, Nb, Ta) may be a group of very promising candidates for thermoelectric applications in the near future.

3. III-Nitride Thermoelectric Devices

As discussed previously, III-Nitride semiconductors have widely used commercially in optical and electronic devices, such as light-emitting diodes (LEDs), ultraviolet (UV) detectors. They are expected to be promising candidates of thermoelectric devices for electricity generation from waste heat harvesting at high temperature regime, due to their excellent thermoelectric properties, low cost, wide bandgap and high thermal stability. This section introduces the fabrication and properties of current III-Nitride thermoelectric power devices, which are mainly based on InN, AlInN, and GaN materials. The design and fabrication of thermoelectric devices trend to be downsized for on-chip integration for power generation. The thermopower properties of the devices are evaluated by the load characteristic (output voltage and output power). We expect to enhance the values of output voltage and power in order to achieve high efficiency of power generation with further improvement in materials quality.

3.1. *InN and AlInN based thermoelectric device*

InN and AlInN are the typical materials used in the thermoelectric devices at high temperature due to their high values of ZT, since the large size of indium atoms can increase phonon scattering and accordingly reduce the thermal conductivity. Thermoelectric devices require scalability for on-chip integration for power generation. III-Nitride films are thin enough to be embedded within integrated devices, with InN and AlInN being easily fabricated down to centimeter-sized scales by designing multi-element structures. Figure 21 (a) gives the typical device structure based on InN or $Al_{0.25}In_{0.75}N$ films prepared by radio-frequency (RF) sputtering, which are composed of 20-pair elements.[73] Chromel elements were sputtered onto the substrate and attached to the films, since its Seebeck coefficient is positive and P-type nitride films are difficult to obtain. Three device samples, $Al_{0.25}In_{0.75}N$-chromel on SiO_2 glass substrate (sample A), InN-chromel on SiO_2 glass (sample B), and InN-chromel on polyimide (sample C) were fabricated to investigated the thermoelectric properties. The area size of the 20-pair devices is 20 mm × 20 mm, and the dimensions of one element are 20 mm (length) × 0.3 mm (width) × 1 μm (thickness). Polyimide is used as the substrate in sample C making the device flexible and appropriate for on-chip integration.

For the purpose of miniaturization and integration of thermoelectric device, a novel roll-type device architecture using InN thin films deposited on polyimide substrate at 100°C using N_2 and Ar gases was demonstrated,[74] as shown in Fig. 21(b). Two device samples, 60-pair (sample D) and 120-pair

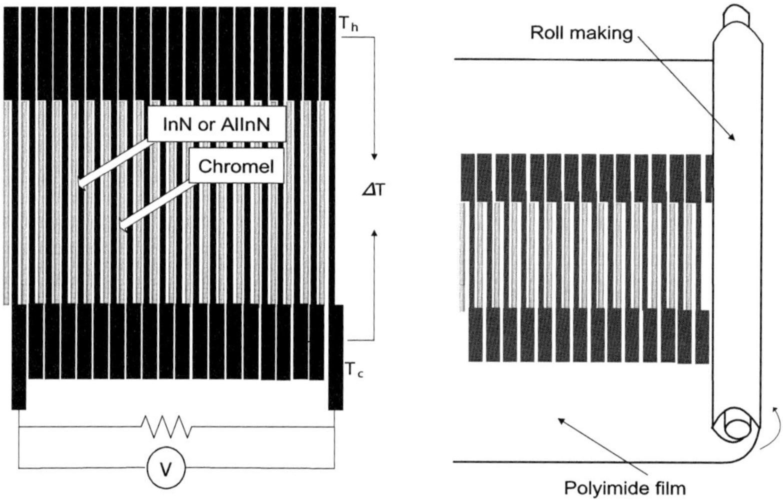

Fig. 21. Fabrication of InN and AlInN based thermoelectric devices. (a) Schematic structure of the typical thermoelectric device. (b) Schematic process of making roll for InN-chromel roll-type device.

InN-chromel (sample E) films, were also prepared by RF sputtering. The area size of the 120-pair devices is 30 mm × 170 mm. However, after being rolled, the length and the diameter of the device are reduced to be 30 mm, 4 mm respectively. The dimensions of one element are 10 mm (length) × 0.3 mm (width) × 1 μm (thickness).

The output voltage (V_{op}) and the output power (P) measurement can be carried out by applying the temperature differences ($\Delta T = T_h - T_c$) between the edges of the devices. T_h was the higher temperature heated at one side of the devices, while T_c was the lower temperature (about 30°C) at another side of the device. Figure 22 shows as an example, the current dependence of output voltage and output power as a parameter of temperature difference ΔT, of sample B. With increasing ΔT, both output voltage and output power increase. It can be attributed to the fact that the Seebeck coefficient increases with increasing ΔT. The output voltage can be expressed by using Eq. (1),[75] and the maximum output power P_{max} can be estimated by using Eq. (2)[76]:

$$V_{\text{op}} = \Delta T S N \tag{1}$$

$$P_{\max} = V_{\text{op}}^2 / 4R_{\text{in}} \tag{2}$$

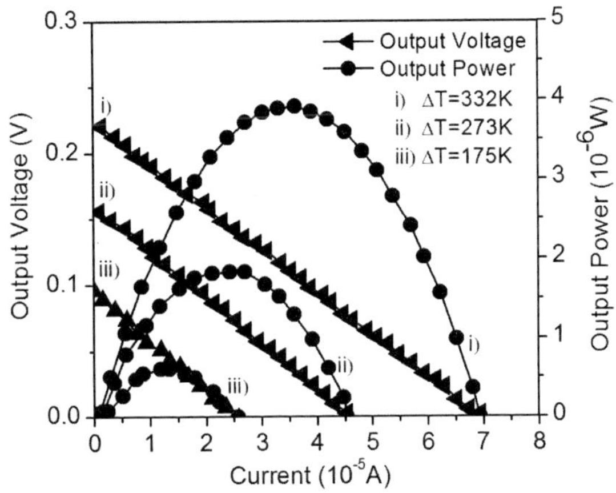

Fig. 22. Current dependence of output voltage V_{op}, and output power P at various ΔT, for InN-chromel of 20-pairs devices on SiO_2 glass (sample B).

Where S is the Seebeck coefficient, N is the number of the elements designed in the device structure, R_{in} is the internal resistances derived from the slopes of the I–V curves, which can also be obtained by Eq. (3):

$$R_{\text{in}} = N_{\rho} \frac{l}{wd} \tag{3}$$

Here d is the thickness of the films, l and w are the line length and line width of the structure element, ρ is the electrical resistivity. Combined with Eq. (1) and Eq. (2), P_{max} can also be expressed by

$$P_{\text{max}} = \frac{N \Delta T^2 S^2}{4\rho(l/dw)} \tag{4}$$

As we can see from the above equations, the output voltage and the maximum power can be enhanced by increasing the element number N, the size of w and d and decreasing the line length l in the designing processes.

Table 1 gives the summarized load characteristic of samples A-E. The maximum output power and the open output voltage of every sample as a parameter of ΔT are illustrated. The table also gives the load characteristic of a 20-pair InON-chromel on SiO_2 glass[77] namely sample F, which is the III-Oxynitride based thermoelectric device. With the same design size, same substrate and same element number, the internal resistance R_{in} of

Table 1. Summary of characteristic values of sample A-F.[73,74,77]

Sample name	Fabrication details	ΔT(K)	V_{op}(V)	P_{max} (10^{-6} W)	R_{in}(kΩ)
A	20-pair $Al_{0.25}In_{0.75}N$-chromel on SiO_2 glass	345	0.21	1.6	6.5
B	20-pair InN-chromel on SiO_2 glass	332	0.22	3.9	3.2
C	20-pair InN-chromel on polyimide	280	0.089	0.33	6.1
		163	0.048	0.087	
D	Roll-type 60-pair InN-chromel on polyimide	160	0.118	0.023	258
E	Roll-type 120-pair InN-chromel on polyimide	168	0.21	0.065	171
F	20-pair InON-chromel on SiO_2 glass	214	0.049	0.065	8.3

sample B is smaller than that of sample A and sample F, indicating that InN may be more advantageous than $Al_{0.25}In_{0.75}N$ and InON for use as a power generation device. The maximum output power and open output voltage of sample C are smaller than those of sample B, which is attributed to the different thermal expansion between SiO_2 glass and the polyimide film. Comparing the samples C, D and E at ΔT around 163 K, sample E has the largest value of open output voltage due to the largest number of elements (N = 120). Although the maximum output power of Roll-type sample D and E are lower than sample C, the roll-type structures will have more potential for the downsized and higher integration thermoelectric devices.

3.2. *GaN based thermoelectric devices*

GaN is one of the most widely used III-nitride thermoelectric materials, having high prospects for the applications in thermoelectric power devices. Three kinds of GaN based thermoelectric device structures with centimeter-size and sub-millimeter size dimensions are introduced in this sub-section. A GaN based thermoelectric power device was first designed and fabricated to be a 4-pair single-crystalline free-standing GaN (fr-GaN) (sample 1).[76] Free-standing GaN devices with bulk-sizes are also important for power generation applications due to high electrical conductivity with large thickness d. A 3-pair epitaxial GaN layers (epi-GaN) device structure (sample 2) was developed and fabricated for the purpose of on-chip integration.[78] The fr-GaN and the epi-GaN were grown by hydride vapor phase epitaxy (HVPE) method, and were doped by Si to achieve n-type dopants. Chromel was sputted onto thc SiO_2 glass substrate and soldered onto fr-GaN in sample 1 or epi-GaN in sample 2. The element dimensions of epi-GaN are 5 mm (length) × 2 mm (width) × 35 μm (thickness), while the element dimensions of fr-GaN are 5 mm (length) × 2 mm (width) × 390 μm (thickness). The structure of sample 1 is presented in Fig. 23.

Figure 24 shows the measured dependence of output voltage V_{op} and output power P on current with $T_c \sim 30°C$ at various ΔT for sample 1 and 2. Table 2 gives the summarized load characteristics of samples 1 and 2, including the maximum output power P_{max} and open output voltage

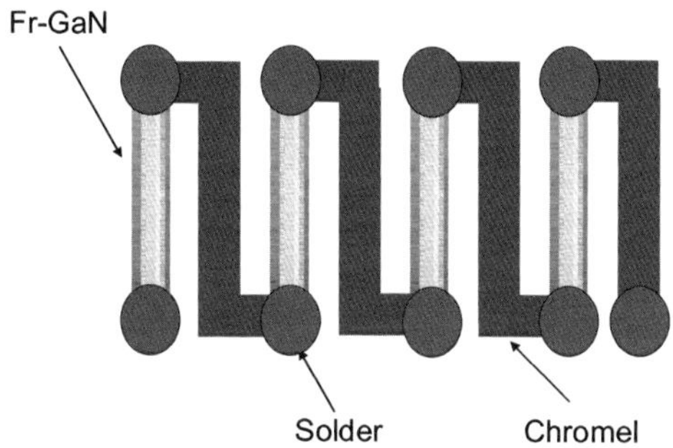

Fig. 23. The Schematic structure of a 4-pair fr-GaN device.

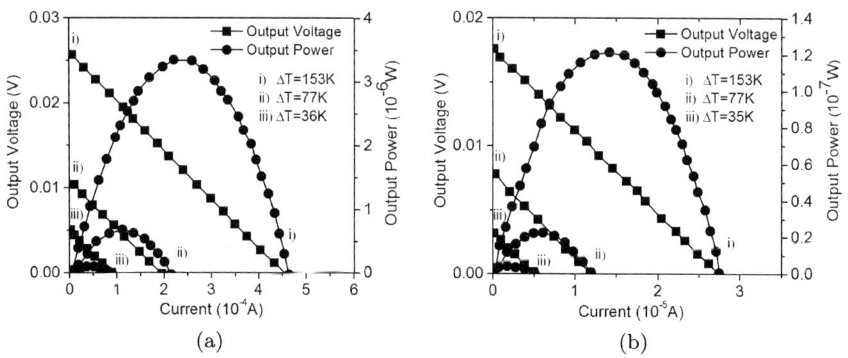

Fig. 24. Current dependences of V_{op} and P as a parameter of various ΔT. (a) 4-pair fr-GaN device on SiO_2 glass (sample 1). (b) 3-pair epi-GaN device on SiO_2 glass (sample 2).

Table 2. Summary of characteristic values of sample 1, 2, 3.[75,76,78]

Sample name	Fabrication details	Growth	T_{avg}(K)	ΔT(K)	V_{op}(V)	P_{max} (10^6W)
1	4-pair fr-GaN device on SiO_2 glass	HVPE	92	153	0.028	3.35
2	3-pair epi-GaN device on SiO_2 glass	HVPE	92	153	0.017	0.12
3	25-pair epi-GaN l on polyimide	MOCVD	508	30	0.3	2.1

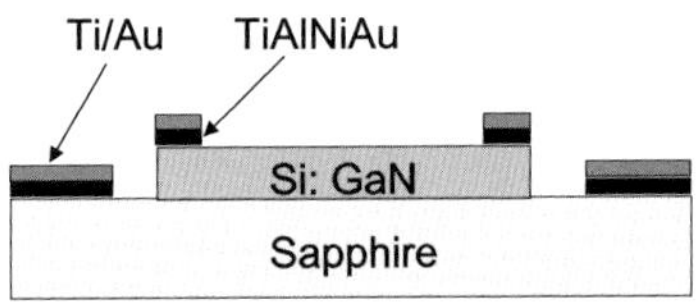

Fig. 25. Cross-section of sample 3.

of every sample at $\Delta T = 153$ K at an average temperature of 92 K, where the average temperature $T_{avg} = (T_c + T_h)/2$. With the same temperature condition, both the maximum output power and open output voltage of sample 2 are smaller than those of sample 1, which is attributed to the smaller element number and smaller thickness in sample 2. However sample 2 is more appropriate for the on-chip integration. Sztein *et al.*[75] reported a GaN-based integrated lateral thermoelectric device deposited on a sapphire substrate by MOCVD (sample 3). As shown in Fig. 25, a contact metal of Ti/Al/Ni/Au was deposited to ensure thermally stable Ohmic contacts, and a pad metal of Ti/Au interconnects was used to achieve p–n pairs. Then the series of 1, 5, 10 and 25 pairs GaN layers devices were fabricated, and the dimensions of one element are 1 mm (length) × 0.1 mm (width) × 3.5 μm (thickness). Figure 26 shows the dependence of load characteristics on element number N. It is observed that both V_{op} and P_{max} increase linearly with the number of elements, and V_{op} reaches a maximum of 0.22 V for the 25 element device, which is in accordance with the Eq. (1) and Eq. (4). Figure 27 shows the effect of ΔT on load characteristics. P_{max} and V_{op} of sample 3 at $\Delta T = 30$ K and $T_{avg} = 508$ K are also listed in Table 2.

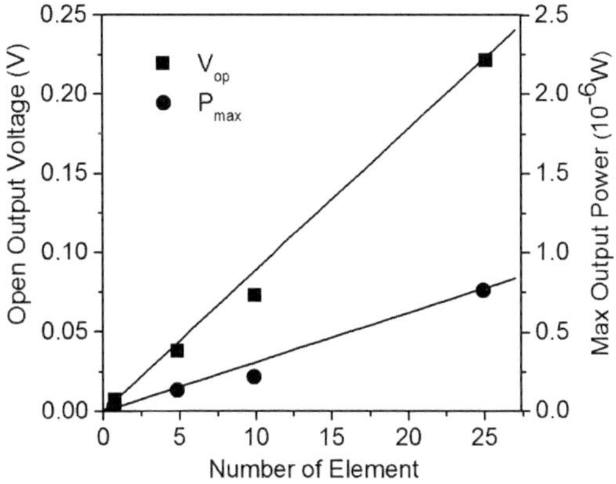

Fig. 26. Dependence of load characteristics on number of elements at $\Delta T = 34.5\,K$ and $T_{avg} = 510\,K$, for sample 3.

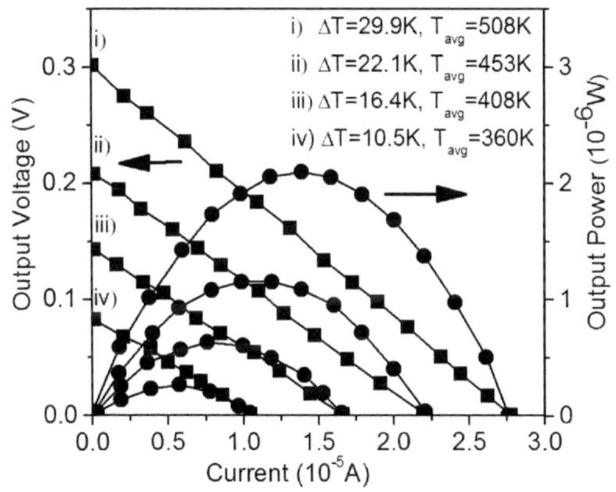

Fig. 27. Current dependences of V_{op} and P as a parameter of various ΔT and T_{avg} for sample 3, N = 25.

Comparing with sample 1 and 2, sample 3 displays the highest open output voltage. The reason is that the high average temperature of sample 3 enhances the Seebeck coefficient, then the open output voltage can be enhanced with the increasing Seebeck coefficient and element number although the decreased ΔT of sample 3. On the other hand, P_{max} can be further increased by adjusting the values of the parameters in Eq. (4), thus to fabricate highly customizable, downsized, low cost, micro power generators for excellent thermoelectric applications.

In summary, the development of the thermoelectric devices is to enhance the load characteristic by choosing the appropriate materials, advanced processing and design parameters. III-Nitride films are considered to be the most appropriate material for on-chip integrated thermoelectric devices since they can be grown thinly enough. With the current development, centimeter sized multi-element structure of InN-chromel on SiO_2 glass or roll-type InN-chromel on polyimide, and sub-millimeter sized multi-element structure of epitaxial GaN-TiAlNiAu on sapphire may be the best choices of the III-Nitride thermoelectric devices for the purpose of improving the thermoelectric properties and the on-chip power supply. However, the output power is not high enough and the fabrication of thermoelectric devices using nitride system has just started up to now. In the future, the highly customizable minimization sized on-chip device with improved processes and more elements could be a major development direction for the III-Nitride thermoelectric power applications.

4. General Discussion of Band structure on Materials' Thermoelectric Performance

III-Nitrides can maintain a high thermoelectric figure of merit at high temperatures due to the wide band-gap inherent in these materials that prevents the excitation of minority carriers and the loss of the n- or p-type character of the material at high temperatures.[7] Thus III-nitrides offer a unique niche to fill because the viable operational temperature of most semiconductors is limited to $<800\,K$.[79] The conversion efficiency of a thermoelectric device relies on the thermoelectric figure of merit at a certain temperature. An overall increase in ZT is a challenging task and requires at times and optimization between the different quantities.[80,81] For typical thermoelectric materials, κ_{Ph} is dominated by the lattice thermal conductivity and has been vigorously analyzed,[82] therefore the maximum ZT is then close to the maximum of the parameter S^2/ρ, namely the power factor. One of the approach to increase the power factor and thus ZT is to achieve high Seebeck coefficient, which requires a wide-bandgap energy, high charge carrier effective masses, and high temperatures as can be seen from the Mott formula in Ref. 82. The second approach is to increase the electrical conductivity of the material, requiring high carrier densities, increasing in the degenerate limit. Moreover, phonon scattering plays an essential role to achieve low thermal conductivity and results in higher ZT values[83,84]; however, the benefits of wide-bandgap materials can most readily be seen in the power factor, which will be the focus of the following sections.

4.1. *Band gap, Seebeck coefficient, and electrical conductivity*

TE materials, assuming parabolic energy bands, have a power factor that is well understood. High power factors necessitate high mobility to increase electrical conductivity, but high effective masses to increase the Seebeck value leading to a contradiction. Mahan and Sofo concluded that materials with a delta function transport distribution function (TDF) lead to the best compromise between conductivity and Seebeck value.[85] This is in itself directly related to the applicability of wide-bandgap materials to increase the overall power factor performance. In theory, by creating a delta function transport function right at the conduction band ($E_F - E_C = 0$, or $E_F - E_V = 0$) one maximizes the overall power factor, assuming that the MFP of carriers is constant.[82] Mahan and Sofo amongst others. showed that the best band gap must be greater than $\sim 6\,k_B T$ with T being the operating temperature. Others have shown similar calculations using variety of extended assumptions from indirect band gap materials to direct.[86,87]

Figure 28 shows the Seebeck coefficient versus the Fermi level. One can see as the operating temperature increases, so to must the band gap as a general trend. This overall trend directly relates to the Fermi level and band gap stability at different temperatures. As the temperature is increased, more carriers are promoted to the conduction ban contributing to conduction, for small band gap materials, this can be dominant at relatively low temperatures $\sim <500$ K. With this increase in carrier concentration, our optimal transport distribution function loses its delta-function optimization, lowering the Seebeck coefficient while increasing electrical conductivity, degrading the overall power factor. However, larger effective mass materials correspond to larger Seebeck coefficient values. These larger effective masses are directly related to the position of the chemical potential, depending on the material's effective masses through their density of states D(E). Figure 29 shows the Seebeck coefficient versus carrier concentration for some thermoelectric materials. This general trend is observed for

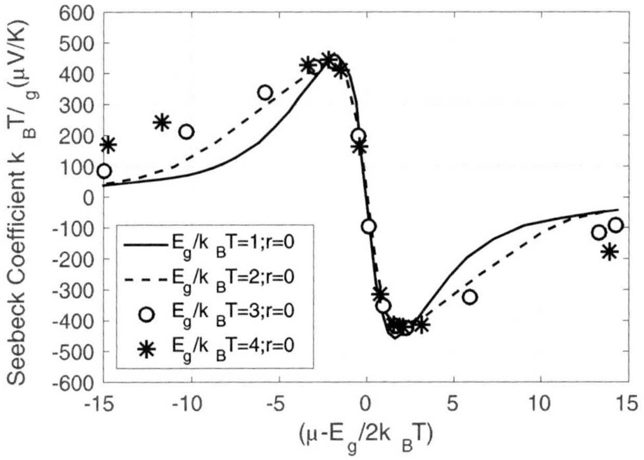

Fig. 28. Seebeck coefficient vs Fermi level position calculated for a two-band parabolic model.

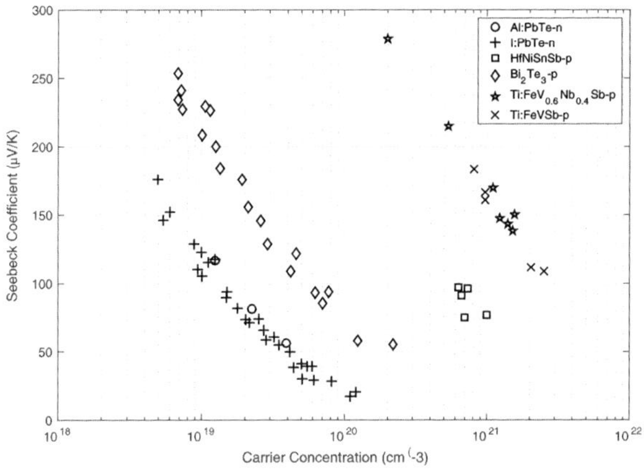

Fig. 29. Seebeck coefficient versus carrier density using a single parabolic band model. Data from for p-type Bi_2Te_3, for Al-doped PbTe, and for p-type half-Heuslers from for $(Hf_{0.6}Zr_{0.4})_{1-x}(Yb,Sb)_xNiSn_{0.98}Sb_{0.02}$, for Ti-doped FeVSb, and for Ti-doped $FeV_{0.6}Nb_{0.4}Sb$.

materials with larger effective masses having larger Seebeck values. Directly relating to the bandgap value, with larger bandgap materials in general having larger effective masses. Unfortunately, this lowers the electrical conductivity, so again a tradeoff is needed.

4.2. *Strategies for power factor optimization*

Good thermoelectric materials have been relegated to doped semiconductors leading from low electrical conductivity in insulators and metals having limited Seebeck coefficients and high thermal conductivity of phonons as well as electrons.[81,94] III-Nitrides, as discussed, have been considered for TE type applications for several reasons, such as high temperature stability, high mechanical strength, and radiation hardness.[6] Nitride materials also allow alloying and nano-structuring within the fabrication process allowing tuning of thermoelectric properties.[6,75]

Several different strategies have been explored to enhance the Seebeck coefficient as well as the electrical conductivity independently. Quantum confinement, resonance levels, and electron energy filtering have been discussed as ways to enhance the Seebeck coefficient of thermoelectric materials.[88] The principle of quantum confinement allows for a more delta-function shape of the density of states with benefits as previously discussed in the above section. Energy dependent scattering resonance levels have been discussed since the inception of thermoelectrics as way to enhance the overall power factor as pointed out by Ioffe in the early 1950s and 60s. However at high temperatures, applicable to wide-bandgap thermoelectrics, these scattering mechanisms would become dominated by phonon scattering, eliminating much of the designed benefit. Electron energy filtering in theory could have applications at higher temperatures. Initially proposed by Shakouri *et al.*[95] for thermionics, by band engineering, heterostructures, or band offsets, low energy electrons are filtered out leaving only high energy electrons thereby increasing the Seebeck coefficient. This in turn will reduce the electrical conductivity, so a tradeoff again must be considered. The fabrication of these types of structures to be implemented at high temperatures will leave much to address, in theory however, energy filtering should be feasible. Enhancement of electrical conductivity has also been addressed in many papers.[88] With large effective masses, wide-bandgap materials are already at a disadvantage for carrier mobility, directly related to electrical conductivity. Modulation doping with other types of band engineering have been proposed as ways to increase electrical conductivity. Tradeoffs with the Seebeck coefficient must be again analyzed in turn to ascertain whether an overall power factor increase would be attainable.

Acknowledgement

The authors are grateful for US National Science Foundation's support through the NSF CMMI-1560834 CAREER Research Program.

References

1. Energy, U.S.D.o., *Waste Heat Recovery: Technology and Opportunities in U.S. Industry* (2008).
2. R.W., G.H.J.a.D., *The use of semiconductors in thermoelectric refrigeration. Brit. J. Appl. Phys.*, **54**(11), 5 (1954).
3. S. Chhajed, Lee, Wonseok Cho, Jaehee., Schubert, E Fred., and Kim, Jong Kyu, *Strong light extraction enhancement in GaInN light-emitting diodes by using self-organized nanoscale patterning of p-type GaN. Applied Physics Letters*, **98**(7), 071102 (2011).
4. C. J. Neufeld, Toledo, Nikholas G., Cruz, Samantha C., Iza, Michael., DenBaars, Steven P., Mishra, Umesh K. *High quantum efficiency InGaN/GaN solar cells with 2.95 eV band gap. Applied Physics Letters*, **93**(14), 3502 (2008).
5. P. Waltereit, *et al.*, *Nitride semiconductors free of electrostatic fields for efficient white light-emitting diodes. Nature*, **406**(6798), 865–868 (2000).
6. B. N. Pantha, R. Dahal, J. Li, J. Y. Lin, H. X. Jiang, and G. Pomrenke, *Thermoelectric properties of* $InxGa_{1-x}N$ *alloys. Applied Physics Letters*, **92**(4), 042112 (2008).
7. N. Lu and I. Ferguson, *III-nitrides for energy production: photovoltaic and thermoelectric applications.* Semiconductor Science and Technology, **28**(7), 074023 (2013).
8. G. Streetman Ben, S. B., *Solid State electronic Devices.* 5 ed. Vol. 1., New Jersey: Prentice Hall (2000).
9. V. Bougrov, L. M. E., S. L. Rumyantsev, and A. Zubrilov, *Properties of Advanced Semiconductor Materials GaN, AlN, InN, BN, SiC, SiGe.* 1 ed. Vol. 1., New York: John Wiley & Sons, Inc (2001).
10. W. Y. Wu ChaoWei, *Thermoelectric characteristic of the rough InN/GaN core-shell nanowires. Journal of Applied Physics*, **116**(10), 3 (2014).
11. B. Kucukgok, W. B., A. G. Melton, N. Lu, and I. T. Ferguson, *Comparison of thermoelectric properties of GaN and ZnO samples. Physica Status Solidi*, **11**(3–4), 4 (2014).
12. M. A. Elisa n. Hurwitz, Andrew Melton, Bahadir Kucukgok, Liqin Su, Mateusz Orocz, Muhammad Jamil, Na Lu, and Ian t. Ferguson, *Thermopower Study of GaN-Based Materials for Next-Generation Thermoelectric Devices and Applications. Journal of Electronic Materials*, **40**(5), 5 (2011).
13. A. Sztein, H. Ohta, J. E. Bowers, S. P. DenBaars, and S. Nakamura, *High temperature thermoelectric properties of optimized InGaN. Journal of Applied Physics*, **110**(12), 123709 (2011).
14. X. W. Bahadir Kucukgok, Xiaojia Wang, Zhiqiang Liu, Ian T. Ferguson and Na Lu, *The structural properties of InGaN alloys and the interdependence on the thermoelectric behavior.* AIP Advances, **6**(2), 7 (2016).
15. A. Sztein, *Calculated thermoelectric properties of InxGa1-xN, InxAl1-xN, and AlxGa1-xN. Journal of Applied Physics*, **113**(18), 183707–183718 (2013).

16. W. Liu, *Thermal conduction in AlGaN alloys and thin films. Journal of Applied Physics*, **97**(7), 073710–073716 (2005).
17. J. Callaway, *Model for Lattice Thermal Conductivity at Low Temperatures. Physical Review*, **113**(4), 1046–1051 (1959).
18. B. Abeles, *Lattice Thermal Conductivity of Disordered Semiconductor Alloys at High Temperatures. Physical Review*, **131**(5), 1906–1911 (1963).
19. H. Tong, *et al.*, *Analysis of Thermoelectric Characteristics of AlGaN and InGaN Semiconductors*. 2009.
20. A. Sztein, *Polarization field engineering of GaN/AlN/AlGaN superlattices for enhanced thermoelectric properties. Applied Physics Letters*, **104**(4), 042106-1 (2014).
21. B. Pantha, *Epitaxial growth of III-nitride nanostructures and applications for visible emitters and energy generation*, J. Lin, *et al.*, Editors. ProQuest Dissertations Publishing (2009).
22. A. Alshaikhi and G. P. Srivastava, *Theoretical investigations of phonon intrinsic mean-free path in zinc- blende and wurtzite AlN. Phys. Rev. B*, **76**(19), 195205 (2007).
23. F. M. Morales, M. J. M., R. García, B. Reuters, H. Kalisch, and A. Vescan, *Evaluation of interpolations of InN, AlN and GaN lattice and elastic constants for their ternary and quaternary alloys. Journal of Physics: D Applied Physics*, **46**(24), 8 (2013).
24. P. D. Borges, S. L., *Electronic and thermoelectric properties of InN studied using ab initio density functional theory and Boltzmann transport calculations. Journal of Applied Physicas*, **116**(22), 5 (2014).
25. I.-L.L. Hung-Hsun Huang, Yuh-Renn Wu, *Study of thermoelectric properties of indium nitride nanowire. Phys Status Solidi A*, **208**(7), 4 (2011).
26. D. Gall, *et al.*, *Growth of poly-and single-crystal ScN on MgO (001): Role of low-energy* $N^{\wedge +} \sim 2$ *irradiation in determining texture, microstructure evolution, and mechanical properties. Journal of Applied Physics*, **84**(11) (1998).
27. V. Rawat, *et al.*, *Thermal conductivity of (Zr, W) N/ScN metal/semiconductor multilayers and superlattices. Journal of Applied Physics*, **105**(2), 024909 (2009).
28. S. W. King, R. F. Davis, and R. J. Nemanich, *Gas source molecular beam epitaxy of scandium nitride on silicon carbide and gallium nitride surfaces. Journal of Vacuum Science & Technology A*, **32**(6), 061504 (2014).
29. J. Dismukes, *et al.*, *Vapor deposition of semiconducting mononitrides of Sc, Y, and the rare earth elements. Journal of Crystal Growth*, **9**, 295–295 (1971).
30. J. P. Dismukes, W. M. Y., J. J. Tietjen, and R. E. Novak, R.C.A. Review., **31**(680) (1970).
31. D. Gall, *et al.*, *Electronic structure of ScN determined using optical spectroscopy, photoemission, and ab initio calculations. Physical Review B*, **63**(12), 125119 (2001).
32. C. Stampfl, *et al.*, *Electronic structure and physical properties of early transition metal mononitrides: Density-functional theory LDA, GGA, and screened-exchange LDA FLAPW calculations. Physical Review B*, **63**(15), 155106 (2001).
33. M. Moreno-Armenta and G. Soto, *Electronic structure of scandium nitride with nitrogen and scandium deficits. Computational materials science*, **40**(2), 275–281 (2007).
34. W. R. Lambrecht, *Electronic structure and optical spectra of the semimetal ScAs and of the indirect-band-gap semiconductors ScN and GdN. Physical Review B*, **62**(20), 13538 (2000).

35. H. A. Al-Brithen, A. R. Smith, and D. Gall, *Surface and bulk electronic structure of ScN (001) investigated by scanning tunneling microscopy/spectroscopy and optical absorption spectroscopy. Physical Review B*, **70**(4), 045303 (2004).
36. H. A. Al-Brithen, *et al.*, *Phase stability, nitrogen vacancies, growth mode, and surface structure of ScN (001) under Sc-rich conditions. Journal of Crystal Growth*, **242**(3), 345–354 (2002).
37. M. Moram, Z. Barber, and C. Humphreys, *The effect of oxygen incorporation in sputtered scandium nitride films.* Thin Solid Films, **516**(23), 8569–8572 (2008).
38. G. Travaglini, *et al.*, *Electronic structure of ScN. Physical Review B*, **34**(6), 3876 (1986).
39. J. M. Gregoire, *et al.*, *High mobility single crystalline ScN and single-orientation epitaxial YN on sapphire via magnetron sputtering. Journal of Applied Physics*, **104**(7), 074913 (2008).
40. J. Dismukes, *et al.*, *Vapor deposition of semiconducting mononitrides of scandium, yttrium, and the rare-earth elements. RCA Rev*, **31**, 680–691 (1970).
41. G. J. Snyder and E. S. Toberer, *Complex thermoelectric materials. Nature materials*, **7**(2), 105–114 (2008).
42. S. Kerdsongpanya, *et al.*, *Anomalously high thermoelectric power factor in epitaxial ScN thin films. Applied Physics Letters*, **99**(23), 232113 (2011).
43. C. X. Quintela, *et al.*, *Epitaxial CrN thin films with high thermoelectric figure of merit. Advanced Materials*, **27**(19), 3032–3037 (2015).
44. P. V. Burmistrova, *et al.*, *Effect of deposition pressure on the microstructure and thermoelectric properties of epitaxial ScN (001) thin films sputtered onto MgO (001) substrates. Journal of Materials Research*, **30**(05), 626–634 (2015).
45. L. Goncalves, *et al.*, *Optimization of thermoelectric properties on Bi 2 Te 3 thin films deposited by thermal co-evaporation.* Thin Solid Films, **518**(10), 2816–2821 (2010).
46. J. R. Sootsman, H. K., C. Uher, J. J. D'Angelo, C.-I. Wu, T. P. Hogan, T. Caillat, and M. G. Kanatzidis, *Angew. Chem., Int.* Ed. **47**(8618) (2008).
47. P. V. Burmistrova, *et al.*, *Thermoelectric properties of epitaxial ScN films deposited by reactive magnetron sputtering onto MgO (001) substrates. Journal of Applied Physics*, **113**(15), 153704 (2013).
48. A. Shakouri, in *nternational Conference on Thermoelectronics.* Clemson, SC (2005).
49. D. Vashaee and A. Shakouri, *Improved thermoelectric power factor in metal-based superlattices. Physical Review Letters*, **92**(10), 106103 (2004).
50. M. Zebarzadi, Z. B., R. Singh, A. Shakouri, R. Wortman, V. Rawat, and T. D. Sands, *J. Electron. Mater.*, **38**, 4 (2009).
51. V. Rawat and T. Sands, *Growth of TiN/GaN metal/semiconductor multilayers by reactive pulsed laser deposition. Journal of Applied Physics*, **100**(6), 064901 (2006).
52. M. Zebarjadi, *et al.*, *Thermoelectric transport in a ZrN/ScN superlattice. Journal of Electronic Materials*, **38**(7), 960–963 (2009).
53. B. Saha, T. D. Sands, and U. V. Waghmare, *First-principles analysis of ZrN/ScN metal/semiconductor superlattices for thermoelectric energy conversion. Journal of Applied Physics*, **109**(8), 083717 (2011).
54. G. Farault, *et al.*, *Crystal chemistry and electronic structure of the metallic ternary nitride, SrTiN2. Chemistry of Materials*, **15**(20), 3922–3929 (2003).

55. O. Seeger, *et al.*, *Synthesis and structure of Ba [ZrN2] and Ba2 [NbN3].* Zeitschrift für anorganische und allgemeine Chemie, **620**(11), 2008–2013 (1994).
56. D. Gregory, *et al.*, *Synthesis, Structure, and Magnetic Properties of the New Ternary Nitride BaHfN 2 and of the BaHf 1- x Zr x N 2 Solid Solution. Journal of Solid State Chemistry*, **137**(1), 62–70 (1998).
57. J. Bednorz, *et al.*, *Jap. JAP letters to appear. z. Phys. B*, **64**, 188 (1986).
58. Y. Kamihara, *et al.*, *Iron-Based Layered Superconductor La [O1-x F x] FeAs (x = 0.05-0.12) with T c = 26 K. Journal of the American Chemical Society*, **130**(11), 3296–3297 (2008).
59. K. Mizushima, *et al.*, *LixCoO2* (0 < *x* < −1)*: A new cathode material for batteries of high energy density. Materials Research Bulletin*, **15**(6), 783–789 (1980).
60. K. Koumoto and T. Mori, *Thermoelectric nanomaterials*. Springer (2015).
61. D. H. Gregory, M. G. B., D. J. Siddons, Mrs Online Proceeding Library (2011).
62. I. Ohkubo and T. Mori, *Two-Dimensional Layered Complex Nitrides as a New Class of Thermoelectric Materials. Chemistry of Materials*, **26**(8), 2532–2536 (2014).
63. I. Ohkubo and T. Mori, *Origin of Projected Excellent Thermoelectric Transport Properties in d0-Electron AMN2 (A = Sr or Ba; M = Ti, Zr, Hf) Layered Complex Metal Nitrides. European Journal of Inorganic Chemistry*, **2015**(22), 3715–3722 (2015).
64. R. A. R. Al Orabi, *et al.*, *Prediction of high thermoelectric potential in AMN 2 layered nitrides: electronic structure, phonons, and anharmonic effects. Journal of Materials Chemistry A*, **3**(18), 9945–9954 (2015).
65. I. Terasaki, Y. Sasago, and K. Uchinokura, *Large thermoelectric power in NaCo 2 O 4 single crystals. Physical Review B*, **56**(20), R12685 (1997).
66. K. Takada, *et al.*, *Superconductivity in two-dimensional CoO2 layers. Nature*, **422**(6927), 53–55 (2003).
67. D. J. Singh, *Electronic structure of NaCo 2 O 4. Physical Review B*, **61**(20), 13397 (2000).
68. K. Sugiura, *et al.*, *Anisotropic carrier transport properties in layered cobaltate epitaxial films grown by reactive solid-phase epitaxy. Applied Physics Letters*, **94**(15), 152105-1 (2009).
69. I. Ohkubo and T. Mori, *Anisotropic thermoelectric properties in layered complex nitrides with* α*-NaFeO2-type structure. APL Materials*, **4**(10), 104808 (2016).
70. I. Ohkubo and T. Mori, *Three-dimensionality of electronic structures and thermoelectric transport in SrZrN2 and SrHfN2 layered complex metal nitrides. Inorganic Chemistry*, **53**(17), 8979–8984 (2014).
71. T. Okuda, *et al.*, *Large thermoelectric response of metallic perovskites: Sr 1- x La x TiO 3* (0 <∼ *x* <∼ 0.1). *Physical Review B*, **63**(11), 113104 (2001).
72. S. Ohta, *et al.*, *High-temperature carrier transport and thermoelectric properties of heavily La-or Nb-doped SrTiO3 single crystals. Journal of Applied Physics*, **97**(3), 034106 (2005).
73. S. Yamaguchi, *et al.*, *Thermoelectric devices using InN and Al[sub 1-x]In[sub x]N thin films prepared by reactive radio-frequency sputtering. Applied Physics Letters*, **84**(26), 5344 (2004).
74. R. Izaki, *et al.*, *Roll-type thermoelectric devices with InN thin films. Applied Physics Letters*, **87**(24), 243508 (2005).

75. A. Sztein, *et al.*, *GaN-based integrated lateral thermoelectric device for micro-power generation. Applied Physics Express*, **2**(11), 111003 (2009).
76. S. Yamaguchi, *et al.*, *Thermoelectric properties of and devices based on free-standing GaN. Applied Physics Letters*, **86**(25), 252102 (2005).
77. P. Richard and M. Cheyrezy, *Composition of reactive powder concrete. Cement and Concrete Research*, **25**(7), 1501–1511 (1995).
78. N. Kaiwa, *et al.*, *Thermoelectric properties and thermoelectric devices of free-standing GaN and epitaxial GaN layer. Thin Solid Films*, **515**(10), 4501–4504 (2007).
79. L.-P.H. Zhao, Tie-Jun, Zhu., Ya-Guang, Wang., Han-Hui, Xie., Zhao-Jun, Xu., Xin, Bing.,, *Shifting up the optimum figure of merit of p-type bismuth telluride-based thermoelectric materials for power generation by suppressing intrinsic conduction. NPG Asia Materials*, **6**(2) (2014).
80. Z. Chen, G. Han, L. Yang, L. Cheng, and J. Zou, *Nanostructured thermoelectric materials: Current research and future challenge.* Progress in Natural Science: Materials International, **22**(6), 15 (2012).
81. T. M. Tritt and M. A. Subramanian, *Thermoelectric materials, phenomena, and applications: a bird's eye view. MRS Bulletin*, **31**, 7 (2006).
82. C. Jeong, Kim, Raseong., Lundstrom, Mark S., *On the best bandstructure for thermoelectric performance: A Landauer perspective. Journal of Applied Physics*, **111**(11), 113707 (2012).
83. S. H. Lo, He, Jiaqing., Biswas, Kanishka., Kanatzidis, G. Mercouri, Dravid, Vinayak, P. *Phonon Scattering and Thermal Conductivity in p-Type Nanostructured PbTe-BaTe Bulk Thermoelectric Materials. Advanced Functional Materials*, **22**(24), 5175–5184 (2012).
84. O. Delaire, Ma, Jie., Marty, Karol., May, Andrew F., McGuire, Michael A., Du, Mao-Hua., Singh, David J., Podlesnyak, A., Ehlers, G., Lumsden, MD., *Giant anharmonic phonon scattering in PbTe. Nature materials*, **10**(8), 614–619 (2011).
85. G. Mahan, Sofo, JO, *The best thermoelectric. Proceedings of the National Academy of Sciences*, **93**(15), 7436–7439 (1996).
86. G. Mahan, *Good thermoelectrics. Solid state physics*, **51**, 81–157 (1997).
87. G. S. Nolas, Sharp, Jeffrey., Goldsmid, Julian, *Thermoelectrics: basic principles and new materials developments*. Vol. 45. Springer Science & Business Media (2013).
88. A. M. Dehkordi, Zebarjadi, Mona., He, Jian., Tritt, Terry M., *Thermoelectric power factor: Enhancement mechanisms and strategies for higher performance thermoelectric materials.* Materials Science and Engineering: R: Reports, **97**, 1–22 (2015).
89. B. P. Ren, *et al.*, *High-Thermoelectric Performance of Nanostructured Bismuth Antimony Telluride Bulk Alloys.* **320**(5876), 634–638 (2008).
90. A. D. LaLonde, Y. Pei, and G. J. Snyder, *Reevaluation of PbTe 1- x I x as high performance n-type thermoelectric material. Energy & Environmental Science*, **4**(6), 2090–2096 (2011).
91. C. M. Jaworski and J. P. Heremans, *Thermoelectric transport properties of the n-type impurity Al in PbTe. Physical Review B*, **85**(3), 033204 (2012).
92. T. Zhu, *et al.*, *Effects of yttrium doping on the thermoelectric properties of Hf0. 6Zr0. 4NiSn0. 98Sb0. 02 half-Heusler alloys. Journal of Applied Physics*, **108**(4), 044903 (2010).

93. C. Fu, *et al.*, *High Band Degeneracy Contributes to High Thermoelectric Performance in p-Type Half-Heusler Compounds. Advanced Energy Materials*, **4**(18) (2014).
94. A. Shakouri, *Recent Developments in Semiconductor Thermoelectric Physics and Materials.* http://dx.doi.org/10.1146/annurev-matsci-062910-100445 (2011).
95. J. Bowers E. and Ali Shakouri, *Heterostructure integrated thermionic coolers.* 1997.

CHAPTER 3

III-NITRIDE NANOSTRUCTURES FOR INTERSUBBAND OPTOELECTRONICS

C. B. Lim, A. Ajay, J. Lähnemann, D. A. Browne and E. Monroy*

CEA-Grenoble, INAC-PHELIQS
17 av. des Martyrs, 38000 Grenoble, France
eva.monroy@cea.fr

In intersubband optoelectronics, transitions between confined electron (or hole) levels in the conduction (or valence) band of heterostructures are utilized to emit or detect infrared radiation. III-nitride semiconductors have emerged as promising materials for new intersubband technologies due to their large conduction band offset and sub-picosecond intersubband relaxation time. Furthermore, the large energy of longitudinal-optical phonons in GaN opens up prospects for ultra-high-speed devices for fiber-optic telecommunication, or for the operation at room temperature of THz quantum cascade lasers. Additionally, GaN-based intersubband devices could cover the 5–10 THz band, inaccessible to GaAs-based technologies.

1. Introduction

1.1. *Intersubband transitions*

Intersubband (ISB) or intraband transitions are electronic transitions between confined levels in either the conduction band or the valence band of semiconductor heterostructures. Quantum wells (QWs) or quantum dots (QDs) are nanostructures that can be engineered to present ISB transitions in the infrared (IR) spectral range. The desired operation wavelength can be selected by varying the size of the nanostructures. The first observation of ISB transitions was reported in 1974, in the surface accumulation layer of n-doped silicon.[1] Since then, ISB absorption has been observed in a variety of configurations, including GaAs[2] and GaN[3] QWs. This is at the base of a tremendous technological development which resulted in commercially-available devices such as quantum cascade lasers (QCLs)[4] or quantum well infrared photodetectors (QWIPs).[5,6]

ISB transitions are governed by certain polarization selection rules. The necessity for momentum conservation implies that photons can only effectively couple to electrons with a negligible momentum. Therefore, ISB electron-photon interactions are only possible when the light has an electric field component parallel to the direction of carrier confinement. In the case of QWs, this selection rule imposes the use of waveguide configurations or surface gratings to improve light coupling. Furthermore, due to the inversion symmetry, only transitions between wavefunctions of opposite parity are allowed. A comprehensive introduction to ISB physics can be found in Bastard[7] or Liu and Capasso.[8]

1.2. *Some notes on many-body effects*

A prerequisite for the observation of ISB absorption is the population of the first electronic level in the nanostructures. The necessarily high doping levels lead to shifts of the ISB transitions due to many-body effects. In general, two kind of many-body effect must be considered, namely those modifying the energy levels (exchange interaction) and those modifying the intraband absorption energy (plasmon screening and excitonic shift).[9]

The exchange interaction is due to the repulsion between electrons with the same spin, following the Pauli exclusion principle. The shift induced by the exchange interaction, E_{exch}, can be estimated following the approximation of Bandara *et al.*[10,11] using the Hartree-Fock method in a one-dimensional approach:

$$E_{exch}(k) = \frac{-e^2 k_F}{4\pi\,\varepsilon\varepsilon_0}\left[\frac{2}{\pi}\boldsymbol{E}\,(k/k_F) - 0.32\,(k_F/k_L)\right] \tag{1}$$

where e is the electron charge, ε the dielectric constant, ε_0 the vacuum permittivity, $k_F = \sqrt{2\pi n_S}$ with n_S being the surface charge density in the wells, $k_L = \pi/L$ with L being the QW thickness, and $\boldsymbol{E}(k/k_F)$ is the complete elliptical integral of the second kind.

The ISB absorption energy can be corrected for the shift induced by plasmon screening (depolarization shift) and for the Coulomb interaction between the excited electron and the quasi-hole left in the ground state (excitonic or final-state interaction) by the following equation[9]

$$E_{21} = (e_2 - e_1)\sqrt{1 + \alpha - \beta} \tag{2}$$

where e_1 and e_2 are the energies of the first and second electronic levels in the well, and α and β represent the depolarization shift and the exciton interaction, respectively.

The depolarization shift is due to the interaction between external radiation and the electron plasma, which leads to a modulation of the carrier density and an increase of the transition energy. The value of α can be estimated from[9,12]

$$\alpha = \frac{2e^2 n_S}{\varepsilon\varepsilon_0(e_2 - e_1)} \int_{-\infty}^{\infty} dz \left[\int_{-\infty}^{z} \Psi_2(z')\Psi_1(z')dz'\right]^2 \tag{3}$$

with $\Psi_1(z')$ and $\Psi_2(z')$ being the wave functions associated to e_1 and e_2, respectively.

Finally, β can be calculated as[9]

$$\beta = \frac{-2n_s}{(e_2 - e_1)} \int_{-\infty}^{\infty} dz\, |\Psi_1(z)|^2\, |\Psi_2(z)|^2 \frac{\partial}{\partial n(z)} \times \left(\left(\frac{9\pi}{4}\right)^{\frac{1}{3}} \frac{-e^2 m^*}{4\pi^2 r_s \varepsilon^2 \varepsilon_0 a_B}\left[1 + B\frac{r_s}{A}\ln\left(1 + \frac{A}{r_s}\right)\right]\right) \tag{4}$$

where $A = 21$, $B = 0.7734$,[9,13] m^* is the effective mass, $n(z)$ is the electron density in volume, $a_B = 4\pi\varepsilon_0\hbar^2/(m_0 e^2)$ is the Bohr radius with $\hbar$ being the reduced Planck constant and m_0 being the electron mass, and

$$r_s = \left[\frac{4\pi}{3}\left(\frac{\varepsilon a_B}{m^*}\right)^3 n(z)\right]^{-\frac{1}{3}} \tag{5}$$

It is important to keep in mind the limitations associated with the calculations of many-body effects. The models of the exchange interaction and the depolarization shift described above are approximations that consider these effects as a perturbation of the Hartree-Fock equation.[9] Equation (1) is an approximation that accounts for the deformation of the ground state assuming square QWs,[10,11] and Eq. (3) is extracted from a two-level model which requires that there is significant oscillator strength for transitions only to one excited state.[9] When the spectral shift of the ISB transition becomes comparable to the $e_2 - e_1$ energy, the validity of these approximations is arguable. They provide a valid qualitative description, but a rigorous solution of the Hartree-Fock equation would be required for a full quantification of the effects.

1.3. *Motivation for III-nitride ISB studies*

In contrast to other materials used for ISB applications, III-nitride semiconductors present a large conduction band offset (≈1.8 eV for GaN/AlN[14–16]). Therefore, it becomes feasible to fabricate ISB devices operating in the near-IR spectral range, particularly in the 1.3–1.55 μm wavelength window for fiber-optic communications.[17–19] Even at these high energies, III-nitrides do not exhibit problems of inter-valley scattering, since the L and X points of the Brillouin zone are much higher in energy (>2 eV) than the Γ point. Additionally, GaN is transparent for most of the IR spectral region, namely for wavelengths longer than 360 nm, except for the Reststrahlen band (phonon absorption band) located from 9.6 to 19 μm. Also, absorption in the range of 7.3 to 9 μm has been observed in bulk GaN substrates,[20–22] and is attributed to the second harmonic of the Reststrahlen band.

Additionally, there is an interest to develop III-nitride ISB devices for the THz frequency range (far IR). This spectral region has attracted attention due to potential applications in medical diagnostics, security screening and quality control. GaAs-based QCLs have already demonstrated potential as THz sources; however, their operating temperature and spectral range are limited by the longitudinal optical (LO) phonon energy at 36 meV (8.7 THz). Thanks to its large LO-phonon energy (91 meV), GaN has the potential to operate at higher temperature than GaAs and to cover the whole THz domain.

The use of indium containing alloys (InGaN, AlInN) has been explored,[23–26] since it would offer and even larger conduction band offset and a reduction of the electron effective mass. However, the huge lattice mismatch (11.0% and 13.8% in plane lattice mismatch between InN and GaN, and between InN and AlN, respectively), and the chemical inhomogeneity of these ternary compounds have hindered this research line.

2. ISB Characterization of GaN/AlGaN Quantum Wells

2.1. *Modeling*

III-nitride semiconductors are preferentially grown in the wurtzite crystallographic structure. Because of the low symmetry of the crystal, the positive and negative gravity centers of the dipoles do not coincide. Consequently, there is a spontaneous polarization field along the <0001> axis. The magnitude of the polarization depends on the ideality of the crystal, the cation-nitrogen bond length, and the chemical properties of the cation, and it is therefore smaller for GaN ($P_{sp} = -0.029\,\mathrm{C/m^2}$ in Ref. 27 or $-0.034\,\mathrm{C/m^2}$ in Ref. 28) than for AlN ($P_{sp} = -0.081\,\mathrm{C/m^2}$ in Ref. 27 or $-0.090\,\mathrm{C/m^2}$ in Ref. 28). Superimposed to the spontaneous polarization, a piezoelectric polarization is present in GaN/AlN heterostructures that is linearly dependent on the strain field within the material. In the case of two-dimensional (0001)-oriented structures like QWs, the strain-induced piezoelectric polarization can be expressed as:

$$P_{pz} = 2\epsilon_x \left(e_{31} - e_{33}\frac{c_{13}}{c_{33}} \right) \tag{6}$$

where ϵ_x is the in-plane strain, c_{ij} are elastic constants from the elastic tensor, and e_{ij} are piezoelectric constants from the piezoelectric tensor.

The electronic properties of (0001)-oriented nitride QWs are strongly affected by the polarization discontinuity between the well and barrier materials. Figure 1(a) presents the band diagram of a GaN/AlN (2 nm/3 nm) superlattice calculated using the nextnano^3 8-band k.p Schrödinger-Poisson solver[29] with the material parameters described in Ref. 30. The structure was considered pseudomorphic on an AlGaN substrate whose Al mole fraction matches the average Al content of the MQW. Due to the built-in electric field, the electron wavefunctions are shifted toward the [0001] direction and the hole wavefunctions toward the [000–1] direction, leading to the so-called quantum-confined Stark effect (QCSE), i.e. redshift and reduction of the oscillator strength of the band-to-band transition. For QWs larger than 2 nm, $e_2 - e_1$ is mostly determined by the QCSE, since both e_1 and e_2 lie in the triangular part of the QW potential profile. A description of the evolution of the ISB energy $e_2 - e_1$ as a function of the QW thickness is presented in Fig. 1(d).

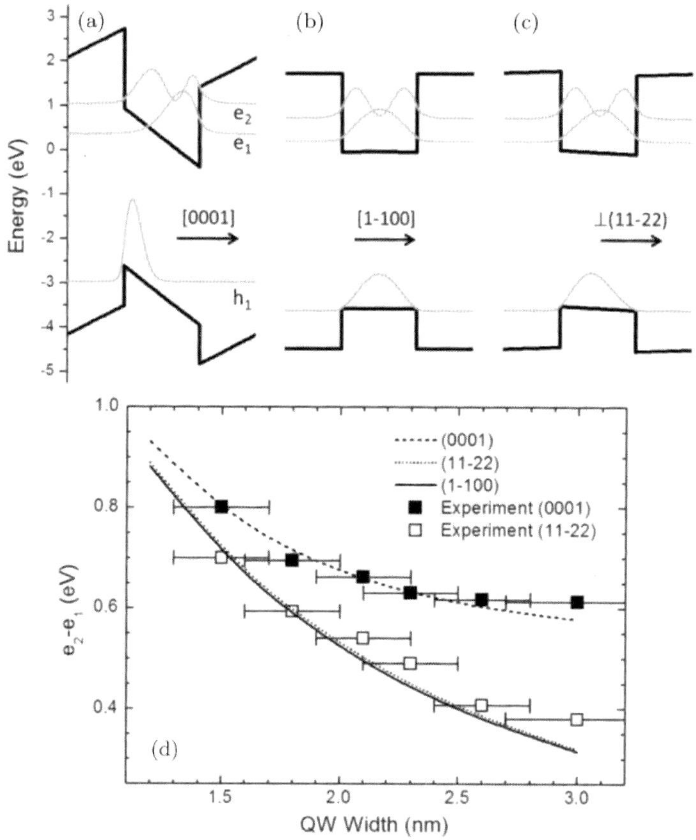

Fig. 1. Band diagram of GaN/AlN QWs in an infinite superlattice with 3-nm-thick AlN barriers and 2-nm-thick GaN QWs, grown on the (a) (0001), (b) (1-100) and (c) (11-22) planes. The structure is considered pseudomorphic on an AlGaN substrate with 60% Al. The squared electron wave functions of the ground hole state, h_1, the ground electron state, e_1, and the excited electron state, e_2 are presented. (d) Variation of $e_2 - e_1$ a function of the QW thickness in GaN/AlN MQW structures with 3-nm-thick barriers, for the above mentioned crystallographic orientations. All simulations assume the lattice at room temperature. Experimental results for planes (0001) and (11-22) are superimposed.

Another consequence of the polarization in polar heterostructures is that the charge distribution depends not only on the doping level in the QWs, but also on the carrier redistribution due to the internal electric field. In general, the polarization discontinuity between the active region and the contact or cladding layers leads to the formation of a depletion layer on one side of the active region and an accumulation layer on the other side. Therefore, a realistic view of the charge distribution in a device is only achieved by extending the electronic modelling to the complete structure.

The presence of internal electric fields in polar materials increases the design complexity of ISB devices, which has motivated research in alternative crystallographic orientations. Nonpolar planes like the $m - \{1{-}100\}$ or the $a - \{11{-}20\}$ planes are a simple alternative to obtain heterostructures without internal electric field,[31] as illustrated in Fig. 1(b). However, due to the strong surface anisotropy, nonpolar growth on foreign substrates results in a high density of crystalline defects. The unavailability of free-standing nonpolar GaN until recently has delayed the development of nonpolar ISB technologies, and has supported instead the growth on semipolar planes,[31] i.e. those (*hkil*) planes with at least two non-zero h, k, or i Miller indices and a nonzero l Miller index. Semipolar planes allow a considerable reduction in the internal electric field[32] while presenting a lower in-plane anisotropy than non-polar surfaces.[33,34] The most common example is the $(11{-}22)$ plane, for which smooth two-dimensional layers can be synthesized on m-sapphire both by metalorganic vapor phase epitaxy and by molecular-beam epitaxy. In nonpolar/semipolar orientations, Eq. (6) is no longer valid. An analysis of the strain and piezoelectric polarization in two-dimensional nonpolar and semipolar heterostructures can be found in Ref. 35. Considering both the effect of the spontaneous and the piezoelectric polarization, semipolar structures exhibit quasi-square potential band profiles with relatively symmetric wavefunctions [see Fig. 1(c)].[34,36]

2.2. *Polar GaN/AlGaN quantum wells*

2.2.1. *GaN/AlN quantum wells*

The ISB absorption of AlN/GaN MQWs has been studied by a number of groups, which have shown the tunability of $e_2 - e_1$ in the 1.0–3.5 μm wavelength range.[16,30,37–43] Figure 2(b,c) shows the ISB absorption of structures with various well thicknesses measured by Fourier transform IR spectroscopy (FTIR) in the waveguide geometry described in Fig. 2(a), using polarizers to discern between the transverse-electric (TE) and transverse-magnetic (TM) polarized light. The samples show an absorption

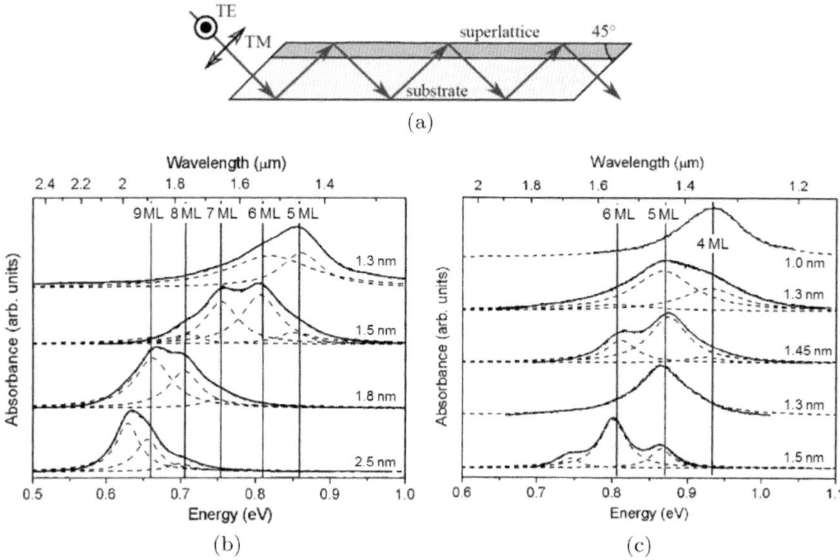

Fig. 2. (a) Schematic description of the waveguide geometry generally used in FTIR transmission measurements. (b,c) Absorbance spectra for TM polarized light in samples consisting of 20 periods of GaN/AlN QWs with 3-nm-thick AlN barriers and various QW thicknesses (indicated in the figure, with ML = monolayer). The curves are vertically shifted for clarity. Vertical lines mark the mean energies of structures. Dashed lines represent Lorentzian fitting curves. (Reprinted, with permission, from Tchernycheva, *et al.*, *Phys. Rev. B* **73**, 125347 (2006).[16] © 2006 The American Physical Society.)

peak for TM-polarized light assigned to the $e_1 - e_2$ transition, while no absorption is observed for TE-polarized light. For large QWs ($\geq$2 nm), an additional absorption line associated to the $e_1 - e_3$ transition is often observed (not shown). This transition is initially forbidden,[44] but it is allowed in polar QWs since the internal electric field breaks the symmetry of the potential in the well. A set of experimental values of $e_2 - e_1$ are presented in Fig. 1(d) as a function of the QW width, showing a good fit with theoretical calculations.

The full width at half maximum (FWHM) of the absorption remains in the 70–100 meV range for QWs doped at $5 \times 10^{19}\,\text{cm}^{-3}$. A record small line width of 40 meV has been measured by photo-induced absorption in non-intentionally doped structures.[45] The spectra present either a Lorentzian shape or are structured with 2–3 well-defined Lorentzian-shaped peaks. The energy location of such multiple peaks corresponds to theoretical values of $e_2 - e_1$ in QWs whose thickness is equal to an integer number of GaN monolayers (ML in Fig. 2).

Unlike band-to-band transitions, ISB transitions are expected to be insensitive to temperature. However, in polar III-nitrides the huge piezoelectric fields introduce an additional temperature-dependent effect because of the different thermal expansions of AlN and GaN. In this sense, it has been verified experimentally that the ISB absorption energy in GaN/AlN decreases only by ≈6 meV at 400°C relative to its room temperature value.[42]

Most studies of the ISB properties of GaN/AlN QWs are performed on material synthesized by molecular-beam epitaxy (MBE), due to the requirements in terms of interface sharpness and layer thickness control. GaN/AlN QWs displaying ISB transitions in the near-IR can also be synthesized by metalorganic vapor phase epitaxy (MOVPE).[43,46–48] In this case, a critical parameter to attain devices in the telecommunication spectral range (1.3 μm, 1.55 μm) is a drastic reduction of the growth temperature far below the 1050–1100°C optimum range for GaN, in order to minimize the GaN-AlN interdiffusion. Furthermore, deposition under compressive strain is recommended, since it prevents the redshift of the ISB transitions due to instabilities of the GaN/AlN interface.[49]

2.2.2. *Absorption saturation and relaxation time*

Saturation of the ISB absorption at ≈1.55 μm with recovery times in the 140–400 fs range has been demonstrated by several groups.[50–55] This result has motivated research on GaN QWs to develop ISB all-optical switches at telecommunication wavelengths, as first proposed by Suzuki *et al.*[56,57] In general, such devices consist of GaN/AlN MQWs embedded in a ridge waveguide.[54,58–63] A strong control light pulse can be used to saturate the $e_1 - e_2$ absorption, while the absorption of a weaker signal pulse is used to read out the saturation state of the MQW. A 10 dB transmittance modulation depth can be achieved with control switching energies of ≈24 pJ with 150 fs pulses in a 5-μm-wide (multimode) GaN-on-AlN ridge waveguide containing 3 periods of GaN/AlN QWs.[63]

2.2.3. *ISB photoluminescence*

In spite of the high oscillator strength of ISB transitions, ISB photoluminescence is an inefficient process due to the competition with even faster nonradiative relaxation channels, particularly the interaction with LO phonons on sub-picosecond time scales. Nevertheless, there are reports of room-temperature ISB luminescence obtained by optical pumping GaN/AlN MQWs.[64–66] The QWs used for these experiments present three confined levels. TM-polarized emission at 2–2.3 μm assigned to the $e_3 - e_2$ ISB transition was observed under TM-polarized excitation at wavelengths corresponding to the $e_1 - e_3$ ISB transition.

2.2.4. *Towards longer wavelengths*

Using binary compounds (GaN/AlN), $e_2 - e_1$ can be tuned in the 1.0–3.5 μm wavelength range by changing the QW thickness from 1 nm to 7 nm, with AlN barrier thicknesses in the 1.5–5.1 nm range.[16,30,37–43] For large QWs, the first two confined electron levels get trapped in the triangular section of the QW, which results in a saturation of $e_2 - e_1$ for further increasing QW thickness. To shift the absorption towards longer wavelengths, it is hence necessary to reduce the internal electric field in the QWs. A first approach consisted of employing ternary AlGaN barriers with reduced Al mole fraction. By changing the geometry and composition, the ISB absorption can be tailored to cover the near-IR range above 1.0 μm and mid-IR region up to 10 μm.[3,68–79]

Reducing the ISB transition energy below 60 meV (wavelength $>20\,\mu$m) requires further band engineering to compensate the internal electric field in the QWs. A first proposed architecture consisted of a 3-layer well (step-QW) with a virtually flat potential profile.[67,81–83] Figures 3(a) and (b) show a schematic description of the structure and the conduction band profile of one period, in this case consisting of an $Al_{0.1}Ga_{0.9}N/GaN/Al_{0.05}Ga_{0.95}N$ (3 nm/3 nm/10 nm) sequence. The design can be split into two portions, namely the "barrier", which comprises the high-Al-content $Al_xGa_{1-x}N$ layer and the GaN layer, and the "well", which is the low-Al-content $Al_xGa_{1-x}N$ layer. The design creates a flat band in the "well" by having the "barrier" balanced at the same average Al percentage, so that the average polarization in the "barrier" is approximately equal to

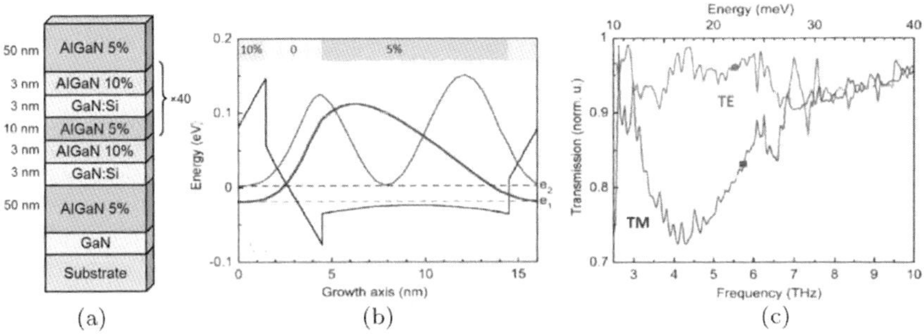

Fig. 3. (a) Schematic description of an $Al_{0.05}Ga_{0.95}N/Al_{0.1}Ga_{0.9}N/GaN$ (10 nm/3 nm/3 nm) step-QW structure. (b) Conduction band profile of a single period of the structure, including the squared wavefunctions of levels e_1 and e_2. (c) Transmission spectra for TM- (square) and TE- (circle) polarized light at T = 4.7 K [Reprinted, with permission, from Machhadani, *et al.*, *Appl. Phys. Lett.* **97**, 191101 (2010).[67] © 2010 American Institute of Physics].

the polarization in the "well". Samples following the step-QW design have shown ISB absorption at ≈2 THz (≈70 μm) and ≈13 THz (≈22 μm).[67,82] The weakness of the step-QW design lies in the fact that a deviation in the Al mole fraction or thickness of any of the layers has a drastic effect on the ISB transition energy. The limitations of the step-QW configuration can be surmounted by the insertion of an additional AlGaN layer to separate the GaN layer from the low-Al-content $Al_xGa_{1-x}N$ well,[81] at the price of enlarging the barrier, thus hindering tunneling transport.

As an alternative, a pseudo-square QW was proposed, consisting of a 4-layer architecture where the compensation of the polarization-induced internal electric field is obtained by creating a gradual increase in polarization along the potential "well" formed by three low-Al-content layers [see schematic description and band diagram in Figs. 4(a,b)].[80] This design has single-layer barriers which allow tunneling transport. Experimentally, it has been shown that the ISB wavelength can be varied from 150 to 75 μm by changing the size of the "well", and from 100 to 50 μm by changing the doping level, as illustrated in Fig. 4(c).

2.2.5. *Many-body effects*

In order to observe ISB absorption, the first electronic level must be populated. As a result of the high doping density, many-body effects are

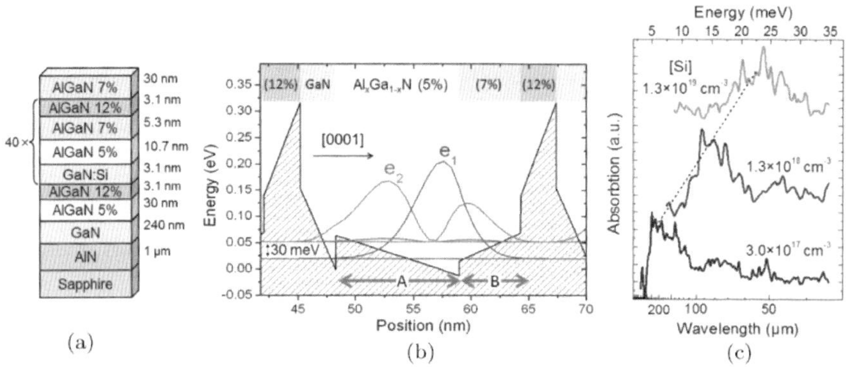

Fig. 4. (a) Schematic description of a pseudo-square QW structure. (b) Band profile of a pseudo-square QW including the squared wavefuctions of e_1 and e_2. The nominal compositions of the layers are indicated above. (c) Spectral absorption of TM-polarized light in samples with different doping levels. [Reprinted, with permission, from Beeler, *et al.*, *Appl. Phys. Lett.* **105**, 131106 (2014).[80] © 2014 American Institute of Physics].

important phenomena for ISB design; they can strongly affect the subband structure and drastically change the absorption spectra. In heavily Si doped GaN/Al(Ga)N MQWs, a blueshift and broadening of the ISB absorption has been observed,[39,80,84,85] as illustrated in Fig. 5. Theoretical calculations explain it as mostly related to the exchange interaction.[39,80] In mid- and far-IR structures, the magnitude of the shift is comparable to the value of $e_2 - e_1$.

2.3. *Semipolar GaN/AlGaN quantum wells*

Near-IR ISB absorption in the range of 1.5–3.75 μm has been reported on semipolar (11–22)-oriented GaN/AlN MQWs heteroepitaxially grown on m-sapphire.[34,36] The reduction in the internal electric field results in a redshift of the ISB energy in comparison with polar QWs with the same dimensions, as shown in Fig. 6(b) and Fig. 1(d).

In semipolar QWs, the ISB absorption shows a FWHM around 80–190 meV, and can be well fitted by a Gaussian function (or a sum of several Gaussian functions), in contrast to the Lorentzian shape of the absorption in polar QWs. This difference can be explained based on a structural analysis. Polar GaN/AlN QWs exhibit chemically abrupt interfaces containing monoatomic steps with a characteristic in-plane distance above 25 nm. For

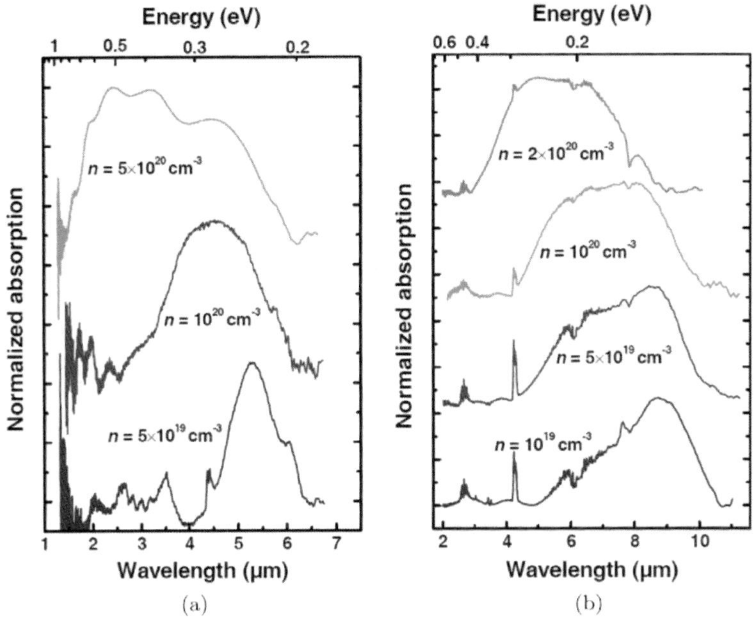

Fig. 5. IR absorption spectra for TM-polarized light measured from (a) $GaN/Al_{0.2}Ga_{0.8}N$ (3 nm/3 nm) and (b) $GaN/Al_{0.1}Ga_{0.9}N$ (7 nm/4 nm) MQWs grown on GaN-on-Si(111) templates with different doping levels. Spectra are vertically shifted for clarity. [Reprinted, with permission, from Kandaswamy, *et al.*, *Appl. Phys. Lett.* **96**, 141903 (2010).[84] © 2010 American Institute of Physics].

semipolar QWs, shown in Fig. 6(a), a higher interface roughness is observed with a characteristic in-plane step distance of $\approx$5 nm, i.e. smaller than the De Broglie wavelength. Therefore, averaged over their coherence length, the electrons feel an effective thickness which is not necessarily equal to an integer number of atomic layers. As a consequence, the total ISB absorption arising from different QW regions exhibits an inhomogeneous broadening and can be fitted by a Gaussian function.

2.4. *Nonpolar GaN/AlGaN quantum wells*

The first experimental results obtained by heteroepitaxy of a-plane GaN/AlN MQWs on r-sapphire showed ISB optical absorption at $\lambda \approx 2.1\,\mu$m with a FWHM $= 120$ meV.[86] Nowadays, the availability of high-quality free-standing nonpolar GaN substrates has encouraged research

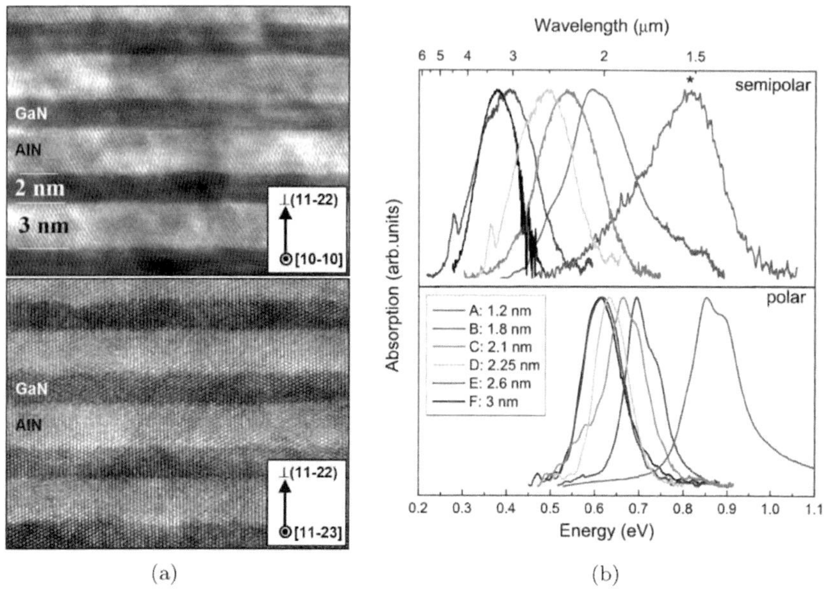

Fig. 6. (a) Cross-section high-resolution transmission electron microscopy image of semipolar GaN/AlN QWs viewed along the [10–10] and [11–23] zone axis. (b) ISB absorption spectra for semipolar (top) and polar (bottom) GaN/AlN QWs with different well thicknesses. The spectrum labeled with a star has been obtained by photoinduced absorption. [Reprinted, with permission, from Machhadani, *et al.*, *J. Appl. Phys.* **113**, 143109 (2013).[36] © 2013 American Institute of Physics].

on nonpolar orientations.[87–95] A comparative analysis of a- vs m-oriented GaN/AlN MQWs, grown on free-standing GaN has shown that the m-plane is the most promising nonpolar orientation for ISB application, based on several criteria. Structurally, the growth of m-oriented MQWs systematically resulted in smoother surfaces and better mosaicity in comparison with a-oriented samples with similar structure. In terms of ISB transitions, m-plane GaN/AlN MQWs showed room-temperature ISB absorption in the 1.5–2.9 μm range with linewidth similar to those of polar structures (see Fig. 7) whereas absorption in a-plane MQWs was either broader or non-existent.[92,95] Similar to the semipolar case, the $e_2 - e_1$ transition energy is redshifted with respect to polar MQWs due to the absence of internal electric fields.

By incorporating Ga in the MQW barriers and enlarging the QW width, heterostructures for ISB absorption in the mid-IR region can be

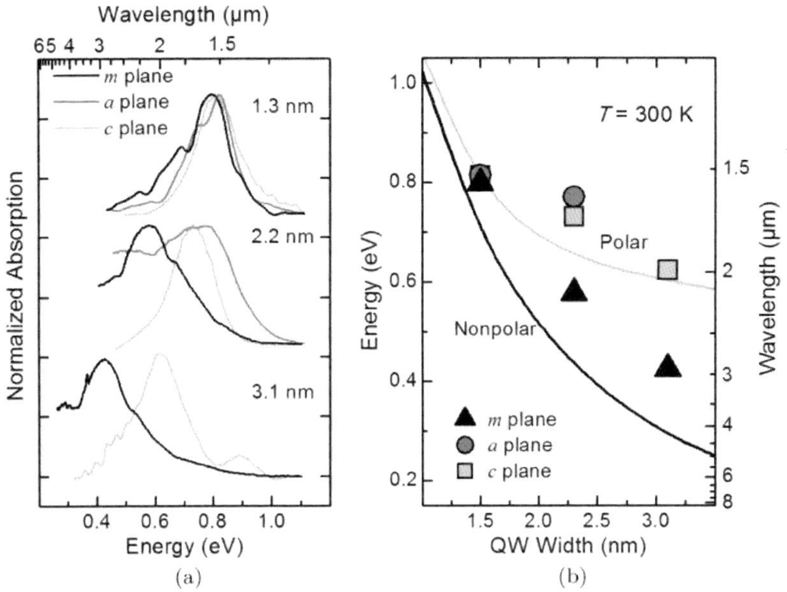

Fig. 7. (a) Spectral absorption of TM-polarized light in m-, a- and c-plane GaN/AlN MQWs with $\approx$3.5 nm AlN barriers and the QW thickness indicated in the figure. (b) ISB transition energy as a function of the QW width. Solid lines correspond to theoretical simulations assuming that the in-plane lattice parameters of the MQWs correspond to those of a relaxed AlGaN alloy with the average Al composition of the structure. [Reprinted, with permission, from Lim, *et al.*, *J. Appl. Phys.* **118**, 014309 (2015).[92] © 2015 American Institute of Physics].

designed. Room-temperature ISB absorption in the 4.0–5.8 μm range has been observed in m-plane GaN/AlGaN MQWs.[92,96] The long wavelength limit in this range is set by the substrate absorption associated with the second order of the GaN substrate Reststrahlen band.

In spite of these encouraging results, the lattice mismatch in these near- and mid-IR structures (4% along the c axis for GaN/AlN) leads to the formation of extended defects (dislocations, stacking faults). Furthermore, cracks develop during the cooling down process as a result of the temperature dependence of the lattice mismatch between GaN and AlN. In contrast, when reducing the Al mole fraction of the QW barriers below 10%, epitaxy-induced defects are no longer observed [see Fig. 8(a)]. This Al content is feasible when aiming at ISB transitions in the THz spectral range. Using m-GaN/AlGaN MQWs, low temperature ($T = 5 - 10$ K) ISB absorption in the 1.5–9 THz range has been demonstrated,[93,97] which confirms that most

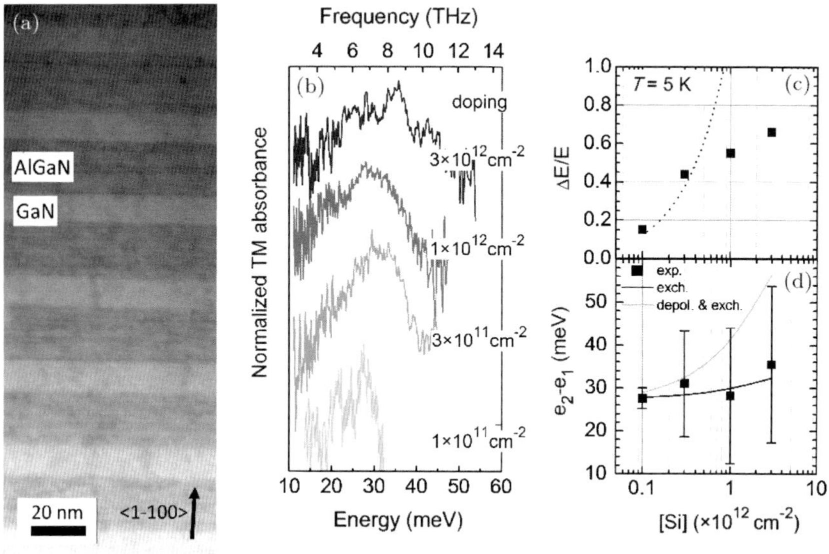

Fig. 8. (a) Cross-section high-angle annular dark-field scanning transmission electron microscopy image of an *m*-GaN/AlGaN MQW viewed along the <11-20> zone axis. (b) Normalized TM-polarized ISB absorption of GaN/$Al_{0.075}Ga_{0.925}N$ (10 nm/18.5 nm) MQWs with different doping levels (indicated in the figure). (c) Normalized linewidth of the absorbance as a function of the doping density in the MQWs. The dashed line indicates its theoretical trend. (d) ISB transition energy as a function of the doping density in the MQWs and theoretical evolution considering the exchange interaction (gray) and both the exchange interaction and the depolarization shift (black). The error bars illustrate the absorption linewidth. [Reprinted, with permission, from Lim, *et al.*, *Nanotechnology* **27**, 145201 (2016).[94] © 2016 IOP Publishing Ltd].

of the 7–10 THz band forbidden to GaAs-based technologies can be covered by GaN.

The effect of the doping density in the QWs has been studied in the mid- and far-IR. As illustrated in Fig. 8(b), increasing the silicon concentration in the wells results in a blueshift of the ISB absorption due to the exchange interaction and depolarization shift, and also causes an enhanced broadening due to scattering by ionized impurities.[91,94,97] The magnitude of the ISB absorption first increases with Si incorporation, and then saturates for doping densities around $1 \times 10^{12}\,cm^{-2}$. Also, for the samples with doping levels higher than $10^{12}\,cm^{-2}$, the blueshift and broadening are lower than theoretically predicted, as illustrated in Figs. 8(c)–(d). This can be explained by the presence of free carriers in the excited electron level e_2 when the Fermi level is increased.

3. Intraband Characterization of GaN/AlN Quantum Dots

3.1. *Intraband absorption*

An alternative approach to MQWs for the fabrication of intraband devices is the use of quantum dot (QD) superlattices.[98,99] Quantum dot IR photodetectors (QDIPs) are expected to ultimately outperform QWIPs in terms of low dark current, high photoelectric gain and sensitivity.[100] Furthermore, under certain conditions, intraband absorption in QDs can be almost independent of the polarization of the light.[101–103]

In the case of (0001)-oriented GaN, QDs can be synthesized by plasma-assisted MBE through GaN deposition under compressive strain and under N-rich conditions.[45,104] They can also be induced by a growth interruption in vacuum in NH_3-MBE[105] or in plasma-assisted MBE when growing under Ga-rich conditions,[106] or they can be generated by MOVPE under very low V/III ratios.[107] GaN QDs are well-defined hexagonal truncated pyramids with $\{1-103\}$ facets, which form on top of a 2-ML-thick GaN wetting layer.[108] The QD size can be tuned by modifying parameters like the deposition time, the growth temperature, or the growth interruption time after deposition of the QDs. By adjusting the growth conditions, QDs with height (diameter) in the range of 1–5 nm (10–40 nm), and density between $10^9\,cm^{-2}$ and $10^{12}\,cm^{-2}$ can be synthesized.[109] Silicon dopants can be incorporated into the QDs without perturbation of the morphology.

Si-doped polar GaN/AlN QD superlattices exhibit TM-polarized intraband absorption at room-temperature, which can be tuned from 1.38 to 1.68 μm as a function of the QD size.[109,110] This absorption line is attributed to transitions from the ground state of the conduction band, s, to the first excited electronic state confined along the growth axis, p_z. The broadening of the absorption peak is typically 80–100 meV, and it presents a Gaussian shape, which is explained by the energy dispersion associated to the diameter distribution of the QDs. The homogeneous line width of the $s - p_z$ transition in GaN/AlN QDs absorbing at 1.55 μm was estimated to be 15 meV at 5 K from nonlinear spectral hole-burning experiments.[111]

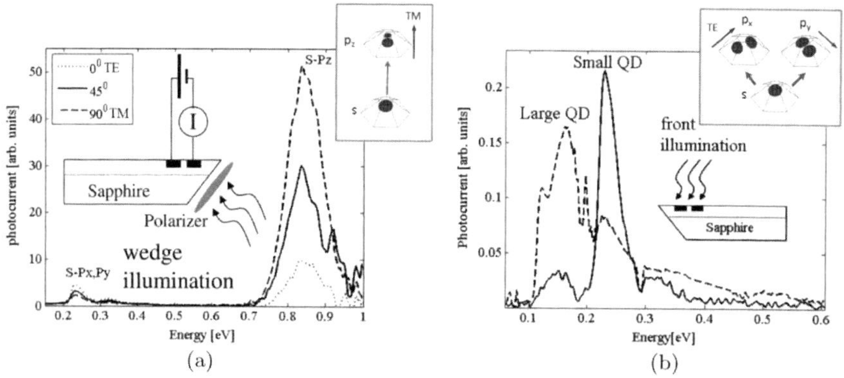

Fig. 9. (a) Polarization-dependent spectral photocurrent of a 20-period GaN/AlN QD sample measured in wedge illumination. Normalized response of both $s - p_z$ peak at near IR and $s - p_x$, p_y peak at mid-IR; The $s - p_z$ peak is 10 times larger than $s - p_x$, p_y peak. (b) Mid-IR spectral response of two 20-period GaN/AlN QD samples measured at T = 12 K under front illumination and 10 V bias. [Reprinted, with permission, from Vardi, *et al.*, *Phys. Rev. B* **80**, 155439 (2009).[112] © 2009 The American Physical Society].

The lateral confinement in the QDs gives rise to additional intraband transitions that can be observed under TE-polarized excitation. The observation of the $s - p_{x,y}$ transition in GaN/AlN QDs was first reported by Vardi *et al.*, using in-plane electronic transport at low temperatures,[112] as illustrated in Fig. 5. The measured $s - p_{x,y}$ energy (0.1–0.3 eV) is significantly larger than the equivalent transition energy in InGaAs/GaAs QDs due to the strong internal electric field in the QDs, which results in stronger in-plane confinement of the electrons at the QD top facet.

TM-polarized IR absorption in the 1.6–2 μm wavelength range, attributed to $s - p_z$ intraband transitions, has also been reported in ternary $Al_xGa_{1-x}N$/AlN QDs (x = 0–0.42) measured at room temperature. For increasing Al mole fraction in the QWs, the $s - p_z$ transition is redshifted as a result of the reduction of the band offset, which is in good agreement with theoretical calculations.[113,114]

3.2. *Absorption saturation and relaxation time*

Saturation of the intraband absorption in GaN/AlN QDs was first demonstrated by Nevou *et al.*[115] in samples containing 200 periods of dots, obtaining saturation for a control power in the range of 15–137 MW/cm^2

(0.03–0.27 pJ/μm^2). In spite of the large signal variation (a consequence of the focusing uncertainty in the sample), even the upper estimate of the saturation intensity for QDs is smaller than the corresponding value for GaN/AlN QWs (9.46 W/μm^2).[60] Time-resolved measurements revealed a multiexponential decay of the differential transmission, with an ultrafast (160 fs) and a slower (1.5 ps) component. The time constants were attributed to the excited state lifetime and to the thermalization time in the ground state, respectively. Based on these results, Monteagudo-Lerma *et al.* fabricated saturable absorbers consisting of 3 periods of GaN/AlN QDs inserted in a GaN-on-AlN waveguide structure.[116] In the case of 2-μm-wide (monomode) waveguides, a 10 dB change in the transmittance was achieved for input energies of $\approx$8 pJ with 150 fs pulses.

3.3. *Intraband photoluminescence*

Room-temperature intraband emission has been observed in optically-pumped GaN/AlN QDs.[117] The $p_z - s$ intraband luminescence was observed at $\lambda = 1.48\,\mu$m under optical excitation at $\lambda = 1.34\,\mu$m perpendicular to the [0001] growth axis. The population of the p_z state arises from Raman scattering by A_1 longitudinal optical phonons in GaN.

4. Intraband Absorption in GaN/AlN Nanowire Heterostructures

4.1. *Intraband absorption*

Three-dimensional nanostructures with a high aspect ratio, commonly referred to as nanowires, have received significant attention in the past decade. As a consequence of the proximity of free surfaces, heterostructures embedded in nanowires offer some basic advantages for devices requiring a low density of structural defects in the active region, and for the integration of materials with large lattice mismatch. Particularly the latter is advantageous for the combination of GaN and AlN binary compounds for near-IR applications. Additionally, the three-dimensional strain distribution in nanowire heterostructures results in an attenuation of the internal electric fields.[118–120] Thereby, the wavelength range accessible using simple GaN/AlGaN heterostructures in the mid-IR spectral range can be extended. For sufficiently small nanowire diameters, embedded nanodisks can be considered as QDs that should exhibit both the TM-polarized $s - p_z$ and the TE-polarized $s - p_{x,y}$ intraband transitions.

Intraband transitions have been observed in GaN/AlN heterostructures within self-assembled nanowires grown on Si(111). Micrographs of such samples are shown in Figs. 10(a,c), and a schematic description of the sample structure is given in Fig. 10(b).[121] To populate the electron ground state, the GaN nanodisks were doped n-type using germanium. The thickness of the nanodisks was varied in the 2–8 nm range. As revealed by temperature-dependent and time-resolved photoluminescence measurements, the carrier confinement in the nanodisks is very efficient.[118] The carrier lifetimes for the band-to-band transition are almost independent of temperature between 10 and 300 K, which confirms the insensitivity to non-radiative recombination. The IR transmission measured by FTIR for a series of samples with GaN/AlN (4 nm/4 nm) nanodisks with varied doping levels is shown in Fig. 10(d). Absorption is observed only for TM-polarized light, and consequently assigned to the $s - p_z$ intraband transition in the nanodisks. This absorption line is blueshifted for increasing Ge doping, whereas a redshift is observed with increasing nanodisk thickness. Note that due to many-body effects, namely the exchange interaction and depolarization shift, the intraband transitions are strongly blueshifted, as described in Fig. 10(e).[121]

4.2. *Absorption saturation and relaxation time*

Tanaka *et al.*[122] studied the intraband relaxation dynamics at 1.55 μm in an ensemble of GaN/AlN nanodisks in self-assembled GaN nanowires by pump-and-probe experiments. An absorption modulation with a FWHM of $\approx$230 fs was observed for an effective pump beam intensity of $\approx$1.4 pJ/mm^2 per pulse. The saturation intensity for the Brewster configuration was estimated at 67 pJ/mm^2 for 100 fs pulses.

5. Resonant Tunneling Transport

A prerequisite for certain types of electrically driven ISB devices is resonant tunneling transport through the quantum barriers. Studies of electron transport through single GaN/AlN/GaN barriers confirmed electron tunneling in 0.5- and 1-nm-thick AlN barriers.[123,124] For AlN barriers in the 1–3 nm range, conductive atomic force microscopy measurements

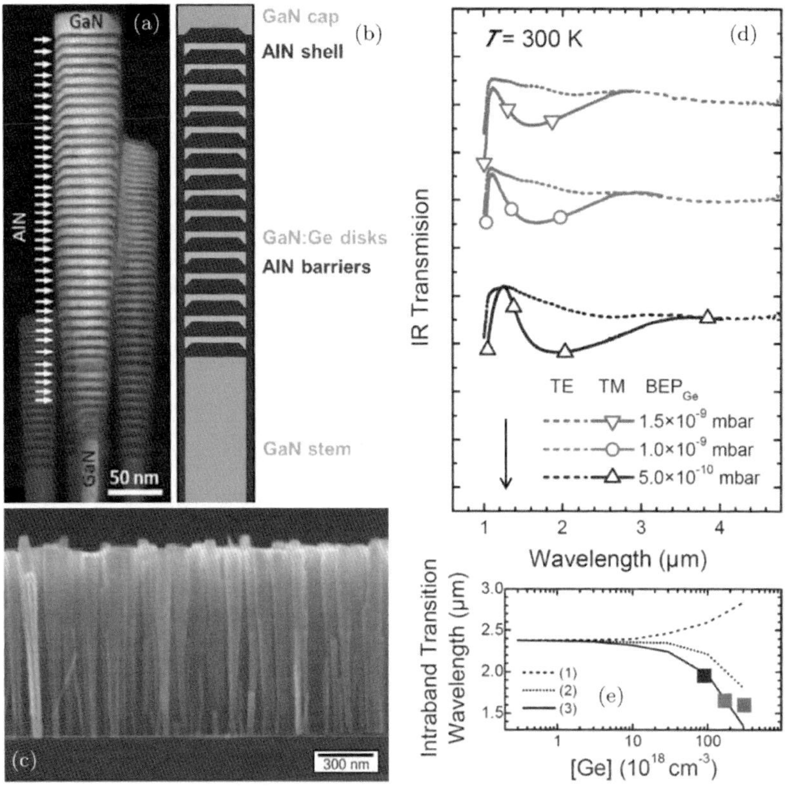

Fig. 10. (a) High-angle annular dark field scanning transmission electron microscopy image of Ge-doped GaN/AlN nanowire heterostructures. (b) Schematic description of the structure. (c) Cross-section scanning electron microscopy image of the same sample. (d) Room-temperature IR transmission spectra for TE-(dashed) and TM-polarized (solid) light measured for Ge-doped GaN/AlN (4 nm/4 nm) nanowire heterostructure with different doping levels in the GaN disks. The spectra are vertically shifted for clarity. (e) Variation of the intraband transition wavelength as a function of the estimated Ge concentration. Dots are experimental values from (d); the dashed line labeled (1) is a 1D calculation of the intraband transition accounting for the screening of the internal electric field; the dotted line (2) incorporates corrections associated to both screening and exchange interaction; the solid line (3) accounts for screening, exchange interaction, and depolarization shift. [Reprinted, with permission, from Beeler, *et al.*, *Nano Lett.* **14**, 1665 (2014).[121] © 2014 American Chemical Society].

revealed efficient current blockage, with pure screw dislocations being the main source of leakage current.[124,125] Finally, leakage due to AlN relaxation and interband tunneling was observed for 5 nm thick barriers.[124,126]

Resonant tunneling diodes (RTDs) consisting of a double-barrier heterostructure are the simplest devices to explore resonant tunneling transport. In spite of the promising results in single barriers, the first reports of Al(Ga)N/GaN RTDs were highly controversial regarding the nature of the observed negative differential resistance (NDR).[124,127–139] NDR was only observed in forward bias[a] and when scanning from negative towards positive bias. The asymmetry introduced by polarization could explain the asymmetric current-voltage characteristics,[140–144] but not the hysteresis, which was attributed to the presence of trap-like defects and impurities in the heterostructures.[145,146] Better reproducibility was achieved by decreasing the substrate dislocation density,[132,134,135] decreasing the device mesa size,[124,147] or using low-aluminum-content AlGaN/GaN double-barriers.[137,148] The reproducibility was also drastically improved when scanning the voltage from a sufficiently negative starting bias due to the detrapping of charges via the Poole-Frenkel effect.[124,147] Finally, NDR has also been observed in GaN/Al(Ga)N MQWs,[148–151] where it was interpreted as resonant tunneling transport between the fundamental and excited states of adjacent QWs.

Nanowire heterostructures offer an alternative approach that provides dislocation-free material with attenuated piezoelectric effects. Resonant tunneling transport was reported in single GaN nanowires containing closely-spaced AlN double barriers.[152] The NDR features were reproducible for both bias sweeping directions, their bias location was tunable with a back-gate bias, and no degradation was found after repeated measurements. NDR associated to resonant tunneling has also been observed in the case of GaN nanowires containing multiple AlN barriers,[153] and for GaN/AlN double barriers in the core-shell geometry.[154]

6. GaN-Based ISB Devices

6.1. *Photodetectors*

Demonstration of near-IR photoconductive GaN/AlN QWIPs have been reported early on.[155,156] However, these devices suffer from a low yield

[a]It is conventionally considered that the positive electrode is the top contact in samples grown along the [0001] axis.

due to the structural defects originating from the large lattice mismatch (2.5% along the *a* direction). An alternative approach to bypass the leakage problem consists of designing devices that can operate at zero bias, which will guarantee a minimum dark current.[157–161] A polar GaN/AlN MQW can provide a photovoltaic signal based on resonant optical rectification.[159,162] Due to the asymmetric potential profile in polar QWs, the excitation of an electron into the upper quantized level is accompanied by a charge displacement along the growth direction, creating an electrical dipole moment. Stacking a large number of QWs, the microscopic dipoles add up to a significant external photovoltage. It was also shown that the responsivity of such devices could be enhanced by a factor of 60 by using QDs instead of QWs as active elements.[163]

The photovoltaic response of QW-based ISB detectors can be further improved by extracting the photoexcited electron from the upper level of the active QW towards an extractor region where it experiences multiple relaxations towards the ground level of the next active QW. This is the operating principle of quantum cascade detectors (QCDs).[165,166] In near-IR GaN/AlGaN QCDs, the extractor region can consist of an AlN/AlGaN short-period superlattice with its confined levels forming a phonon ladder towards the next active QW.[167–169] A simplified QCD design replacing the extractor superlattice by an AlGaN layer has also been demonstrated.[170]

Pump and probe measurements of GaN-based QCDs point to an ISB scattering time of 0.1 ps in the active QWs, and a 1 ps transit time through the extractor.[171] This corresponds to an expected intrinsic frequency bandwidth above 160 GHz. Detectors with the band structure described in Fig. 11 and containing a 40 period (active QW + extractor) cascade have demonstrated room temperature responsivities of $\approx$10 mA/W at 1.5 μm, with a -3 dB bandwidth of $\approx$40 GHz.[164,167,172] The photoresponse of such detectors at normal incidence can be increased by a factor of 30 by using a nanohole array integrated on top of the detector.[173] In the mid-IR, GaN/$Al_{0.5}Ga_{0.5}N$ QCDs have been demonstrated with a peak responsivity of 100 μA/W at 4 μm when operated at 140 K.[174]

In the mid- and far-IR spectral regions, the reduction of the lattice mismatch in the structure makes it more accessible to fabricate photoconductive QWIPs. Devices operating in the 3–5 μm atmospheric window[175] and in the THz domain[176] have been demonstrated operating at low temperature (5 K

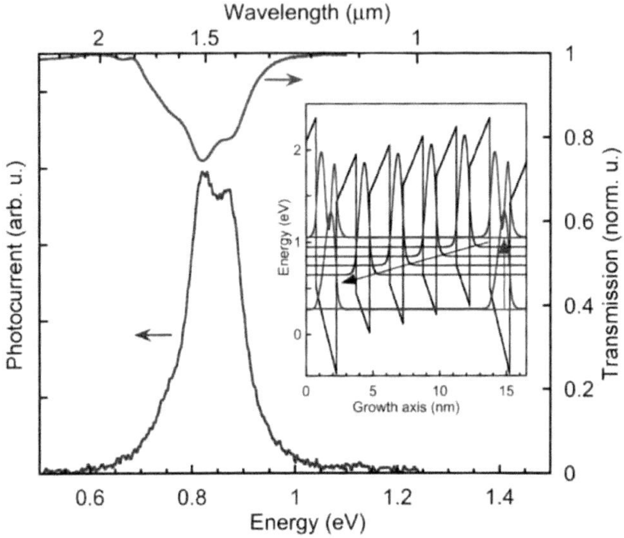

Fig. 11. Room-temperature photocurrent spectrum and transmission spectrum in multi-pass waveguide configuration of the same QCD structure. The inset shows the conduction band diagram and squared envelope functions of one QCD period, which is integrated by a 1.6 ± 0.26 nm GaN active QW followed by an extractor consisting of 4 periods of 1.0 ± 0.15 nm $Al_{0.3}Ga_{0.7}N$ QWs with 1.5 ± 0.15 nm AlN barriers. [Reprinted, with permission, from Sakr, *et al.*, *Appl. Phys. Lett.* **102**, 011135 (2013).[164] © 2013 American Institute of Physics].

and up to 50 K, respectively). The active region of such detectors presents a bound-to-quasi-bound configuration following the step-QW design.

More recently, InGaN/(Al)GaN QWIPs have been fabricated on free-standing non-polar *m*-plane GaN substrates.[177] These devices, consisting of $In_{0.095}Ga_{0.905}N/Al_{0.07}Ga_{0.93}N$ (2.5 nm/56.2 nm) and $In_{0.1}Ga_{0.9}N/GaN$ (3 nm/50 nm) MQWs, displayed photocurrent peaks at 7.5 μm and 9.3 μm, respectively, when characterized at low temperature (14 K).

6.2. *Electro-optical modulators*

Electro-optical amplitude and phase modulators allow tuning the amplitude, phase, and/or polarization state of an optical beam as a function of the control voltage. The first electro-absorption ISB modulation experiments on AlN/GaN QWs were based on the electrical depletion of a 5-period AlN/GaN MQW structure grown on a thick GaN buffer layer.[178] The ISB absorption was modulated by applying an external field, which modifies the charge distribution. To increase the modulation depth, the interaction of light with the active medium should be enhanced, which can be achieved when embedding the QWs in an optical waveguide.[18] By inserting 3 active GaN/AlN QWs operating at $\lambda = 1.55\,\mu$m in a 1-μm-thick $Al_{0.5}Ga_{0.5}N$ waveguiding layer deposited on an AlN-on-sapphire template, a modulation depth of 10 dB was observed for a 5 V voltage swing.

The modulation speed can be improved by emptying active QWs into a local reservoir, instead of transferring carriers over the whole active region. This is the principle of the coupled-QW modulator: The electro-modulation originates from electron tunneling between a wide well (reservoir) and a narrow well separated by an ultrathin (~1 nm) AlN barrier. Room-temperature ISB electro-modulated absorption at telecommunication wavelengths in GaN/AlN coupled QWs has been demonstrated.[179–181] Devices consisting of $15 \times 15\,\mu m^2$ mesas displayed a −3 dB bandwidth which was limited by the device capacitance to 3 GHz.

All the above-described electro-optical modulators rely on a modulation of the light intensity via ISB absorption. Based on the Kramers-Kronig relations, the ISB absorption should also translate into a variation of the refractive index at wavelengths close to the transition, which can be used for phase modulation.[182] Using a depletion modulator consisting of 3 GaN/AlN QWs inserted in an $Al_{0.5}Ga_{0.5}N$/AlN ridge waveguide on sapphire, Lupu *et al.*[183] reported a variation of the refractive index at 1.5 μm of $\Delta n = -5 \times 10^{-3}$ as the population evolved from complete depletion to full population of the QWs. This result is in close agreement with the observation of a refractive index variation from -5×10^{-3} to 6×10^{-3} in 100-period Si-doped GaN/AlN (1.5 nm/3 nm) MQWs using a free-space Mach-Zehnder interferometer configuration.[184] These results open the way for the realization of ISB Mach-Zehnder interferometer phase modulators in the wavelength range used for optical communications.

6.3. *Quantum cascade lasers*

QCLs rely on optical transitions from electron transitions down potential steps where population inversion is maintained via control of tunneling in a suitably designed semiconductor MQW structure.[4] Due to the polarization selection rules of ISB transitions, these devices are in-plane emitters, with their electric-field vector perpendicular to the plane of the layers.

An electron injected into the *active QWs* first undergoes a radiative ISB transition and is rapidly extracted from the lower laser state by a fast nonradiative transition, which is vital for maintaining the population inversion. Then, the electron tunnels through the *injector region* towards the upper level of the next active QW. By stacking a number of periods of active region + injector in a series (a cascade), multiple photons per electron are obtained.

Due to the large lattice mismatch of the GaN/AlN system, the fabrication of GaN-based near-IR QCLs does not appear feasible, despite several theoretical proposals.[17,187,188] However, using a GaN/AlGaN structure, Song *et al.* achieved quantum cascade electroluminescence in the mid-IR ($\lambda = 4.9\,\mu$m peak wavelength, with FWHM of 110 meV) under pulsed operation at 80 K.[189]

There is increasing interest for the fabrication of GaN QCLs in the far-IR, where the lattice mismatch requirements are lowered, and the exploitation of the large LO phonon energy to realize room temperature operation should be possible. Various designs for a GaN THz QCL have been proposed,[163,190–201] all focusing on the resonant-phonon architecture.[202] Terashima and Hirayama have presented THz QCL designs based on 4-well resonant-phonon GaN/AlGaN structures.[194,203,204] Electroluminescence at 1.37 THz ($\lambda = 219\,\mu$m) has been reported in a first structure[203] grown both on a GaN-on-sapphire template and on bulk GaN, and polarization-dependent electroluminescence at 2.82 THz ($\lambda = 106\,\mu$m), slightly tunable by changing the driving voltage in the 20–21 V range, has been reported using a second design grown on an AlN-on-sapphire template.[205] Most recently Terashima *et al.* have reported functioning QCLs utilizing the pure 3-level design described in Fig. 12(a) grown by both MBE as well as MOVPE.[185] The MBE laser was grown on a GaN-on-sapphire template, and it consisted of 100 periods of 1.5 nm $Al_{0.2}Ga_{0.8}N$/ 4 nm GaN/ 1.5 nm $Al_{0.2}Ga_{0.8}N$/ 6 nm GaN. As illustrated in Fig. 12(b), lasing was observed at 5.37 THz ($\lambda = 55.8\,\mu$m) for pulsed injection (14.5 V, 1.75 kA/cm^2) at a temperature of 5.8 K with a peak optical power observed at 2.8 kA/cm^2. Low pressure MOVPE was used to fabricate another QCL with an identical QW structure but with 200 periods and a slightly lower Al concentration ($Al_{0.15}Ga_{0.85}N$) than the MBE laser. The reported lasing occurs for pulsed measurements at 5.2 K at a frequency of 6.97 THz ($\lambda = 43.0\,\mu$m)

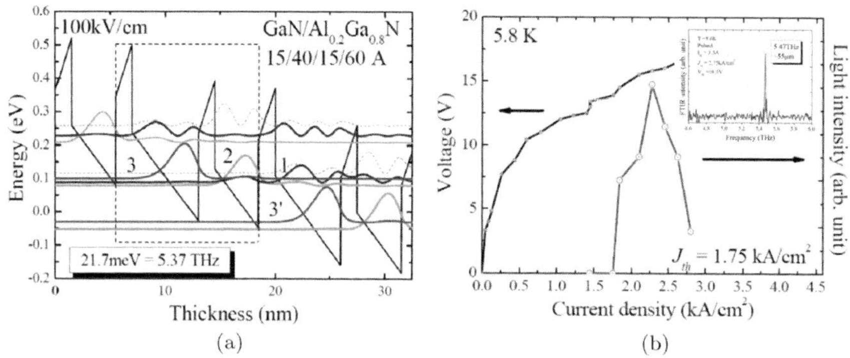

Fig. 12. (a) Conduction band profile and squared wavefunctions of the pure 3-level c-plane QCL structure under an external electric field of 100 kV/cm with GaN wells and $Al_{0.2}Ga_{0.8}N$ barriers. (b) The light intensity and voltage vs. current density characteristics for the QCL is shown along with a frequency spectrum in the inset. [Reprinted, with permission, from W. Terashima and H. Hirayama, *SPIE Proc.* **9483**, 948304 (2015) and *SPIE Newsroom* (DOI: 10.1117/2.1201507.006058) (2015).[185,186] © 2015 Society of Photo-Optical Instrumentation Engineers (SPIE).]

with a corresponding threshold current density of 0.75 kA/cm^2. These recent results are promising first steps for the development of III-Nitride based QCLs.

7. Conclusions

In this chapter, we summarized recent research on III-nitride nanostructures for the fabrication of ISB optoelectronic devices. The large conduction band offset of the GaN/AlN system and its sub-picosecond ISB scattering rate make them promising materials for high-speed devices for fiber-optics telecommunications. Furthermore, GaN is transparent in a large spectral region, notably for wavelengths longer than 360 nm, except for the Reststrahlen band (9.6–19 μm). One exciting challenge for ISB optoelectronics is their extension towards the far-IR spectral range. Beyond 20 μm, the large energy of GaN longitudinal-optical phonons opens prospects for high-temperature THz quantum cascade lasers and ISB devices covering the 5–10 THz band, inaccessible to GaAs due to phonon absorption.

ISB absorption in polar GaN/AlGaN QWs can be tuned from 1.0 μm to 10 μm. However, when decreasing the energy between electronic levels towards the far infrared (>20 μm), the internal electric field associated to the spontaneous and piezoelectric polarization poses a major hurdle for device design. For polar crystallographic orientations, sophisticated band engineering is required to achieve a pseudo-square potential distribution, while the use of nonpolar orientations allows much easier device design and operation without the influence of this polarization-related electric field. Going beyond planar heterostructures, QDs can lift the constraints on the light polarization with additional TE-polarized ISB transitions, and NWs have benefits for the integration of the lattice mismatched binaries GaN and AlN aimed at near-IR applications.

A variety of GaN-based ISB devices have been demonstrated. The feasibility of GaN/AlN all-optical switches and electro-optical modulators operating at 1.55 μm was verified by several groups. Photovoltaic QWIPs and QCDs in the near-IR, and photoconductive QWIPs in the 3–5 μm atmospheric window and in the THz domain have also been reported. For both all-optical switches and photovoltaic detectors, QDs were demonstrated to be a good alternative to QWs to improve the device performance. Despite the low efficiency of ISB photoluminescence, near-IR emission was observed in optically-pumped GaN/AlN QWs and QDs. Mid-IR ISB electroluminescence was measured on chirped AlGaN/GaN MQWs, and in quantum cascade GaN/AlGaN superlattices. Great effort was made to propose different designs of a GaN-based THz QCL based on the resonant-phonon architecture. From the experimental viewpoint, lasing was reported in the 1.37–6.97 THz range for pulsed injection at 5 K, opening prospects for the future development of room-temperature III-Nitride based QCLs.

Acknowledgements

This work is supported by the EU ERC-StG project “TeraGaN” (grant #278428). A. A. acknowledges financial support by the French National Research Agency via the GaNEX program (ANR-11-LABX-0014).

References

1. A. Kamgar, P. Kneschaurek, G. Dorda, and J. Koch, *Phys. Rev. Lett.*, **32**, 1251 (1974).
2. L. C. West and S. J. Eglash, *Appl. Phys. Lett.*, **46**, 1156 (1985).
3. N. Suzuki and N. Iizuka, *Jpn. J. Appl. Phys.*, **38**, L363 (1999).

4. J. Faist, F. Capasso, D. L. Sivco, C. Sirtori, A. L. Hutchinson, and A. Y. Cho, *Science*, **264**, 553 (1994).
5. B. F. Levine, K. K. Choi, C. G. Bethea, J. Walker, and R. J. Malik, *Appl. Phys. Lett.*, **50**, 1092 (1987).
6. B. F. Levine, *J. Appl. Phys.* **74**, R1 (1993).
7. G. Bastard, *Wave Mechanics Applied to Semiconductor Heterostructures* (Les Editions de Physique, Les Ulis, France, 1988).
8. H. C. Liu and F. Capasso, *Intersubband Transitions in Quantum Wells?: Physics and Device Applications I* (Academic Press, San Diego, CA, 2000).
9. M. Helm, in *Intersubband Transit. Quantum Wells Phys. Device Appl. Ed H C Liu F Capasso* (Acad. Press, San Diego, 2000).
10. K. M. S. V. Bandara, D. D. Coon, B. O, Y. F. Lin, and M. H. Francombe, *Appl. Phys. Lett.*, **53**, 1931 (1988).
11. K. M. S. V. Bandara, D. D. Coon, B. O, Y. F. Lin, and M. H. Francombe, *Appl. Phys. Lett.*, **55**, 206 (1989).
12. S. J. Allen, D. C. Tsui, and B. Vinter, *Solid State Commun.*, **20**, 425 (1976).
13. W. L. Bloss, *J. Appl. Phys.*, **66**, 3639 (1989).
14. N. Binggeli, P. Ferrara, and A. Baldereschi, *Phys. Rev. B*, **63**, 245306 (2001).
15. D. Cociorva, W. G. Aulbur, and J. W. Wilkins, *Solid State Commun.*, **124**, 63 (2002).
16. M. Tchernycheva, L. Nevou, L. Doyennette, F. Julien, E. Warde, F. Guillot, E. Monroy, E. Bellet-Amalric, T. Remmele, and M. Albrecht, *Phys. Rev. B*, **73**, 125347 (2006).
17. D. Hofstetter, E. Baumann, F. R. Giorgetta, R. Théron, H. Wu, W. J. Schaff, J. Dawlaty, P. A. George, L. F. Eastman, F. Rana, P. K. Kandaswamy, F. Guillot, and E. Monroy, *Proc. IEEE*, **98**, 1234 (2010).
18. H. Machhadani, P. Kandaswamy, S. Sakr, A. Vardi, A. Wirtmüller, L. Nevou, F. Guillot, G. Pozzovivo, M. Tchernycheva, A. Lupu, L. Vivien, P. Crozat, E. Warde, C. Bougerol, S. Schacham, G. Strasser, G. Bahir, E. Monroy, and F. H. Julien, *New J. Phys.*, **11**, 125023 (2009).
19. M. Beeler, E. Trichas, and E. Monroy, Semicond. *Sci. Technol.*, **28**, 74022 (2013).
20. M. Hao, S. Mahanty, R. S. Qhalid Fareed, S. Tottori, K. Nishino, and S. Sakai, *Appl. Phys. Lett.*, **74**, 2788 (1999).
21. J. Yang, G. J. Brown, M. Dutta, and M. A. Stroscio, *J. Appl. Phys.*, **98**, 43517 (2005).
22. M. Welna, R. Kudrawiec, M. Motyka, R. Kucharski, M. Zając, M. Rudziński, J. Misiewicz, R. Doradziński, and R. Dwiliński, *Cryst. Res. Technol.*, **47**, 347 (2012).
23. G. Cywiński, C. Skierbiszewski, A. Fedunieiwcz-Żmuda, M. Siekacz, L. Nevou, L. Doyennette, M. Tchernycheva, F. H. Julien, P. Prystawko, M. Kryśko, S. Grzanka, I. Grzegory, A. Presz, J. Z. Domagała, J. Smalc, M. Albrecht, T. Remmele, and S. Porowski, *J. Vac. Sci. Technol. B*, **24**, 1505 (2006).
24. G. Chen, X. Q. Wang, X. Rong, P. Wang, F. J. Xu, N. Tang, Z. X. Qin, Y. H. Chen, and B. Shen, *Sci. Rep.*, **5**, 11485 (2015).
25. C. Edmunds, L. Tang, D. Li, M. Cervantes, G. Gardner, T. Paskova, M. J. Manfra, and O. Malis, *J. Electron. Mater.*, **41**, 881 (2012).
26. O. Malis, C. Edmunds, M. J. Manfra, and D. L. Sivco, *Appl. Phys. Lett.*, **94**, 161111 (2009).
27. F. Bernardini, V. Fiorentini, and D. Vanderbilt, *Phys. Rev. B*, **56**, R10024 (1997).

28. A. Zoroddu, F. Bernardini, P. Ruggerone, and V. Fiorentini, *Phys. Rev. B*, **64**, 45208 (2001).
29. S. Birner, T. Zibold, T. Andlauer, T. Kubis, M. Sabathil, A. Trellakis, and P. Vogl, *IEEE Trans. Electron Devices*, **54**, 2137 (2007).
30. P. K. Kandaswamy, F. Guillot, E. Bellet-Amalric, E. Monroy, L. Nevou, M. Tchernycheva, A. Michon, F. H. Julien, E. Baumann, F. R. Giorgetta, D. Hofstetter, T. Remmele, M. Albrecht, S. Birner, and L. S. Dang, *J. Appl. Phys.*, **104**, 93501 (2008).
31. J. S. Speck and S. F. Chichibu, *MRS Bull.*, **34**, 304 (2011).
32. A. E. Romanov, T. J. Baker, S. Nakamura, J. S. Speck, and ERATO/JST UCSB Group, *J. Appl. Phys.*, **100**, 23522 (2006).
33. L. Lahourcade, E. Bellet-Amalric, E. Monroy, M. Abouzaid, and P. Ruterana, *Appl. Phys. Lett.*, **90**, 131909 (2007).
34. L. Lahourcade, P. K. Kandaswamy, J. Renard, P. Ruterana, H. Machhadani, M. Tchernycheva, F. H. Julien, B. Gayral, and E. Monroy, *Appl. Phys. Lett.*, **93**, 111906 (2008).
35. A. E. Romanov, T. J. Baker, S. Nakamura, J. S. Speck, and ERATO/JST UCSB Group, *J. Appl. Phys.*, **100**, 23522 (2006).
36. H. Machhadani, M. Beeler, S. Sakr, E. Warde, Y. Kotsar, M. Tchernycheva, M. P. Chauvat, P. Ruterana, G. Nataf, P. De Mierry, E. Monroy, and F. H. Julien, *J. Appl. Phys.*, **113**, 143109 (2013).
37. N. Iizuka, K. Kaneko, and N. Suzuki, *Appl. Phys. Lett.*, **81**, 1803 (2002).
38. K. Kishino, A. Kikuchi, H. Kanazawa, and T. Tachibana, *Appl. Phys. Lett.*, **81**, 1234 (2002).
39. A. Helman, M. Tchernycheva, A. Lusson, E. Warde, F. H. Julien, K. Moumanis, G. Fishman, E. Monroy, B. Daudin, D. Le Si Dang, E. Bellet-Amalric, and D. Jalabert, *Appl. Phys. Lett.*, **83**, 5196 (2003).
40. T. G. Andersson, X. Y. Liu, T. Aggerstam, P. Holmström, S. Lourdudoss, L. Thylen, Y. L. Chen, C. H. Hsieh, and I. Lo, *Microelectron. J.*, **40**, 360 (2009).
41. X. Y. Liu, P. Holmström, P. Jänes, L. Thylén, and T. G. Andersson, *Phys. Status Solidi B*, **244**, 2892 (2007).
42. K. Berland, M. Stattin, R. Farivar, D. M. S. Sultan, P. Hyldgaard, A. Larsson, S. M. Wang, and T. G. Andersson, *Appl. Phys. Lett.*, **97**, 43507 (2010).
43. C. Bayram, N. Péré-laperne, and M. Razeghi, *Appl. Phys. Lett.*, **95**, 201906 (2009).
44. R. Yang, J. Xu, and M. Sweeny, *Phys. Rev. B*, **50**, 7474 (1994).
45. F. Guillot, B. Amstatt, E. Bellet-Amalric, E. Monroy, L. Nevou, L. Doyennette, F. H. Julien, and L. S. Dang, *Superlattices Microstruct.*, **40**, 306 (2006).
46. E. Baumann, F. R. Giorgetta, D. Hofstetter, S. Golka, W. Schrenk, G. Strasser, L. Kirste, S. Nicolay, E. Feltin, J. F. Carlin, and N. Grandjean, *Appl. Phys. Lett.*, **89**, 41106 (2006).
47. H. Sodabanlu, J.-S. Yang, M. Sugiyama, Y. Shimogaki, and Y. Nakano, *Appl. Phys. Lett.*, **95**, 161908 (2009).
48. J.-S. Yang, H. Sodabanlu, M. Sugiyama, Y. Nakano, and Y. Shimogaki, *Appl. Phys. Lett.*, **95**, 162111 (2009).
49. S. Nicolay, E. Feltin, J.-F. Carlin, N. Grandjean, L. Nevou, F. H. Julien, M. Schmidbauer, T. Remmele, and M. Albrecht, *Appl. Phys. Lett.*, **91**, 61927 (2007).

50. N. Iizuka, K. Kaneko, N. Suzuki, T. Asano, S. Noda, and O. Wada, *Appl. Phys. Lett.*, **77**, 648 (2000).
51. C. Gmachl, S. V. Frolov, H. M. Ng, S.-N. G. Chu, and A. Y. Cho, *Electron. Lett.*, **37**, 378 (2001).
52. J. D. Heber, C. Gmachl, H. M. Ng, and A. Y. Cho, *Appl. Phys. Lett.*, **81**, 1237 (2002).
53. R. Rapaport, G. Chen, O. Mitrofanov, C. Gmachl, H. M. Ng, and S. N. G. Chu, *Appl. Phys. Lett.*, **83**, 263 (2003).
54. N. Iizuka, K. Kaneko, and N. Suzuki, *Opt. Express*, **13**, 3835 (2005).
55. J. Hamazaki, H. Kunugita, K. Ema, A. Kikuchi, and K. Kishino, *Phys. Rev. B*, **71**, 165334 (2005).
56. N. Suzuki and N. Iizuka, *Jpn. J. Appl. Phys.*, **36**, L1006 (1997).
57. N. Suzuki, N. Iizuka, and K. Kaneko, *Proc. SPIE*, **3940**, 127 (2000).
58. N. Iizuka, K. Kaneko, and N. Suzuki, *Electron. Lett.*, **40**, 962963 (2004).
59. N. Iizuka, K. Kaneko, and N. Suzuki, *IEEE J. Quantum Electron.*, **42**, 765 (2006).
60. Y. Li, A. Bhattacharyya, C. Thomidis, T. D. Moustakas, and R. Paiella, *Opt. Express*, **15**, 17922 (2007).
61. H. Sodabanlu, J.-S. Yang, T. Tanemura, M. Sugiyama, Y. Shimogaki, and Y. Nakano, *Appl. Phys. Lett.*, **99**, 151102 (2011).
62. N. Iizuka, H. Yoshida, N. Managaki, T. Shimizu, S. Hassanet, C. Cumtornkittikul, M. Sugiyama, and Y. Nakano, *Opt. Express*, **17**, 23247 (2009).
63. L. Monteagudo-Lerma, S. Valdueza-Felip, F. B. Naranjo, P. Corredera, L. Rapenne, E. Sarigiannidou, G. Strasser, E. Monroy, and M. González-Herráez, *Opt. Express*, **21**, 27578 (2013).
64. L. Nevou, F. H. Julien, R. Colombelli, F. Guillot, and E. Monroy, *Electron. Lett.*, **42**, 1308 (2006).
65. L. Nevou, M. Tchernycheva, F. H. Julien, F. Guillot, and E. Monroy, *Appl. Phys. Lett.*, **90**, 121106 (2007).
66. K. Driscoll, Y. Liao, A. Bhattacharyya, L. Zhou, D. J. Smith, T. D. Moustakas, and R. Paiella, *Appl. Phys. Lett.*, **94**, 81120 (2009).
67. H. Machhadani, Y. Kotsar, S. Sakr, M. Tchernycheva, R. Colombelli, J. Mangeney, E. Bellet-Amalric, E. Sarigiannidou, E. Monroy, and F. H. Julien, *Appl. Phys. Lett.*, **97**, 191101 (2010).
68. C. Bayram, *J. Appl. Phys.*, **111**, 13514 (2012).
69. G. Chen, Z. L. Li, X. Q. Wang, C. C. Huang, X. Rong, L. W. Sang, F. J. Xu, N. Tang, Z. X. Qin, M. Sumiya, Y. H. Chen, W. K. Ge, and B. Shen, *Appl. Phys. Lett.*, **102**, 192109 (2013).
70. C. Edmunds, L. Tang, D. Li, M. Cervantes, G. Gardner, T. Paskova, M. J. Manfra, and O. Malis, *J. Electron. Mater.*, **41**, 881 (2012).
71. C. Gmachl, H. M. Ng, S.-N. George Chu, and A. Y. Cho, *Appl. Phys. Lett.*, **77**, 3722 (2000).
72. C. Gmachl, H. M. Ng, and A. Y. Cho, *Appl. Phys. Lett.*, **79**, 1590 (2001).
73. C. C. Huang, F. J. Xu, X. D. Yan, J. Song, Z. Y. Xu, L. B. Cen, Y. Wang, J. H. Pan, X. Q. Wang, Z. J. Yang, B. Shen, B. S. Zhang, X. S. Chen, and W. Lu, *Appl. Phys. Lett.*, **98**, 132105 (2011).
74. P. K. Kandaswamy, H. Machhadani, C. Bougerol, S. Sakr, M. Tchernycheva, F. H. Julien, and E. Monroy, *Appl. Phys. Lett.*, **95**, 141911 (2009).

75. H. M. Ng, C. Gmachl, T. Siegrist, S. N. G. Chu, and A. Y. Cho, *Phys. Status Solidi A*, **188**, 825 (2001).
76. H. M. Ng, C. Gmachl, J. D. Heber, J. W. P. Hsu, S. N. G. Chu, and A. Y. Cho, *Phys. Status Solidi B*, **234**, 817 (2002).
77. N. Péré-Laperne, C. Bayram, L. Nguyen-The, R. McClintock, and M. Razeghi, *Appl. Phys. Lett.*, **95**, 131109 (2009).
78. B. Sherliker, M. Halsall, I. Kasalynas, D. Seliuta, G. Valusis, M. Vengris, M. Barkauskas, V. Sirutkaitis, P. Harrison, V. D. Jovanovic, D. Indjin, Z. Ikonic, P. J. Parbrook, T. Wang, and P. D. Buckle, Semicond. *Sci. Technol.*, **22**, 1240 (2007).
79. W. Tian, W. Y. Yan, X. Hui, S. L. Li, Y. Y. Ding, Y. Li, Y. Tian, J. N. Dai, Y. Y. Fang, Z. H. Wu, C. H. Yu, and C. Q. Chen, *J. Appl. Phys.*, **112**, 63526 (2012).
80. M. Beeler, C. Bougerol, E. Bellet-Amalric, and E. Monroy, *Appl. Phys. Lett.*, **105**, 131106 (2014).
81. M. Beeler, C. Bougerol, E. Bellet-Amalric, and E. Monroy, *Appl. Phys. Lett.*, **103**, 91108 (2013).
82. M. Beeler, C. Bougerol, E. Bellet-Amalaric, and E. Monroy, *Phys. Status Solidi A*, **211**, 761 (2014).
83. F. Wu, W. Tian, W. Y. Yan, J. Zhang, S. C. Sun, J. N. Dai, Y. Y. Fang, Z. H. Wu, and C. Q. Chen, *J. Appl. Phys.*, **113**, 154505 (2013).
84. P. K. Kandaswamy, H. Machhadani, Y. Kotsar, S. Sakr, A. Das, M. Tchernycheva, L. Rapenne, E. Sarigiannidou, F. H. Julien, and E. Monroy, *Appl. Phys. Lett.*, **96**, 141903 (2010).
85. D.-F. Liu, J.-G. Jiang, Y. Cheng, and J.-F. He, *Phys. E Low-Dimens. Syst. Nanostructures*, **54**, 253 (2013).
86. C. Gmachl and H. M. Ng, *Electron. Lett.*, **39**, 567 (2003).
87. J. Shao, D. N. Zakharov, C. Edmunds, O. Malis, and M. J. Manfra, *Appl. Phys. Lett.*, **103**, 232103 (2013).
88. J. Shao, L. Tang, C. Edmunds, G. Gardner, O. Malis, and M. Manfra, *J. Appl. Phys.*, **114**, 23508 (2013).
89. A. M. Armstrong, K. Kelchner, S. Nakamura, S. P. DenBaars, and J. S. Speck, *Appl. Phys. Lett.*, **103**, 232108 (2013).
90. T. Kotani, M. Arita, and Y. Arakawa, *Appl. Phys. Lett.*, **105**, 261108 (2014).
91. T. Kotani, M. Arita, and Y. Arakawa, *Appl. Phys. Lett.*, **107**, 112107 (2015).
92. C. B. Lim, M. Beeler, A. Ajay, J. Lähnemann, E. Bellet-Amalric, C. Bougerol, and E. Monroy, *J. Appl. Phys.*, **118**, 14309 (2015).
93. C. B. Lim, A. Ajay, C. Bougerol, B. Haas, J. Schörmann, M. Beeler, J. Lähnemann, M. Eickhoff, and E. Monroy, *Nanotechnology*, **26**, 435201 (2015).
94. C. B. Lim, A. Ajay, C. Bougerol, J. Lähnemann, F. Donatini, J. Schörmann, E. Bellet-Amalric, D. A. Browne, M. Jiménez-Rodríguez, and E. Monroy, *Nanotechnology*, **27**, 145201 (2016).
95. C. B. Lim, M. Beeler, A. Ajay, J. Lähnemann, E. Bellet-Amalric, C. Bougerol, J. Schörmann, M. Eickhoff, and E. Monroy, *Jpn. J. Appl. Phys.*, **55**, 05FG05 (2016).
96. T. Kotani, M. Arita, and Y. Arakawa, *Appl. Phys. Lett.*, **105**, 261108 (2014).
97. C. Edmunds, J. Shao, M. Shirazi-HD, M. J. Manfra, and O. Malis, *Appl. Phys. Lett.*, **105**, 21109 (2014).

98. K. W. Berryman, S. A. Lyon, and M. Segev, *Appl. Phys. Lett.*, **70**, 1861 (1997).
99. J. Phillips, K. Kamath, and P. Bhattacharya, *Appl. Phys. Lett.*, **72**, 2020 (1998).
100. V. Ryzhii, *Semicond. Sci. Technol.*, **11**, 759 (1996).
101. D. Pan, E. Towe, and S. Kennerly, *Appl. Phys. Lett.*, **73**, 1937 (1998).
102. Z. Chen, O. Baklenov, E. T. Kim, I. Mukhametzhanov, J. Tie, A. Madhukar, Z. Ye, and J. C. Campbell, *J. Appl. Phys.*, **89**, 4558 (2001).
103. L. Chu, A. Zrenner, M. Bichler, and G. Abstreiter, *Appl. Phys. Lett.*, **79**, 2249 (2001).
104. B. Daudin, F. Widmann, G. Feuillet, Y. Samson, M. Arlery, and J. Rouvière, *Phys. Rev. B*, **56**, R7069 (1997).
105. B. Damilano, N. Grandjean, F. Semond, J. Massies, and M. Leroux, *Appl. Phys. Lett.*, **75**, 962 (1999).
106. N. Gogneau, D. Jalabert, E. Monroy, T. Shibata, M. Tanaka, and B. Daudin, *J. Appl. Phys.*, **94**, 2254 (2003).
107. M. Miyamura, K. Tachibana, and Y. Arakawa, *Appl. Phys. Lett.*, **80**, 3937 (2002).
108. V. Chamard, T. Schülli, M. Sztucki, T. Metzger, E. Sarigiannidou, J.-L. Rouvière, M. Tolan, and C. Adelmann, *Phys. Rev. B*, **69**, 125327 (2004).
109. F. Guillot, M. Tchernycheva, L. Nevou, L. Doyennette, E. Monroy, F. H. Julien, L. S. Dang, T. Remmele, M. Albrecht, T. Shibata, and M. Tanaka, *Phys. Status Solidi A*, **203**, 1754 (2006).
110. M. Tchernycheva, L. Nevou, L. Doyennette, A. Helman, R. Colombelli, F. H. Julien, F. Guillot, E. Monroy, T. Shibata, and M. Tanaka, *Appl. Phys. Lett.*, **87**, 101912 (2005).
111. D. T. Nguyen, W. Wüster, P. Roussignol, C. Voisin, G. Cassabois, M. Tchernycheva, F. H. Julien, F. Guillot, and E. Monroy, *Appl. Phys. Lett.*, **97**, 61903 (2010).
112. A. Vardi, G. Bahir, S. E. Schacham, P. K. Kandaswamy, and E. Monroy, *Phys. Rev. B*, **80**, 155439 (2009).
113. C. Himwas, R. Songmuang, Le Si Dang, J. Bleuse, L. Rapenne, E. Sarigiannidou, and E. Monroy, *Appl. Phys. Lett.*, **101**, 241914 (2012).
114. C. Himwas, M. den Hertog, E. Bellet-Amalric, R. Songmuang, F. Donatini, L. Si Dang, and E. Monroy, *J. Appl. Phys.*, **116**, 23502 (2014).
115. L. Nevou, J. Mangeney, M. Tchernycheva, F. H. Julien, F. Guillot, and E. Monroy, *Appl. Phys. Lett.*, **94**, 132104 (2009).
116. L. Monteagudo-Lerma, S. Valdueza-Felip, F. B. Naranjo, P. Corredera, L. Rapenne, E. Sarigiannidou, G. Strasser, E. Monroy, and M. González-Herráez, *Opt. Express*, **21**, 27578 (2013).
117. L. Nevou, F. H. Julien, M. Tchernycheva, F. Guillot, E. Monroy, and E. Sarigiannidou, *Appl. Phys. Lett.*, **92**, 161105 (2008).
118. M. Beeler, C. B. Lim, P. Hille, J. Bleuse, J. Schörmann, M. de la Mata, J. Arbiol, M. Eickhoff, and E. Monroy, *Phys. Rev. B*, **91**, 205440 (2015).
119. J. Lähnemann, O. Brandt, C. Pfüller, T. Flissikowski, U. Jahn, E. Luna, M. Hanke, M. Knelangen, A. Trampert, and H. T. Grahn, *Phys. Rev. B*, **84**, 155303 (2011).
120. R. Songmuang, D. Kalita, P. Sinha, M. den Hertog, R. André, T. Ben, D. González, H. Mariette, and E. Monroy, *Appl. Phys. Lett.*, **99**, 141914 (2011).
121. M. Beeler, P. Hille, J. Schörmann, J. Teubert, M. de la Mata, J. Arbiol, M. Eickhoff, and E. Monroy, *Nano Lett.* **14**, 1665 (2014).

122. K. Tanaka, K. Ikuno, Y. Kasai, K. Fukunaga, H. Kunugita, K. Ema, A. Kikuchi, and K. Kishino, *J. Lumin.* **128**, 1084 (2008).
123. S. Leconte, F. Guillot, E. Sarigiannidou, and E. Monroy, *Semicond. Sci. Technol.*, **22**, 107 (2007).
124. S. Leconte, L. Gerrer, and E. Monroy, *Microelectron. J.*, **40**, 339 (2009).
125. B. S. Simpkins, E. T. Yu, P. Waltereit, and J. S. Speck, *J. Appl. Phys.*, **94**, 1448 (2003).
126. J. Simon, Z. Zhang, K. Goodman, H. Xing, T. Kosel, P. Fay, and D. Jena, *Phys. Rev. Lett.*, **103**, 26801 (2009).
127. A. Kikuchi, R. Bannai, K. Kishino, C.-M. Lee, and J.-I. Chyi, *Appl. Phys. Lett.*, **81**, 1729 (2002).
128. C. T. Foxon, S. V. Novikov, A. E. Belyaev, L. X. Zhao, O. Makarovsky, D. J. Walker, L. Eaves, R. I. Dykeman, S. V. Danylyuk, S. A. Vitusevich, M. J. Kappers, J. S. Barnard, and C. J. Humphreys, *Phys. Status Solidi C*, **0**, 2389 (2003).
129. A. E. Belyaev, C. T. Foxon, S. V. Novikov, O. Makarovsky, L. Eaves, M. J. Kappers, and C. J. Humphreys, *Appl. Phys. Lett.*, **83**, 3626 (2003).
130. A. Kikuchi, R. Bannai, K. Kishino, C.-M. Lee, and J.-I. Chyi, *Appl. Phys. Lett.*, **83**, 3628 (2003).
131. M. Hermann, E. Monroy, A. Helman, B. Baur, M. Albrecht, B. Daudin, O. Ambacher, M. Stutzmann, and M. Eickhoff, *Phys. Status Solidi C*, **1**, 2210 (2004).
132. S. Golka, C. Pflügl, W. Schrenk, G. Strasser, C. Skierbiszewski, M. Siekacz, I. Grzegory, and S. Porowski, *Appl. Phys. Lett.*, **88**, 172106 (2006).
133. A. M. Kurakin, S. A. Vitusevich, S. V. Danylyuk, A. V. Naumov, C. T. Foxon, S. V. Novikov, N. Klein, H. Lüth, and A. E. Belyaev, *Phys. Status Solidi C*, **3**, 2265 (2006).
134. Z. Vashaei, C. Bayram, and M. Razeghi, *J. Appl. Phys.*, **107**, 83505 (2010).
135. C. Bayram, Z. Vashaei, and M. Razeghi, *Appl. Phys. Lett.*, **97**, 92104 (2010).
136. C. Bayram, Z. Vashaei, and M. Razeghi, *Appl. Phys. Lett.*, **96**, 42103 (2010).
137. C. Bayram, Z. Vashaei, and M. Razeghi, *Appl. Phys. Lett.*, **97**, 181109 (2010).
138. D. Li, L. Tang, C. Edmunds, J. Shao, G. Gardner, M. J. Manfra, and O. Malis, *Appl. Phys. Lett.*, **100**, 252105 (2012).
139. D. Li, J. Shao, L. Tang, C. Edmunds, G. Gardner, M. J. Manfra, and O. Malis, *Semicond. Sci. Technol.*, **28**, 74024 (2013).
140. K. M. Indlekofer, E. Donà, J. Malindretos, M. Bertelli, M. Kocan, A. Rizzi, and H. Lüth, *Phys. Status Solidi B*, **234**, 769 (2002).
141. F. Sacconi, A. Di Carlo, and P. Lugli, *Phys. Status Solidi A*, **190**, 295 (2002).
142. S. Sakr, E. Warde, M. Tchernycheva, and F. H. Julien, *J. Appl. Phys.*, **109**, 23717 (2011).
143. K. Berland, T. Andersson, and P. Hyldgaard, *Phys. Rev. B*, **84**, 245313 (2011).
144. E. Warde, S. Sakr, M. Tchernycheva, and F. H. Julien, *J. Electron. Mater.*, **41**, 965 (2012).
145. M. V. Petrychuk, A. E. Belyaev, A. M. Kurakin, S. V. Danylyuk, N. Klein, and S. A. Vitusevich, *Appl. Phys. Lett.*, **91**, 222112 (2007).
146. L. Yang, H. He, W. Mao, and Y. Hao, *Appl. Phys. Lett.*, **99**, 153501 (2011).
147. M. Boucherit, A. Soltani, E. Monroy, M. Rousseau, D. Deresmes, M. Berthe, C. Durand, and J.-C. De Jaeger, *Appl. Phys. Lett.*, **99**, 182109 (2011).

148. A. Grier, A. Valavanis, C. Edmunds, J. Shao, J. D. Cooper, G. Gardner, M. J. Manfra, O. Malis, D. Indjin, Z. Ikonić, and P. Harrison, *J. Appl. Phys.*, **118**, 224308 (2015).
149. S. Sakr, Y. Kotsar, M. Tchernycheva, E. Warde, N. Isac, E. Monroy, and F. H. Julien, *Appl. Phys. Express*, **5**, 52203 (2012).
150. F. Sudradjat, W. Zhang, K. Driscoll, Y. Liao, A. Bhattacharyya, C. Thomidis, L. Zhou, D. J. Smith, T. D. Moustakas, and R. Paiella, *J. Appl. Phys.*, **108**, 103704 (2010).
151. F. F. Sudradjat, W. Zhang, K. Driscoll, Y. Liao, A. Bhattacharyya, C. Thomidis, L. Zhou, D. J. Smith, T. D. Moustakas, and R. Paiella, *Phys. Status Solidi C*, **9**, 588 (2012).
152. R. Songmuang, G. Katsaros, E. Monroy, P. Spathis, C. Bougerol, M. Mongillo, and S. De Franceschi, *Nano Lett.*, **10**, 3545 (2010).
153. L. Rigutti, G. Jacopin, A. D. L. Bugallo, M. Tchernycheva, E. Warde, F. H. Julien, R. Songmuang, E. Galopin, L. Largeau, and J.-C. Harmand, *Nanotechnology*, **21**, 425206 (2010).
154. S. D. Carnevale, C. Marginean, P. J. Phillips, T. F. Kent, A. T. M. G. Sarwar, M. J. Mills, and R. C. Myers, *Appl. Phys. Lett.*, **100**, 142115 (2012).
155. D. Hofstetter, S.-S. Schad, H. Wu, W. J. Schaff, and L. F. Eastman, *Appl. Phys. Lett.*, **83**, 572 (2003).
156. E. Baumann, F. R. Giorgetta, D. Hofstetter, H. Lu, X. Chen, W. J. Schaff, L. F. Eastman, S. Golka, W. Schrenk, and G. Strasser, *Appl. Phys. Lett.*, **87**, 191102 (2005).
157. D. Hofstetter, E. Baumann, F. R. Giorgetta, M. Graf, M. Maier, F. Guillot, E. Bellet-Amalric, and E. Monroy, *Appl. Phys. Lett.*, **88**, 121112 (2006).
158. E. Baumann, F. R. Giorgetta, D. Hofstetter, S. Golka, W. Schrenk, G. Strasser, L. Kirste, S. Nicolay, E. Feltin, J. F. Carlin, and N. Grandjean, *Appl. Phys. Lett.*, **89**, 41106 (2006).
159. D. Hofstetter, E. Baumann, F. R. Giorgetta, F. Guillot, S. Leconte, and E. Monroy, *Appl. Phys. Lett.*, **91**, 131115 (2007).
160. F. R. Giorgetta, E. Baumann, F. Guillot, E. Monroy, and D. Hofstetter, *Electron. Lett.*, **43**, 185 (2007).
161. D. Hofstetter, E. Baumann, F. R. Giorgetta, R. Théron, H. Wu, W. J. Schaff, J. Dawlaty, P. A. George, L. F. Eastman, F. Rana, P. K. Kandaswamy, S. Leconte, and E. Monroy, *J. Phys. Condens. Matter*, **21**, 174208 (2009).
162. E. Rosencher and P. Bois, *Phys. Rev. B*, **44**, 11315 (1991).
163. D. Hofstetter, J. Di Francesco, P. K. Kandaswamy, A. Das, S. Valdueza-Felip, and E. Monroy, *IEEE Photonics Technol. Lett.*, **22**, 1087 (2010).
164. S. Sakr, P. Crozat, D. Gacemi, Y. Kotsar, A. Pesach, P. Quach, N. Isac, M. Tchernycheva, L. Vivien, G. Bahir, E. Monroy, and F. H. Julien, *Appl. Phys. Lett.*, **102**, 11135 (2013).
165. L. Gendron, M. Carras, A. Huynh, V. Ortiz, C. Koeniguer, and V. Berger, *Appl. Phys. Lett.*, **85**, 2824 (2004).
166. F. R. Giorgetta, E. Baumann, M. Graf, Q. Yang, C. Manz, K. Kohler, H. E. Beere, D. A. Ritchie, E. Linfield, A. G. Davies, Y. Fedoryshyn, H. Jackel, M. Fischer, J. Faist, and D. Hofstetter, *IEEE J. Quantum Electron.*, **45**, 1039 (2009).
167. A. Vardi, G. Bahir, F. Guillot, C. Bougerol, E. Monroy, S. E. Schacham, M. Tchernycheva, and F. H. Julien, *Appl. Phys. Lett.*, **92**, 11112 (2008).

168. S. Sakr, Y. Kotsar, S. Haddadi, M. Tchernycheva, L. Vivien, I. Sarigiannidou, N. Isac, E. Monroy, and F. H. Julien, *Electron. Lett.*, **46**, 1685 (2010).
169. A. Vardi, N. Kheirodin, L. Nevou, H. Machhadani, L. Vivien, P. Crozat, M. Tchernycheva, R. Colombelli, F. H. Julien, F. Guillot, C. Bougerol, E. Monroy, S. Schacham, and G. Bahir, *Appl. Phys. Lett.*, **93**, 193509 (2008).
170. S. Sakr, E. Giraud, M. Tchernycheva, N. Isac, P. Quach, E. Warde, N. Grandjean, and F. H. Julien, *Appl. Phys. Lett.*, **101**, 251101 (2012).
171. A. Vardi, S. Sakr, J. Mangeney, P. K. Kandaswamy, E. Monroy, M. Tchernycheva, S. E. Schacham, F. H. Julien, and G. Bahir, *Appl. Phys. Lett.*, **99**, 202111 (2011).
172. A. Vardi, N. Kheirodin, L. Nevou, H. Machhadani, L. Vivien, P. Crozat, M. Tchernycheva, R. Colombelli, F. H. Julien, F. Guillot, C. Bougerol, E. Monroy, S. Schacham, and G. Bahir, *Appl. Phys. Lett.*, **93**, 193509 (2008).
173. A. Pesach, S. Sakr, E. Giraud, O. Sorias, L. Gal, M. Tchernycheva, M. Orenstein, N. Grandjean, F. H. Julien, and G. Bahir, *Opt. Express*, **22**, 21069 (2014).
174. Y. Song, R. Bhat, T.-Y. Huang, P. Badami, C.-E. Zah, and C. Gmachl, *Appl. Phys. Lett.*, **105**, 182104 (2014).
175. X. Rong, X. Q. Wang, G. Chen, X. T. Zheng, P. Wang, F. J. Xu, Z. X. Qin, N. Tang, Y. H. Chen, L. W. Sang, M. Sumiya, W. K. Ge, and B. Shen, *Sci. Rep.*, **5**, 14386 (2015).
176. F. F. Sudradjat, W. Zhang, J. Woodward, H. Durmaz, T. D. Moustakas, and R. Paiella, *Appl. Phys. Lett.*, **100**, 241113 (2012).
177. A. Pesach, E. Gross, C.-Y. Huang, Y.-D. Lin, A. Vardi, S. E. Schacham, S. Nakamura, and G. Bahir, *Appl. Phys. Lett.*, **103**, 22110 (2013).
178. E. Baumann, F. R. Giorgetta, D. Hofstetter, S. Leconte, F. Guillot, E. Bellet-Amalric, and E. Monroy, *Appl. Phys. Lett.*, **89**, 101121 (2006).
179. L. Nevou, N. Kheirodin, M. Tchernycheva, L. Meignien, P. Crozat, A. Lupu, E. Warde, F. H. Julien, G. Pozzovivo, S. Golka, G. Strasser, F. Guillot, E. Monroy, T. Remmele, and M. Albrecht, *Appl. Phys. Lett.*, **90**, 223511 (2007).
180. N. Kheirodin, L. Nevou, H. Machhadani, P. Crozat, L. Vivien, M. Tchernycheva, A. Lupu, F. H. Julien, G. Pozzovivo, S. Golka, G. Strasser, F. Guillot, and E. Monroy, *IEEE Photonics Technol. Lett.*, **20**, 724 (2008).
181. A. Dussaigne, S. Nicolay, D. Martin, A. Castiglia, N. Grandjean, L. Nevou, H. Machhadani, M. Tchernycheva, L. Vivien, F. H. Julien, T. Remmele, and M. Albrecht, *Proc. SPIE*, **7608**, 76080H (2010).
182. L. B. Cen, B. Shen, Z. X. Qin, and G. Y. Zhang, *J. Appl. Phys.*, **105**, 93109 (2009).
183. A. Lupu, M. Tchernycheva, Y. Kotsar, E. Monroy, and F. H. Julien, *Opt. Express*, **20**, 12541 (2012).
184. E. Gross, A. Nevet, A. Pesach, E. Monroy, S. E. Schacham, M. Orenstein, M. Segev, and G. Bahir, *Opt. Express*, **21**, 3800 (2013).
185. W. Terashima and H. Hirayama, *Proc. SPIE* 948304 (2015).
186. W. Terashima and H. Hirayama, *SPIE Newsroom* (DOI: 10.1117/2.1201507.006058) (2015).
187. V. D. Jovanović, Z. Ikonić, D. Indjin, P. Harrison, V. Milanović, and R. A. Soref, *J. Appl. Phys.*, **93**, 3194 (2003).
188. M. Stattin, K. Berland, P. Hyldgaard, A. Larsson, and T. G. Andersson, *Phys. Status Solidi C*, **8**, 2357 (2011).

189. A. Y. Song, R. Bhat, A. A. Allerman, J. Wang, T.-Y. Huang, C.-E. Zah, and C. F. Gmachl, *Appl. Phys. Lett.*, **107**, 132104 (2015).
190. V. D. Jovanović, D. Indjin, Z. Ikonić, and P. Harrison, *Appl. Phys. Lett.*, **84**, 2995 (2004).
191. G. Sun, R. A. Soref, and J. B. Khurgin, *Superlattices Microstruct.*, **37**, 107 (2005).
192. N. Vukmirović, V. D. Jovanović, D. Indjin, Z. Ikonić, P. Harrison, and V. Milanović, *J. Appl. Phys.*, **97**, 103106 (2005).
193. E. Bellotti, K. Driscoll, T. D. Moustakas, and R. Paiella, *Appl. Phys. Lett.*, **92**, 101112 (2008).
194. W. Terashima and H. Hirayama, *Phys. Status Solidi C*, **6**, S615 (2009).
195. E. Bellotti, K. Driscoll, T. D. Moustakas, and R. Paiella, *J. Appl. Phys.*, **105**, 113103 (2009).
196. H. Yasuda, T. Kubis, and K. Hirakawa, in *36th Int. Conf. Infrared Millim. Terahertz Waves IRMMW-THz* (2011).
197. H. Chou, T. Manzur, and M. Anwar, *Proc. SPIE*, **8023**, 802309 (2011).
198. B. Mirzaei, A. Rostami, and H. Baghban, *Opt. Laser Technol.*, **44**, 378 (2012).
199. G. Sun, J. B. Khurgin, and D. P. Tsai, *Opt. Express*, **21**, 28054 (2013).
200. H. Chou, M. Anwar, and T. Manzur, *Proc. SPIE*, **8268**, 82680O (2012).
201. H. Yasuda, I. Hosako, and K. Hirakawa, in *Conf. Lasers Electro-Opt. CLEO* (2012).
202. B. S. Williams, H. Callebaut, S. Kumar, Q. Hu, and J. L. Reno, *Appl. Phys. Lett.*, **82**, 1015 (2003).
203. W. Terashima and H. Hirayama, *Phys. Status Solidi C*, **8**, 2302 (2011).
204. W. Terashima and H. Hirayama, in *36th Int. Conf. Infrared Millim. Terahertz Waves IRMMW-THz* (2011).
205. Hirayama, H. and Terashima, W., *Proc. SPIE*, **8993**, 89930G (2013).

CHAPTER 4

GAN-BASED PHOTODETECTORS

Ke Jiang, Xiaojuan Sun, Hang Song and Dabing Li*

State Key Laboratory of Luminescence and Applications,
Changchun Institute of Optics, Fine Mechanics and Physics,
Chinese Academy of Sciences,
3888 Dong Nan Hu Road, 130033, P. R. China
**lidb@ciomp.ac.cn*

GaN-based ultraviolet (UV) photodetectors are one of the most important UV photodetectors. Since the 1990s the quality of GaN material improves rapidly and a lot of GaN-based UV photodetectors are reported with different kinds of structures. Here, we introduced the classical structures such as Schottky, MSM, and p-n junction as well as several high gain structures like photoconductors, phototransistors and avalanche photodiodes (APDs). Moreover, some novel GaN-based UV detectors such as surface plasmon resonance enhanced (SPR), graphene/GaN (G/GaN) contact and nanowires (NWs) photodetectors were also involved in our report. Besides, some achievements about AlGaN- and InGaN-based photodetectors were also reported.

1. Introduction

Nowadays, people's requirement for detecting the ultra-violet (UV) region wave becomes more and more intense, especially in the areas of military, medicine, environment monitor and fire accidents warming.[1,2] GaN-based materials own many advantages including high compactness, favorable thermal and chemical stability. More importantly, GaN has wide-direct-band gap of 3.4 eV that can be varied from 0.7 eV to 6.2 eV by altering the Al or In content from 0 to 1, which satisfies the aforementioned applications well. From 1990s, the two-step growth method for GaN in MOCVD system significantly improved the quality of GaN[3,4] which made it possible to be used as the basic materials in the UV device applications. Besides, the realization of GaN-based quantum well structure blue LEDs[5] encouraged the researchers and let them feel full of confidence for GaN materials.

There are many kinds of GaN-based photodetectors, which generally are divided into two types, namely photoconductive and photovoltaic detectors. The mechanism of GaN-based photoconductive UV detectors is that the conductivity of the detectors can be changed by the UV photons which leads to photo-current response. Although the photoconductive detectors hold the advantages of simple structure and great responsivity of 2000 A/W or even higher,[6] the slow response speed and seriously persistent photo-current limit their applications. While, the photovoltaic detectors containing the types of p-n and p-i-n junction, metal-semiconductor (MS) and metal-semiconductor-metal (MSM) can realize fast response speed and low noise. But there isn't gain in photovoltaic detectors, thus the sensitivity needs to be enhanced further. Some novel GaN-based UV detectors like SPR and G/GaN detectors were designed to realize higher sensitivity and speed. Moreover, the phototransistors and the avalanche photodiodes, which can achieve high gain, are also used in GaN-based UV detectors.

2. Classical GaN-based Photodetectors

Because of the simplicity and easy-fabrication of the Schottky and MSM structures, they are studied comparably early and many achievements had been realized. p-n and p-i-n junction based photodetectors have the advantages of low dark current, fast speed and are easy to make array architectures, so they are also widely studied.

2.1. *Schottky structure*

A Schottky type GaN-based UV photodetector holds a comparatively simple structure, consisting a semiconductor with two electrodes of ohmic and Schottky contact, respectively. Because of the different work functions, barrier is generated at the interface and spacing charge region (SCR) comes into being. Otherwise, surface states, dopants related levels and image-force can diminish the barrier height,[7] and the surface states can trap carriers

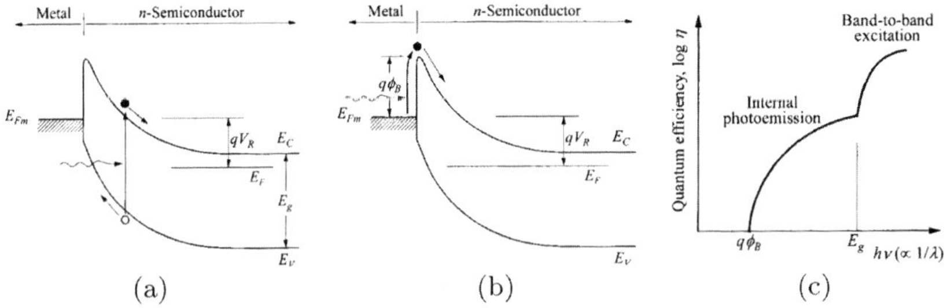

Fig. 1. (a) The band-to-band excitation of electron-hole pair ($h\nu > E_g$); (b) The internal photoemission of excited electrons from metal to semiconductor ($E_g > h\nu > q\varphi$); (c) Quantum efficiency as a function of wavelength showing both processes. [Adapted from Fig. 11 of Chap. 13 of Ref. 7 and Reprinted with permission from John Wiley & Sons Ltd, *Physics of Semiconductor Devices*. Copyright © 2007 John Wiley & Sons, Inc.]

under some certain conditions.[8] The photo-current of Schottky structure GaN-based UV detectors results from three aspects as followings. (1) Photo-generated carriers in SCR and within the neutral region of a diffusion length apart from SCR are swept out by inner electronic field, creating photo-current, as shown in Fig. 1(a). (2) Electrons in metal within an inelastic mean free path from interface absorb energy by photoemission and cross the Schottky barrier, creating photo-current,[9] as shown in Fig. 1(b). (3) When illuminated, photo-generated carriers are swept to interface and trapped by trap centers, lowering the Schottky barrier, which causes a so-called "Secondary current".[8]

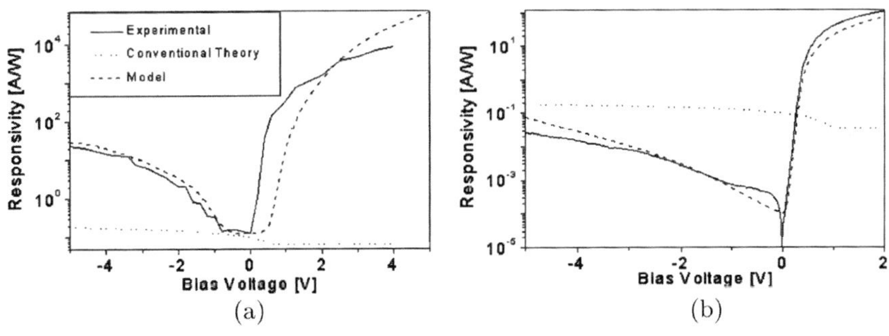

Fig. 2. Simulation of responsivity of a Schottky photodetector considering both three mechanisms. (a) Vertical and (b) Lateral Schottky structure. [Adapted from Figs. 2(a) and (b) of Ref. 9. Reprinted with permission from American Institute of Physics for Applied Physics Letter. Copyright © 2001 American Institute of Physics.]

Surely, those three mechanisms have their own effective working conditions. At relatively long wavelength range, photoemission related photo-current is the main mechanism. Near the band edge, at comparatively low bias, photo-current caused by surface states traps of carriers is obvious. However, at high enough bias, the draft-current of photo-generated carriers contribute most.

Recent years, researchers optimized the performances of Schottky structure GaN-based UV photodetectors by improving GaN material quality, optimizing the structure parameters, and choosing proper metal contacts. The photodetectors based on nanorod epitaxial lateral overgrowth (NRELOG) technology[10] had lower dark currents, higher UV-to-visible rejection ratios, and the noise equivalent power (NEP) decreased about four orders of magnitude with detectivity (D*) increasing approximately four orders of magnitude, respectively.[11] Besides, the intermediate layers sandwiched in the high temperature (HT) epitaxial layer can effectively ameliorate the material quality too. Researchers from *National Cheng Kung University* in *Taiwan* successively took low temperature(LT) GaN,[12] AlGaN[13] and AlInN[14] as intermediate layers to manufacture GaN based Schottky UV detectors, finding that those as-grown GaN materials made great difference on lowering dark current and NEP as well as increasing detectives. Previously, they had used the Mg_xN_y material as buffer layer to replace LT GaN buffer layer,[15] causing about six orders of magnitude dark current decreasing and three orders of magnitude UV-to-visible rejection ratio increasing, respectively. Recently, Japanese scientists adopted N-face GaN material to fabricate a Schottky photodetector whose barrier height reduced approximately 1 eV and response spectrum extended to infrared region.[16]

Schottky AlGaN-based UV detectors were also studied. When the researchers in *Taiwan* studied the influences of insertion layer on Schottky GaN-based UV detectors, they also did something on Schottky AlGaN-based UV detectors, finding the same results.[12,14] Researches on Schottky avalanche photodetectors directly at high voltage also made some progresses. Turgut Tut *et al.*[17] reported a Schottky AlGaN-based solar-blind APD whose repeatable gain was 25 at a bias of 72 V and quantum efficiency reached 55% at a bias of 25 V. Besides, it was not found that the device was seriously brokendown at Geiger mode. Investigators in *Sun Yat-Sen*

University utilized $Al_{0.4}Ga_{0.6}N$ material to make Schottky avalanche solar-blind UV detectors with internal quantum efficiency(IQE) of 46.2% as well as gain of 10^3 both under front and back illumination,[18] which is a promising kind of detector.

Schottky InGaN-based photodetectors are also such a kind of detector seldom studied. In C. Rivera's report,[19] the Schottky InGaN-based photodetector apparently showed a peak responsivity of 0.628 A/W at 1.5 V bias and was obviously affected by polarization effect. Afterword, they reported a vertical Schottky InGaN-based photodetector whose peak responsivity was improved to 1 A/W and detection contrast exceeded 10^4, and the dark current was decreased to some degree.[20]

2.2. *MSM structure*

GaN-based MSM photodetectors are formed by two ohmic or Schottky contact metal electrodes affixing to the opposite ends of a semiconductor. In the first case, i.e. ohmic contact, it is a photoconductive detector. Otherwise, for the Schottky contact, it forms a photovoltaic one. In this section, we focus on the later, i.e. Schottky contact MSM photodetectors. The advantages of this type of photodetectors are simple structure, no need for p-doped GaN layer, weak leakage current and so on, which make it play an important role in GaN-based UV detectors.

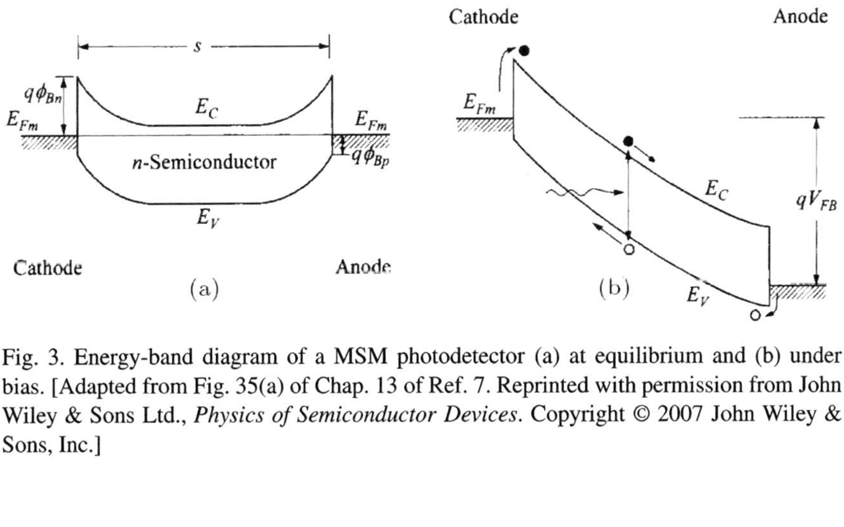

Fig. 3. Energy-band diagram of a MSM photodetector (a) at equilibrium and (b) under bias. [Adapted from Fig. 35(a) of Chap. 13 of Ref. 7. Reprinted with permission from John Wiley & Sons Ltd., *Physics of Semiconductor Devices*. Copyright © 2007 John Wiley & Sons, Inc.]

As early as the end of 20th Century, researchers from *Northwestern University* had demonstrated GaN-based MSM photodetectors with unintentionally doped GaN and Mg-doped GaN whose response speed and noise were extremely fast and low respectively.[21] Afterward, researchers in *Taiwan* found that proper annealed ITO electrodes increased the barrier height and thus led a lower dark current. The as-treated MSM structure photodetector had the responsivity of 7.2A/W at 5 V, demonstrating a further development of GaN-based MSM UV detector.[22] A lot of electrode materials for MSM GaN-based UV detectors had been achieved such as TiW alloy,[23,24] Ni/Au coating[25] and IrO_2 layer.[26] Investigators in *Changchun Institute of Optics, Fine Mechanics and Physics (CIOMP) of Chinese Academy of Sciences (CAS)* took different metals as electrodes, resulting a high response and fast response speed self-driven GaN-based MSM UV detector, as shown in Fig. 4, which provided a novel method to fabricate MSM structure.[27] Inserting LT GaN cap layers whose electron mobility is relatively small into ITO electrodes and GaN epitaxial layers could approach higher resistivity and lower M-S interface states, which significantly reduced dark current and increased UV-to-visible rejection ratio and response speed.[28] However, Hai Lu[29] employed homogeneous GaN substrate to grow GaN epitaxial layers, achieving an excellent GaN film

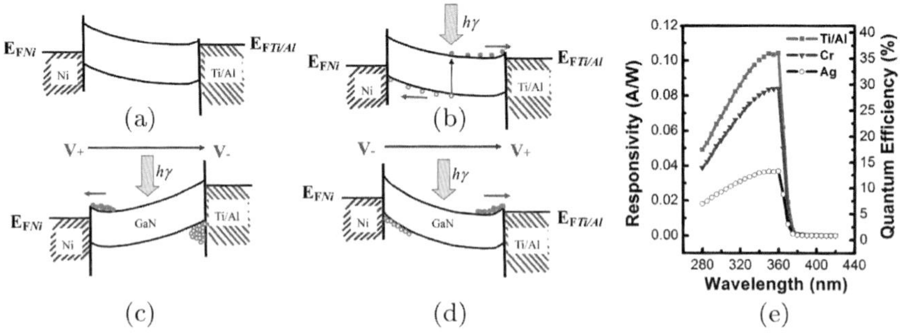

Fig. 4. Energy-band diagram of a MSM photodetector with different metal contacts. (a) Zero bias and darkness; (b) Zero bias and illumination; (c) Above zero bias and illumination; (d) Reverse bias and illumination. (e) Spectral responses of Ni/GaN/Cr, Ni/GaN/Ag and Ni/GaN/Ti/Al detectors under 0 V bias. [Adapted from Figs. 1 and 3(b) of Ref. 27. Reprinted with permission from Nature for scientific reports. License Link: https://creativecommons.org/licenses/by/4.0/legalcode.]

with dislocation density lower than $5 \times 10^6\,cm^{-2}$. The as-fabricated MSM UV detector showed a dark current below 2 pA at 10 V and UV-to-visible rejection ratio as large as 10^5, illustrating the promotions of MSM GaN-based UV detectors. Researchers in *Southern Taiwan University of Science and Technology* applied sputtered AlN nucleation layer technology[30] to MSM GaN-based UV detectors which optimized the performances of dark current, quantum efficiency, UV-to-visible rejection ratio, NEP and detectivity.[31] In a word, improvement of crystal quality reduced the surface states, causing a higher barrier height, which effectively suppressed dark current and resulted in a comparatively better performance.

Some effective methods in MSM GaN-based UV detectors had been transplanted to AlGaN-based deep UV photodetectors. In 1999, *France* scientists[32] used AlGaN materials with 0.25 Al content to fabricate a MSM UV detector whose cut-off wavelength, UV-to-visible rejection ratio, response time and NEP are 310 nm, 10^4, 10 ns and $24\,pW/Hz^{1/2}$ at 28 V respectively, which was the best performance device at that time. In the following years, kinds of methods such as choosing proper electrode materials,[33] appropriate annealing temperature and pattern,[25] insertion of different buffer layers[34,35] were used to achieve solar-blind UV detectors. Deep-UV MSM photodetectors had been reported by scientists in *Bilkent University* utilizing AlGaN material with Al fraction of 0.75.[36] The peak responsivity was 0.53 A/W at 222 nm position, and cut-off wavelength was 229 nm. Notice that the dark current was as low as $5.79 \times 10^{-10}\,A/cm^2$ at a large voltage of 50V. Recently, more approaches had been used to fabricate AlGaN-based MSM solar-blind UV detectors. Researchers in *Taiwan*[37] applied H_2O_2 to a MSM photodetector to passivate surface so as to reduce surface states and thus suppressed surface leakage current. However, the dielectric layer will change the MSM structure to MIS structure and lower the response speed. Dabing Li *et al.*[38,39] in *CIOMP, CAS* studied the influence of the dislocations in GaN based material on the performance of the detectors. They found that the screw dislocations increase the dark current, while the edge dislocations lower the spectral response. Based on these conclusions, they proposed a method of using the SiO_2 nano particles to passivate the dislocations without decrease the speed.

The MSM InGaN-based photodetectors were extensively studied too. In 2001, Yan-Kuin SU *et al.* studied the traits of MSM InGaN-based photodetectors with metals of Au, Ti, Ni, and Pt.[40] Later, Ping-Chuan Chang *et al.*[41,42,43] and Zhengda Huang *et al.*[44] took SiO_2, the un-activated Mg-doped GaN cap layer and β-Ga_2O_3 insulator layer as dielectric layer and attained a photodetector with sufficiently low dark current. In the report of Liwen Sang *et al.*,[45] CaF_2 was inserted into a MSM InGaN-based photodetector, which resulted in a 6 orders of magnitude decrease of dark current compared to the one without dielectric layer. As the best performance MSM InGaN-based photodetector reported up to now, this device exhibited a peak responsivity of 10.4 A/W at 2 V bias, an UV-to-visible ratio of larger than 10^6, a noise-to-signal ratio of over 10^5, and a response time of millisecond magnitude. In addition to inserting dielectric layer, special treatments to InGaN film were also studied. In 2012, Saleh H. Abud *et al.*[46,47] in *Malaysia* etched some porous structures in InGaN film by method of photon-assisted electrochemical etching technique (PEC) and made a MSM structure photodetector. The results showed that with etching time increasing, the response and quantum efficiency improved and the response speed became faster, which provided a new method to make a InGaN-based photodetector.

2.3. *p-n junction structure*

Among all structures, p-n junction structure is one of the earliest studied and simplest structured configurations. This kind of photodetectors consist of heavily doped p-type and n-type materials contacting directly, forming depletion region around the interface. In steady state, the currents crossing depletion region are composed of drift and diffusion current, of which the drift current generates from photo-generated electron-hole pairs in SCR separated by inner electronic field and the other from photo-generated minority carriers whose distance is within a diffusion length apart from SCR boundaries. The basic principles are shown as following Fig. 5.

Such simple structure has its inherent advantages and disadvantages. Firstly, heavily doping narrows the depletion region, and thus reduced the transit time of carriers, leading to a fast response speed. However, extremely thin depletion region means narrow light absorption region and light absorption a diffusion length apart from SCR boundaries contributes nothing to photocurrent, which significantly decrease the sensitivity. To

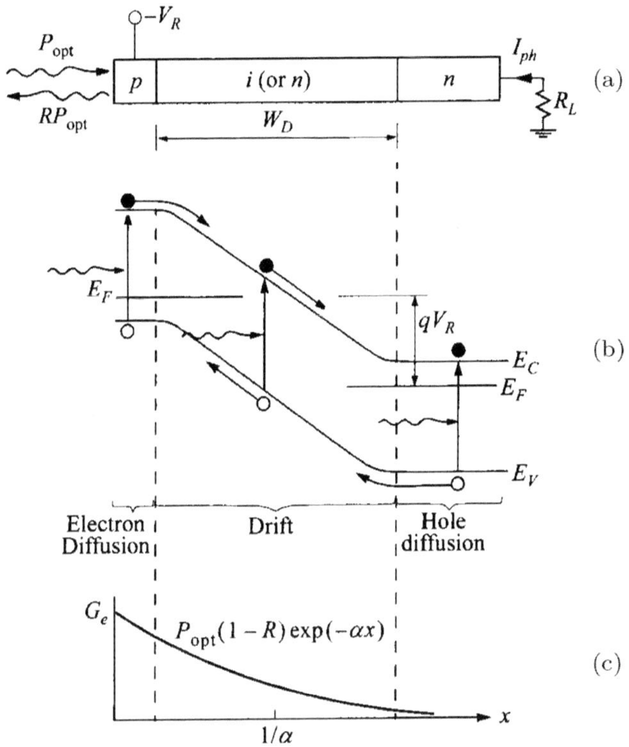

Fig. 5. A typical p-i-n junction structure photodiode. (a) The cross-sectional view; (b) Energy-band diagram under reverse bias; (c) Carrier generation characteristics. [Adapted from Fig. 6 of Chap. 13 of Ref. 7. Reprinted with permission from John Wiley & Sons Ltd, *Physics of Semiconductor Devices*. Copyright © 2007 John Wiley & Sons, Inc.]

solve this problem, a lightly doped or un-doped layer is inserted between p-type and n-type layer, constituting a p-i-n structure. If we optimize the thickness of the i-type layer to reconcile the contradiction between light absorption efficiency and response speed, we can achieve high performance devices. Structures such as p-π-n, p^+-n^+, p^+-i-n^+ deriving from p-n junction all share the same working principle, actually the methods to adjust absorption layers by doping.

A photodiode can be achieved by heterojunction which consists of at least two different kinds of semiconductors contacting with each other. A main merit of heterojunction photodiodes is that the quantum efficiency is not apparently depended on the distance between surface and junction, because the large band-gap material can be transparent and used as a window for the transmission of incoming optical power. In addition, the heterojunction can provide unique material combinations so that the quantum efficiency and response speed can be optimized for a given optical signal wavelength. Another advantage of the heterojunction detectors is a reduced dark current. For a heterojunction photodiode, the lattices of certain materials must match well, otherwise the performances are unfavorable. Usually, we choose AlN, AlGaN, ZnO or SiC[48,49,50,51] as the materials matched to GaN to form heterojunction photodiodes.

Obtaining high quality p-type is a big challenge for p-n junction type detectors. In general, Mg is adopted as the dopants for p-type GaN whose activation energy is as high as 280 meV at room temperature (RT). Furthermore, with the Al content increasing, the activation energy of Mg-doped AlGaN can reach a maximum of 630meV,[52] which causes a sufficiently low doping efficiency and the difficulties to achieve high hole concentration and good quality p-type (Al)GaN. Until now, how to obtain high quality p-type GaN-based layer is still a big problem. The low p-type doping concentration will result in poor ohmic contacts, thus further degrading device performances. Therefore, a lot of work has been done to improve the material quality, especially the p-type GaN layer and optimize the ohmic contact.

In 1995, Q. Chen *et al.*[53] firstly reported a p-n junction visible-blind UV detector whose cut-off wavelength positioned at 370 nm, peak responsivity was 0.09 A/W at 360 nm, and response time was 300 us, respectively. After that, many p-n and p-i-n homojunction or heterojunction GaN-based UV detectors were made successively. M. Rezaghi *et al.*[54] investigated the photovoltaic effect of a GaN-based UV detector with a p-n junction and investigated the influences of carrier concentration, minority carrier diffusion-length and material thickness on response and differential resistance, respectively, providing the concept of optimizing device in aspects of material, thickness, metal contact and geometry configuration.

Meanwhile, other scientists got abundant achievements too. In 1997, J. M. Van Hove *et al.*[55] employed MBE epitaxial technology to fabricate a p-i-n GaN photodetector whose 3dB-bandwidth and peak responsivity were approximately 19 kHz and 0.11 A/W, respectively. In the same year, a p-i-n GaN-based photodetector reached a zero-bias responsivity, response time and NEP of 0.18 A/W, 118 ns and 7×10^{-12} W/Hz$^{1/2}$, respectively, whose noise was even lower than that of a Si-based photodetector.[56] Probably, it was the best performance photodetector then. Later, E. Monroy *et al.*[57] shortened the response time to 105 ns and scientists in *Illinois University*[49] further reduced it to 12 ns by optimizing thickness of p-type layer and the noise was sufficiently low even at −10 V bias. Researchers in *University of California Santa Barbara (UCSB)* adopted the LEOG technology to improve crystal quality, decreasing the response time to 4.5 ns and diminishing the dark current by six orders of magnitude.[58] In 1998, J. C. Carrano *et al.*[59] made a GaN-based p-i-n photodetector with the response time and bandwidth of 90ps and 16GHz, respectively, which was the fastest UV detector at that time. The introduction of graded layer effectively promoted the doping efficiency as well as increased material quality, by which C. J. Collins *et al.*[60] manufactured a p-i-n GaN-based UV detector whose dark current and detectivity were 82 pA/cm^2 at −5 V bias and 2×10^{14} cm.Hz$^{1/2}$/W which was further elevated to 7.5×10^{14} cm.Hz$^{1/2}$/W by scientists in *Bilkent University*.[61] Besides, at an extremely high reverse voltage of −120 V, its dark current was only 1.6 nA.

Device performances more depend on material quality. With the slow advancement of GaN crystal quality, the promotion of GaN-based device performances is suffering a bottleneck. Recently, some beneficial results emerged too. In the year of 2010, James S. Speck *et al.*[62] in *UCSB* applied the NH_3-MBE technology to grow Ga-face p-n junction whose ideal factor was 1.3, which demonstrated the formation of high quality GaN. In 2015, they again reported a GaN/AlN/GaN heterojunction photodiode by the same method.[63] These results made it possible to further improve device performances. Meanwhile, the application of homogeneous GaN single crystal substrates elevated the quality of epitaxial layers, thus reducing dark current and increasing breakdown voltage.[64] The photodiode manufactured by p-Si substrate and GaN nanoparticles was demonstrated to possess detection function.[65] All these studies paved some new paths to research p-n junction GaN-based UV detectors.

AlGaN-based p-n junction photodetectors also made some progresses. In 1999, AlGaN with Al content lower than 0.15 was used to fabricate photodetectors whose cut-off wavelength and UV-to-visible rejection ratio were 338 nm and 10^6, respectively. But their response time exceeded 0.5 us because of the abundant Mg-related defects, hardly reflecting the advantage of high speed.[66] Studies on AlGaN-based p-n photodetectors with much more higher Al fraction were underway at the same time, and one year later they realized a AlGaN-based solar-blind UV detector with cut-off wavelength at 275 nm whose Al content was higher than 0.7.[67] At zero bias, detectivity attained 10^{10} cm.Hz$^{1/2}$/W and at the bias of −5 V, the IQE reached 90%. Besides, the insertion of GaN cap layer increased the responsivity from 0.05 A/W to 0.09 A/W at zero bias. Furthermore, AlN buffer layer as well as Si-In co-doping method in back-illuminated AlGaN-based photodetector improved the response and quantum efficiency further.[68] In 2013, E. Cicek *et al.*[69] achieved a solar-blind UV detector with great high EQE of 89% which is the largest of reported p-n AlGaN-based detectors by utilizing high-quality AlN pattern and optimizing device design and doping.

In conventional structure InGaN-based photodetectors, p-n or p-i-n junction structures were the mostly studied types. In 2002, Yu-Zung Chiou *et al.*[70] made a p-n junction InGaN-based photodetector. At the bias of −0.1 V, −1 V and −3 V, the responsivities were 1.28 A/W, 1.37 A/W and 1.76 A/W, respectively. However, the noise was very strong with the applied bias increase. C. Rivera also reported a p-i-n junction InGaN-based photodetector.[19] At zero bias, a peak responsivity of 0.1 A/W and two cutoffs of 360 nm and 475 nm were found, demonstrating the GaN and InGaN material both contributed to the photocurrent. Later, they studied the relationship between charge collection efficiency and the number of quantum wells in i-layer, finding that it was the well position that affect the performances rather than the quantity.[71] Then they continued to optimize the device and obtained a higher response and a faster response speed in 2009.[72] At the same time, Y. K. SU *et al.*[73] also reported a p-i-n junction InGaN-based photodetector with a responsivity of 0.206 A/W at 380 nm and an IQE of 67%. In the year of 2014, LU Yi-dan *et al.*[6] in *Shanghai Institute of Technical Physics (SITP) of CAS* reported a InGaN-based photodetector with a peak responsivity of 0.22 A/W at 378 nm and an IQE of 88%. The performance of a p-n junction based device was also affected by surface states, which was demonstrated by Yi-Ting

Huang *et al.*[74] whose device with a typical InGaN/GaN LED structure exhibited a phenomenon that when the p-type surface was textured, the dark current rapidly increased. Recently, researchers in *Nanjing University of Posts and Telecommunications* reported some studies about InGaN-based photodetector. Xin Li *et al.*[75] fabricated a p-n junction InGaN-based device on a Si-based GaN substrate and demonstrated the device could function as LED and photodetector on different conditions. On the foundation of the former works, Wei Cai *et al.*[76] further investigated the integration of LEDs and photodetectors, realizing the integration of LEDs, waveguides and photodetectors. The LEDs and photodetectors were strongly coupled through wavegudies, resulting in a signal transmission of high sensitivity and fast speed.

3. High Gain GaN-based Photodetector

GaN-based photoconductors, field effect transistor (FET) photodetectors, heterojunction bipolar transistors(HBT) photodetectors and APDs, all belonging to conventional photodetector structures, are early developed. All of them can attain high gain and thus high responsivity. However, photoconductors, FET photodetectors and HBT photodetectors usually suffer large persistent photocurrent resulting from their carriers capture related photocurrent, which limits their applications and thus results in less and less researches recently. But APDs possess fundamentally different gain mechanism, which exhibit sufficiently high gain as well as low noise, thus they are considered to be one of the most promising structures of GaN-based UV detector and are broadly studied recently.

3.1. *Photoconductors*

A photoconductor consists simply a slab of semiconductor, in bulk or thin-film form, with comic contacts affixed to the opposite ends. As for the high quality semiconductor materials, most kinds of phenomena can be explained by the model of carriers modulation.[7] However, if the model is applied to a GaN-based photoconductor, great deviation will occur. Generally, the GaN-based photoconductors have the following characteristics: (1) Photocurrent response exists for photons with energies below the band gap. (2) Very slow nonexponential photoconductance decays. (3) A highly nonlinear photocurrent response, depending on the illumination irradiance power. (4) Gain dependence on incident power with an exponential correlation. (5) A significant dependence of the photoresponse with temperature. These features can be well interpreted by the model of volume modulation.[77]

In 1992 , M. Asif Khan *et al.*[4] reported a GaN-based photoconductor whose gain and responsivity exceeded 6×10^3 and 2000 A/W at 5 V bias even though the gain mechanism of GaN-based photoconductors was not clear at all. However, this photodetector fully exposed the drawbacks of GaN-based UV photoconductors such as narrow bandwidth (2 kHz) and slow response speed (1 ms or more). After that, K. S. Stevens *et al.*[78] adopted MBE technology to grow Mg doped GaN on Si(111) substrate, achieving a photoconductor with responsivity of 30 A/W at 14 V. Z. C. Huang *et al.*[79] reported a photoconductor with the responsivity of 3200 A/W and response time longer than 0.3 ms. Later, L. B. Flannery *et al.*[80] used p-GaN:Mg materials grown by MBE method to make a photoconductor with response time indeed longer than 5 s. Shortly afterwards, Bo Shen *et al.*[81] in *Nanjing University* fabricated a photoconductor using the undoped GaN materials on 6H-SiC substrate and they studied the relationship between the spectral response and speed to the voltage, as shown in Figs. 6(a) and (b), respectively, which are the important evidences for

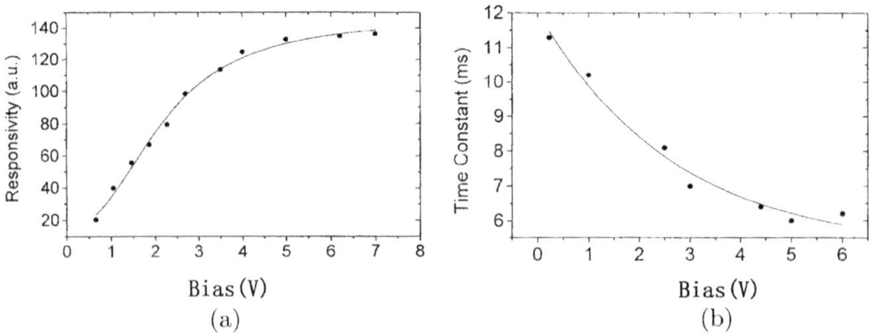

Fig. 6. Typical relationship between (a) Responsivity and bias (b) Time constant and bias of a GaN-based photodetector. [Adapted from Figs. 3 and 6 of Ref. 81. Reprinted with permission from the Institute of Physics for Japanese Journal of Applied Physics. Copyright © 1999 The Japan Society of Applied Physics.]

GaN-based detector researchers to distinguish between photovoltaic and photoconductive response.

For AlGaN-based photoconductor, M. Razeghi attempted to increase Al mole fraction,[82] but the performance was not as good as expected. M. Asif Khan *et al.*[83] increased Al mole fraction from 0.05 to 0.61, and the cut-off wavelength decreased from 350 nm to 240 nm, respectively. However, the response speed was also slow, around 1 ms or 2 ms. They attributed the gain and the slow response speed to the acceptor dopants, which implied that they didn't fully comprehend the photoconductive gain mechanism. InGaN-based photoconductors were comparatively less studied especially for film-based ones except for the report of C. Rivera *et al.*[19] which revealed that a InGaN-based photoconductor held the characteristics of high gain, large dark current, low detectivity and slow response speed. In nano-materials, the same features were exhibited between InGaN and GaN-based photoconductors. In 2011, A. De Luna Bugallo *et al.*[84] adopted the single nanowire with core-shell n-i-n structure to realize a photoconductor with a peak responsivity of 2.5×10^4 A/W at 5 V bias and a gain of 10^3–10^4 magnitude.

3.2. *Phototransistors*

In early stages, transistor-based GaN photodetector were also studied. Compared with diodes, these devices require more complex fabrication process and higher material quality, so there were few reports about these kinds of GaN-based photodetector in recent decades. The principle of a FET photodetector lies on that when the base and gate are biased, a SCR is formed, leading to the conductive channel between source and drain. Under illumination, the photo-generated carriers in SCR are separated by the field between the base and the gate. The photo-generated carriers flowing to channel are rapidly collected and an additional current is produced. Bipolar transistors can be used as a photodetector too, which can achieve a substantially high gain. However, the fabrication of a phototransistor is more complicated than a photodiode, and the inherent large area degrades its high-frequency performance.

In 1995, M. A. Khan *et al.*[85] firstly reported a FET GaN-based photodetector with AlGaN as the gate material. The structure was formed by an un-doped GaN layer following a n-type GaN layer and a n-type AlGaN layer on a sapphire substrate successively. Eventually, this photodetector exhibited a very high peak responsivity of over 3000 A/W. However, the response time of this photodetector was 0.2 ms, which was much longer than photodiodes.[56,57,58]

In 1998, Wei Yang *et al.*[86,87] reported the first GaN/AlGaN HBT UV photodetector whose response spectrum ranged from 320 nm to 360 nm. Under the illumination of 5 nW/cm^2 power, the responsivity reached 5×10^4 A/W with a gain of as high as 1.7×10^5, which was an order of magnitude higher than the largest responsivity GaN-based photoconductor at that time.[79] However, this device also possessed the features that the gain rapidly varied with illumination power and the nonlinearity was very significant, which really limited its applications. Afterward, scientists from *Japan*[88] and *Taiwan*[89] also studied the GaN-based phototransistors, but no improvement was achieved in changing its nonlinearity. From these results, it is not difficult to know that GaN-based phototransistors can easily reach a high gain and thus a larger responsivity and detection contrast, while the poor linearity seriously limited their application areas which results in less and less researches about them.

3.3. *Avalanche photodiodes*

APDs work at a high reverse bias which induces avalanche multiplication effect, resulting in internal current multiplication. Because of high internal gain, APDs can realize large responsivity and thus detect some weak signals, which make it a promising photodetector in recent years. However, at high reverse bias, the devices are easy to breakdown, which makes it necessary to use high quality material.

The avalanche gain for electrons in low-frequency is given by

$$M = \left\{1 - \int_0^{W_D} \alpha_n \exp\left[-\int_x^{W_D} (\alpha_n - \alpha_p) dx'\right]\right\}^{-1} \tag{1}$$

where W_D is the depletion width, and a_n and a_p are the ionization rates for electrons and holes, respectively. The carrier ionization coefficients are the critical parameters affecting avalanche gain. In a practical device, the multiplication for photo-generated carriers can be calculated by

$$M_{ph} = \frac{(I - I_{MD})}{(I_P - I_D)} \tag{2}$$

where I is the total multiplied current, I_p is the primary photocurrent and I_D and I_{MD} are the primary and multiplied dark currents, respectively. From this equation, it can be deduced that the dark current can become very low, so the multiplication factor ($(M_{ph})_{\max}$) is unlimited.

The regeneration avalanche process results in the presence of a large number of carriers in the high-field region long after the primary electrons have traversed through that region. The higher the avalanche gain is, the longer it takes to build up and to quench the avalanche process. These phenomena demonstrate the existence of gain-bandwidth characteristic.

Every electron-hole pair generated at a given distance in the depletion region is independent and does not experience the same multiplication. Since the avalanche gain fluctuates, the mean-square value of the gain ($\langle M^2 \rangle$) is greater than the square of the mean ($\langle M \rangle^2$). The excess noise can be characterized by a noise factor

$$F(M) = \frac{\langle M^2 \rangle}{\langle M \rangle^2} \tag{3}$$

The noise factor $F(M)$ is a measure of the increase in the shot noise compared to an ideal noiseless multiplier and it strongly depends on the ratio of the ionization coefficients a_p/a_n and on the low-frequency multiplication factor M.

The ratio of the ionization coefficients a_p/a_n is the most important physical quantity. In GaN, the gain can reach a maximum when the ionization coefficients of electrons and holes are almost similar to each other and on the contrary the noise can reach a minimum when they are in great difference.[90,91] For GaN, the avalanche breakdown field is approximate 3 MV/cm[92] and the ionization coefficient of holes is larger than that of electrons.[90]

Most common APD structure is p-n junction type including p^+-n, p-n^+ and p-i-n structures. But researchers generally believe that the separated absorption and multiplication region avalanche photodiode (SAM-APD) is one of the best structures with largest potential to achieve high gain and even single photon detection. Recent years, crystal quality of GaN has been improved further, leading to that GaN-based devices can endure higher field.

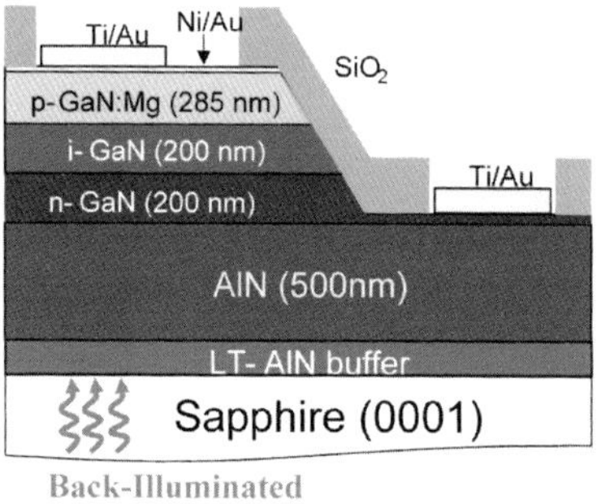

(a)

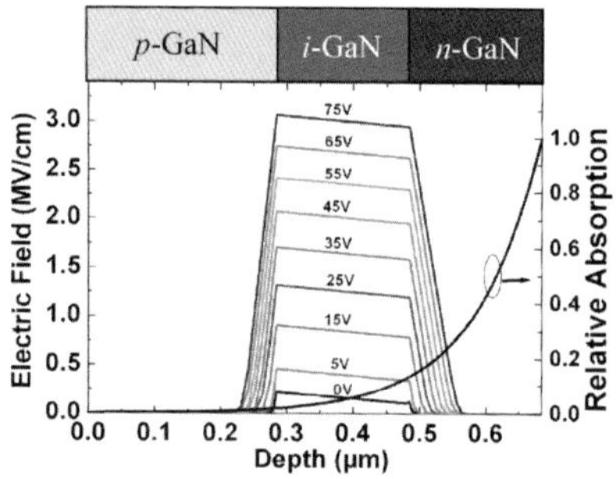

(b)

Fig. 7. (a) A schematic of typical p-i-n APD; [Adapted from Fig. 2 (left) of Ref. 106.] (b) The electric field distribution of the p-i-n APD. [Adapted from Fig. 3 (left) of Ref. 106.] [Reprinted with the permission from SPIE for Quantum Sensing and Nanophotonic Devices VI. Copyright © 2009 SPIE.]

Directly bias both ends of p-n junction, strong field will be built up in depletion region, as shown in Figs. 7(a) and (b). When the bias reaches breakdown voltage, carrier multiplication effect takes place and current increases rapidly. Previous GaN-based APDs are almost p-n junction-based. Several research institutes in which the representatives included *APA Optics*, *the West Point*, *Massachusetts Institute of Technology (MIT)* and *Northwestern University* deeply studied the III-N-based APDs and realized some excellent achievements.

Soon after A. Osinsiky *et al.* reported the GaN-based APD, K. A. McIntosh *et al.*[93] in *MIT* also reported a GaN-based p-n junction APD fabricated on Sapphire substrate by HVPE. Later, K. A. McIntosh *et al.*[94] again fabricated a p-n junction GaN-based APD with lower defect density and no microplasmas in the same method. In 2000, they further studied the photon detection efficiency (PDE) and dark count rate (DCR) of the detector by gated quenching circuit on Geiger Mode, attaining a PDE and minimum DCR of 13.5% and 10 kHz.[95] In the same year, Carrano and Lambert *et al.*[92] reported a p-i-n GaN-based APD with an avalanche gain of 25. Next year, K. A. McIntosh *et al.*[96] also reported a p-i-n GaN-APD whose uniform gain and maximum gain were 20 and 300 on Linear Mode and Geiger Mode, respectively. Meanwhile, Patrick Kung *et al.*[97] fabricated a p-i-n APD on double-polished sapphire substrate by MOCVD. They studied the electric properties, response characteristics and relationship

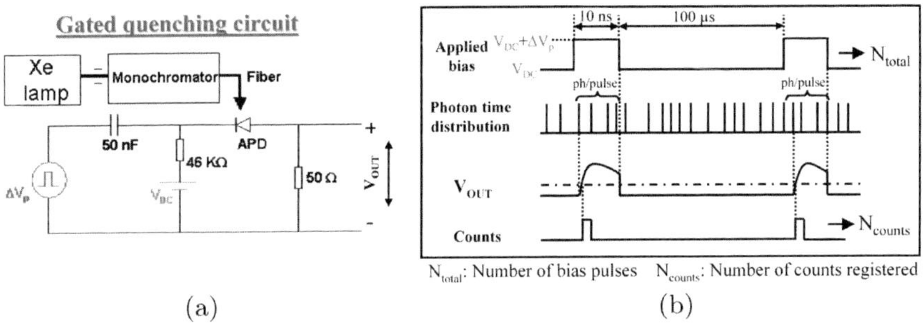

Fig. 8. (a) Geiger mode APD testing circuit. (b) Diagram of an applied bias, and corresponding output for a given input photons (about 3/pulse here). [Adapted from Fig. 8 of Ref. 98. Reprinted withith permission from SPIE for Quantum Sensing and Nanophotonic Devices V. Copyright © 2008 SPIE.]

between breakdown field and i-type layer thickness. The results showed that APD with GaN template had high gain and could bear higher electric field about 2.73 MV/cm. Comparing the performance of the front illumination detectors with the back illumination ones, they found that the hole ionization coefficient was higher than that of electron in GaN.[90] Lastly, they measured the back-illumination I-V characteristics of the AlN template APD in gated quenching circuit, finding an increase of the pulse-deduced current with the increase of voltage.

Based on the previous researches, in 2008, R. A. McClintoch *et al.*[98] manufactured several p-i-n AlN template GaN-APDs with different thicknesses of n-GaN layer. The results showed that when the n-GaN layer was 200 nm, the quantum efficiency was the highest, since the thicker the n-GaN layer was, the more difficultly the photo-generated carriers diffused to multiplication region. On the contrary, excessively thin n-GaN layer leaded to more defects at the interface of AlN template and n-GaN layer. In this case, a large amount of carriers would be trapped, leading to reduce the carriers to the multiplication region. For the sample with 200 nm n-GaN layer, the breakdown voltage was 78V and dark current was 25 nA at low bias. At zero bias, the peak responsivity was 80 mA/W at 360 nm. At breakdown voltage, the responsivity was as high as 550 mA/W at 360 nm. Under Geiger mode, at a certain discrimination voltage, a different photon flux density resulted a corresponding counts distribution, from which we could distinguish the photon flux density. A device of 225 um^2 showed a DCR of 10 kHz and

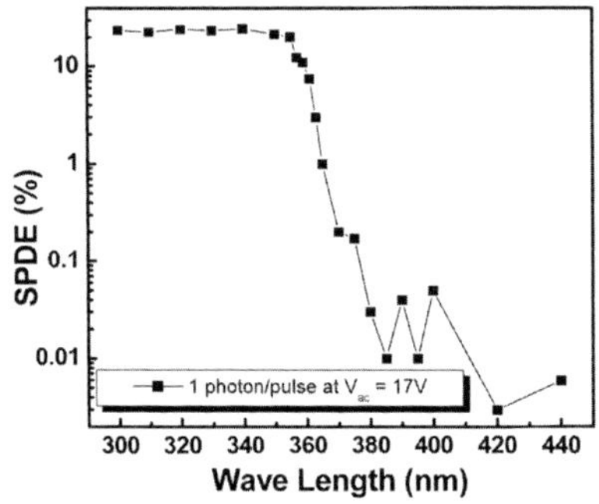

(a)

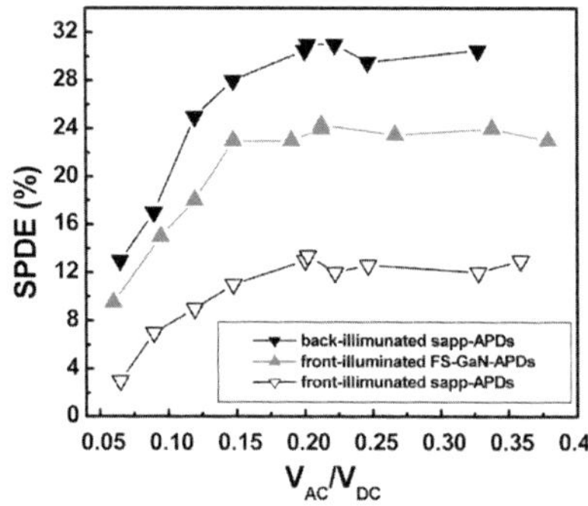

(b)

Fig. 9. (a) Photon counting spectral response of a free-standing-GaN-APD. [Adapted from Fig. 8 of Ref. 101. Reprinted with permission from SPIE for Optical Sensing, Imaging, and Photon Counting: Nanostructured Devices and Applications. Copyright © 2015 SPIE.] (b) SPDE as a function of AC-bias to DC-bias ratio (V_{AC}/V_{DC}) for free-standing-GaN APDs. [Adapted from Fig. 9 of Ref. 101. Reprinted with permission from SPIE for Optical Sensing, Imaging, and Photon Counting: Nanostructured Devices and Applications. Copyright © 2015 SPIE.]

a maximum single photon detection efficiency(SPDE) of 23%. When the photons reached 3 per pulse, the detectivity ran up to 100%. The SPDE spectrum ranged from 360 nm to approximate 200 nm with detectivity above 20%, as shown in Fig. 9.

To improve the performance of GaN-based APD further, the p-i-n APD based on m-plane epitaxial n-GaN was also studied and showed a comparatively low dark current of 8.11 pA and relatively high gain of 8000.[99] A insertion of n-AlGaN layer between n-GaN layer and AlN template weakly increased the responsivity and broadened the response spectrum.[100] Especially in 2015, further optimizing the structure and material quality based on previous works, they obtained a SPDE of 30.5% which was the highest value so far at Geiger mode.[101]

From 2011, Xiaodong Wang and Weida Hu *et al.* in *Shanghai Institute of Technical Physics* theoretically simulated the GaN-based APD. They mainly studied the influence of polarization charge at GaN/AlGaN interface on device performances, and employed the MOS inversion layer theory to explain the results successfully to some degree.[102,103] In 2014 and 2015, researchers from *Tsinghua University* proposed the ideas to employ n-i-p structure to achieve front illumination on which condition the holes trigger

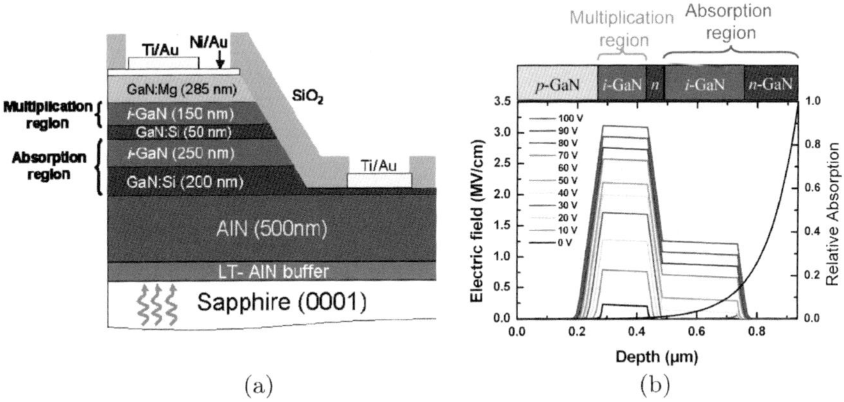

Fig. 10. (a) A typical APD with p-i-n-i-n structure [Adapted from Fig. 2 (right) of Ref. 106. Reprinted with permission from SPIE for Quantum Sensing and Nanophotonic Devices VI. Copyright © 2009 SPIE.] (b) The electric distribution and relative absorption of the APD. [Adapted from Fig. 3 (right) of Ref. 106. Reprinted with permission from SPIE for Quantum Sensing and Nanophotonic Devices VI. Copyright © 2009 SPIE.]

the avalanche multiplication process[104] and to insert AlN layer into i-type layer to form superlattices which increases avalanche gain factor.[105]

The SAM structure is maturely used in other materials based APDs expect for the area of GaN-based UV detection in which researchers from *Northwestern University* did a lot of work. Their studies implied that SAM-APDs have much more advantages than conventional p-n junction APDs in GaN-based UV detection.

R. A. McClintoch *et al.*[106] published a report about III-N APDs in 2009 focusing on the comparison between conventional p-i-n APDs and SAM-APDs (p-i-n-i-n) as well as the traditional Mg-doping and so-called delta Mg-doping. In back illumination, for the p-i-n APD, n-GaN region absorbed 75% of the incident light and the others were absorbed by i-GaN. By contrast, the absorption of the SAM-APD reached 99% in the absorption region, leading the higher gain and lower noise compared with the p-i-n one. The further calculation demonstrated that SAM-APD whose dark current was just 10 nA at breakdown voltage of about 95 V exhibited a gain of 41200, which was an order of magnitude higher than that of a conventional p-i-n APD. Similarly, SAM-APD exhibited low noise, especially in medium and low frequency. Moreover, adopting the delta Mg-doping method to the upper p-type layer of SAM-APD realized a 75 V breakdown voltage

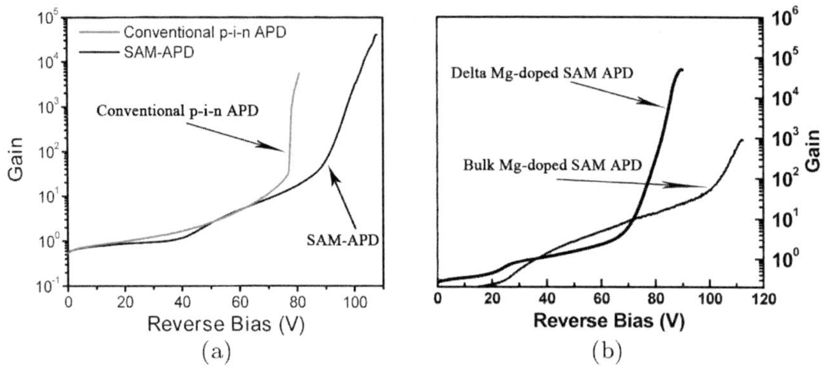

Fig. 11. (a) Comparison of the gain of conventional p-i-n and SAM-APDs. [Adapted from Fig. 6 of Ref. 100. Reprinted with the permission from SPIE for Detectors and Imaging Devices: Infrared, Focal Plane, Single Photon. Copyright © 2009 SPIE.] (b) Comparison of the gain of conventional bulk Mg-doped SAM-APDs and Delta-Mg-doped SAM-APDs. [Adapted from Fig. 7 of Ref. 106. Reprinted with the permission from SPIE for Quantum Sensing and Nanophotonic Devices VI. Copyright © 2009 SPIE.]

which was 35 V lower than that of a traditional doped one. But the gain was 50 times higher, reaching approximately 51000. In 2014, Xiaodong Wang *et al.*[107] investigated the properties of the GaN-based SAM-APD and optimized the performance, achieving higher gain and responsivity. Researchers in *Tsinghua University*[108] designed a novel p-i-p-i-n SAM-APD which triggered avalanche multiplication process by electrons thus conquering the shortage of the hole-triggering avalanche effect device that detectors must be in back-illumination, eventually improving the responsivity and reducing the noise.

The above mentioned SAM-APDs are all homogeneous type, it is possible to utilize heterojunction SAM-APD to realize much lower dark current and higher gain detectors. SiC was adopted as multiplication layer since the quality of SiC is better than that of GaN and GaN or AlGaN is used as the absorption layer. According to some reports,[51,109] gain of SiC/GaN SAM-APD could reach 10^5 magnitude with peak response of 4 A/W. Although the bias came to 90% of the breakdown voltage, it remained an extremely low dark current smaller than 1 pA.

AlGaN-based APDs were also studied by a lot of groups. R. A. McClintoch *et al.*[110] reported a p-i-n type AlGaN-based solar-blind APD with a maximum gain of 700. Later, Hai Lu *et al.*[111] reported a back-illumination

AlGaN-based solar-blind SAM-APD with a comparatively low dark current of 1.06×10^{-8} A/cm^2 at -20 V, reaching a maximum gain larger than 3000 at -91 V. The largest gain of the Schottky AlGaN-based solar-blind APDs is merely 2000, which was fabricated by Zeqiang Huang *et al.*[18] in 2013.

In spite of the encouraging progresses in GaN-APDs, the quality of GaN based material should be improved further to meet the requirement of APDs. When increasing voltage biases to low quality GaN material, premature breakdown will occur at dislocations and crystalline grain boundaries.[112] Premature breakdown makes the dark current increase rapidly, while the avalanche internal field will not have been built-up, leading to the failure of multiplication and gain. Therefore, to improve the crystal quality of GaN is the prevailing research theme now and in the future. Besides, the inherent characteristic that the hole ionization coefficient is larger than that of electron hinders the increase of gain of a GaN APD. Meanwhile, although the noise is effectively suppressed in a SAM-APD, the longer device length undoubtedly results in slow response speed.[7] Currently, it's the emphasis of researchers to balance the gain, noise and response speed to optimize device performances.

Compared to GaN-APDs, researches on AlGaN-APDs face more difficulties. Firstly, it's difficult to grow high quality AlGaN materials. Secondly, we haven't thoroughly understood the avalanche mechanism of AlGaN. Thirdly, there aren't good enough doping methods and electrode contact techniques for AlGaN.

4. Novel GaN-based UV Photodetectors

The quality of the GaN-based material limited the performance of the photodetectors to improve further. Fortunately, some new materials, such as metal plasmon, graphene and nano-materials, offer new opportunities to realize high performance GaN-based photodetectors. In this section, we will introduce the new achievements of GaN-based photodetectors based on surface plasmon, graphene and nano photodetectors.

4.1. *SPR enhanced UV photodetectors*

Near free electron plasmon in metal have their own inherent resonant frequency to which when the oscillation frequency of external electromagnetic field approaches, the effect of resonant enhancement occurs,[113] thus the energy of electromagnetic field transfers to the metal plasmon. If we deposit a certain metal on surface of a semiconductor, the located surface plasmon resonant (LSPR) or surface plasmon propagation (SPP) effect will occur.[114] Moreover, the light scattering between these located plasmon can efficiently increase the utilization ratio of light energy and photo response speed.[115] The position of surface plasmon resonance absorption peak depends on the shapes, sizes and gaps of located plasmons as well as refractive indexes of surrounding materials.

Dabing Li *et al.* in *CIOMP, CAS* firstly achieved a SPR enhanced GaN-based UV detector[116] using Ag nanoparticles. They investigated the photodetectors with different conditions, including photodetector without Ag nanoparticles (A), with as-deposited Ag nanoparticles (B) and with annealed Ag nanoparticles (C). The surface morphologies were charactered by atomic force microscope (AFM), as shown in Fig. 12.

As shown in Fig. 13(a), at room temperature, the response characteristics of these three devices exhibited different behaviors. At −5 V bias, they all peaked at 365 nm where is the absorption peak position of GaN. However, the peak responsivities of these photodetectors were 148 mA/W, 2458 mA/W and 4000 mA/W for samples A, B and C, respectively, which meant that the spectral response was enhanced a lot with the Ag nanoparticles. The response speed property was also studied, as shown in Fig. 13(b), which indicated that the introduction of Ag nanoparticles did not weaken the response speed superiority of MSM photodetectors.

After Dabing Li *et al.* realized the MSM structure SPR-GaN UV photodetector, members in their group also realized a SPR-AlGaN-based MSM solar-blind photodetector by Al nanoparticles, whose peak responsivity reached 0.288A/W at 288 nm. Subsequently, Zhang Wei *et al.*[114] obtained an AlGaN-MSM UV detector by Al nanoparticles with a higher responsivity of 2.34 A/W at 20 V at 269 nm, which was 25 times increased compared to an un-SPR photodetector. Besides, Serkan Butun *et al.*[117,118] utilized nanoantenna structures to realize an Al-SPR-GaN-based MSM UV detector. In 2014, researchers in *Taiwan* inserted Ag nanoparticles into SiO_2 dielectric layer to form a SPR-GaN-based MIS photodetector, which greatly elevated the peak responsivity and UV-to-visible rejection ratio.[119]

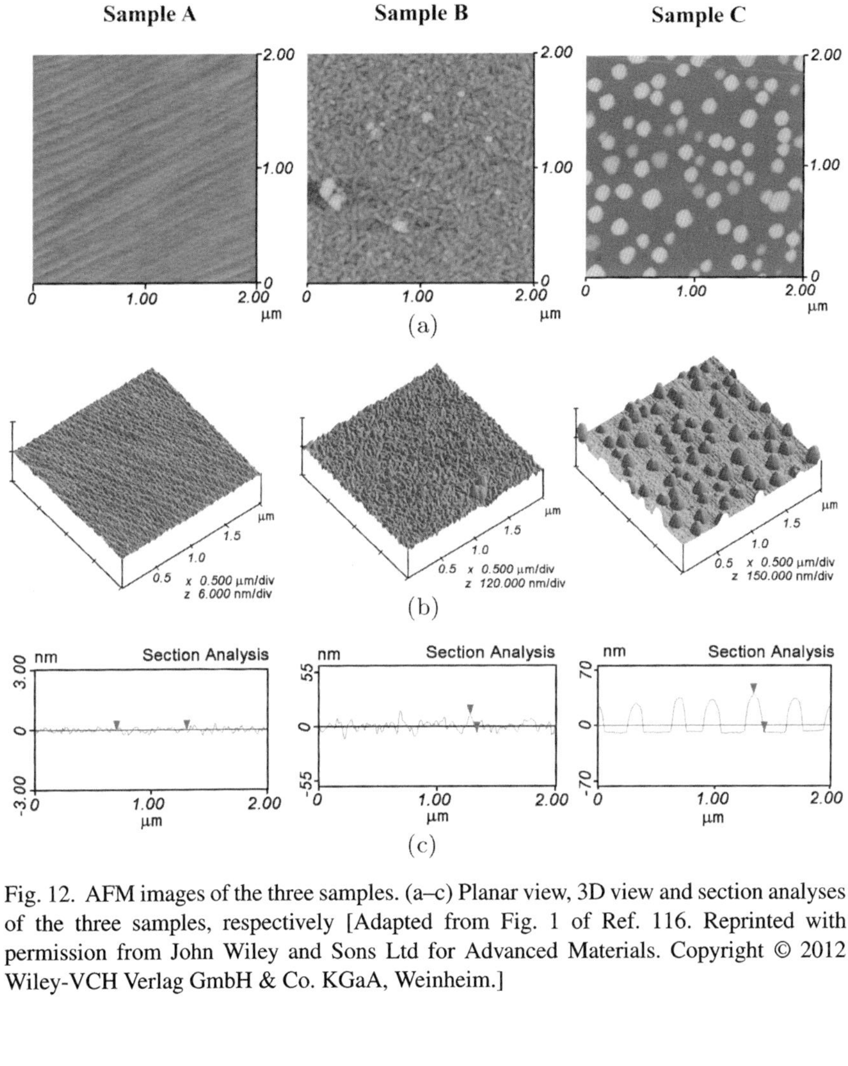

Fig. 12. AFM images of the three samples. (a–c) Planar view, 3D view and section analyses of the three samples, respectively [Adapted from Fig. 1 of Ref. 116. Reprinted with permission from John Wiley and Sons Ltd for Advanced Materials. Copyright © 2012 Wiley-VCH Verlag GmbH & Co. KGaA, Weinheim.]

There exist considerable difficulties in SPR-GaN-based UV detection technology. According to previous reports, we found several blocks as followings. Firstly, deposition of nanoantenna arrays[117] and nanosphere lithography[120] can realize relatively regular shapes, but the sizes are strictly limited by lithography. The technique of electron beam evaporation followed by annealing[116] is simply, while the shapes are irregular and sizes and gaps are nonuniform and hardly effectively controlled. Secondly,

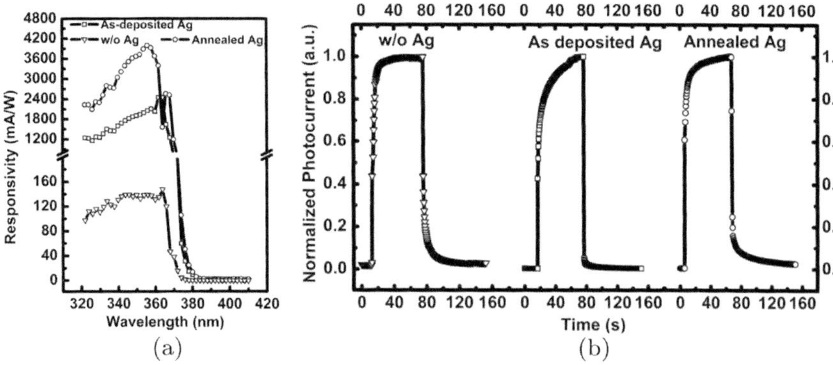

Fig. 13. (a) The spectral response of the three GaN detectors under 5 applied bias; (b) Temporal response of the three detectors measured by shutting on/off the 360 nm light under a 5 V applied bias. [Adapted from Figs. 2 and 5 of Ref. 116. Reprinted with permission from John Wiley and Sons Ltd for Advanced Materials. Copyright © 2012 Wiley-VCH Verlag GmbH & Co. KGaA, Weinheim.]

the commonly used plasmon metal such as Au and Ag are hard to be used in the deep UV region.[120] Even though Al can be used in the UV and deep UV region, the fabrication of Al nanoparticles are challenging. Thirdly, almost all the effective SPR-GaN/AlGaN photodetectors were based on MSM structrue[114,115,116,117,118] and a novel structure design that can simultaneously realize effective light absorption and resonant energy transmission is in urgent need.

4.2. *Graphene/GaN UV photodetectors*

As for photon devices, conductivity and transmittance properties of electrodes directly make difference on quantum efficiency. Series of semitransparent electrode materials such as Ti/W, Ni/Au and ITO[22,23,24,25] were used as the GaN detectors electrodes. However, they are not ideal materials neither in light emitting devices, detectors nor solar-cells. Graphene exhibits outstanding electrical, mechanical, chemical and optical properties. Recently, the research about the combination of graphene and semiconductors became an important content and got a good effect.

The contact of graphene and semiconductor was always considered as an ordinary MS contact and the Fermi level of graphene was believed unchangeable as a common metal,[121,122,123] However, for few layers graphene, the Fermi level could be shifted by the charge carrier transfer,[124,125] which subsequently affected Schottky barrier height and changed the electrical characteristics. It was reported that a transfer of 0.01 electrons per unit cell would shift the Fermi level by 0.47 eV.[126] The significant Fermi level shift obviously influences barrier height, thus has to be taken into account. On the foundation of conventional MS Schottky barrier mode, considering the Fermi level shift of graphene caused by carrier transfer, Haijian Zhong *et al.*[127,128] established a graphene/GaN contact mode by which the barrier height of graphene and other semiconductors can be calculated too.

In addition to the inherent properties of high transmittance and conductivity of graphene, according to known reports, the G/GaN photodetectors have some other characteristics summarized as the following aspects. Firstly, high temperature resistance. S. Tongay *et al.*[129] found that a G/GaN Schottky diode worked stably at 550 K and became non-rectifying above 650 K. When the temperature was elevated to 900 K and then cooled to room temperature, the reverse leakage reduced and rectification recovered. Secondly, radiation hardening. Heather C. Chiamari *et al.*[130] studied the radiation resistance of G/GaN UV detectors. In their report, the responsivity of a directly contacting G/GaN-based MSM photodetector changed only 11% at 750 krad TID radiation intensity compared to 75 krad TID. Thirdly, the responsivity saturates with bias and illumination power increasing. In the studies of Chang-Ju Lee *et al.*,[131] the dark current and photocurrent both slowly changed with bias ranging from 1 V to 15 V, especially when the bias surpassed 4V, the current hardly varied. When A. V. Babichev *et al.*[132] studied a photodetector fabricated by graphene and GaN nanorods, they found the responsivity merely changed within 3 orders of magnitude of illumination power when the power was lower than 10^{-3} W/cm^2. Furthermore, if the illumination power was higher than 10^{-3} W/cm^2, the responsivity decreased instead. Fourthly, there exists noise-like response in a certain wave-range. This phenomenon was always found in reports[131,133,134] of Chang-Ju Lee *et al.* about MSM and MISIM G/GaN photodetectors, and located at the range from 400 nm to 500 nm. Lastly, double-wavelength sensitive. The electric properties of graphene can be transferred by doping,[135] which makes it different from ordinary metals. Therefore, the Schottky barrier height of G/GaN contact changes with graphene doping. Fang Lin *et al.*[136] found the photo-to-dark current ratio of a G/GaN Schottky photodetector exceeded 10^3 at 10 V reverse bias in both UV region and green light of 514 nm.

4.3. *GaN nanowire UV photodetectors*

In recent decades, GaN nano materials obtained great developments. GaN nanostructures are excellent building blocks for the applications in nanophotonics, which have been used to fabricate LEDs,[137] solid state lasers,[138] gas sensoring devices[139] and dye-sensitized solar cells.[140] In the area of detection, researches about nanostructure devices reached many achievements. ZnO NWs[141] and Ge NWs[142] photodetectors were reported in 2002 and 2010, respectively. As one of the most important UV photon device materials, GaN NWs UV photodetectors were studied by a lot of researchers since 2004 when Song Han *et al.*[143] firstly reported this kind of UV photodetector.

Compared to bulk semiconductors, NWs photodetectors possess two outstanding characteristics. First of all, the diameter of a nanowire ranges from tens to hundreds of nanometers in which the stress can be fully released and the dislocations in substrate can't propagate to nanowire structures, thus the nanowires are almost dislocation-free.[144,145,146,147,148] Then, nanowire structures own large surface-to-volume ratio which significantly enlarges the device areas contacting to surroundings such as molecules and photons,[149] making them extremely different from other bulk semiconductors.

Based on characteristics of NWs, the UV detectors fabricated by GaN NWs generally exhibit high gain. In 2007, Reui-San Chen *et al.*[150] reported a GaN NWs UV detector with the gain reaching 5×10^4–1.9×10^5 which is 3 orders of magnitude larger than a film-type photodetector with the similar structure. In 2011, F. Gonzales-Posada *et al.*[151] further enhanced the gain to 10^5–10^8 magnitude, meanwhile the UV-to-visible rejection ratio was higher than 10^6. Considering the feature of large surface-to-volume ratio of a nanowire, researchers attributed such a high gain to surface states.[152,153] A large number of electrons will be trapped by surface states, thus forming depletion region near surface where no free charges exist and photo-generated carriers are immediately swept out by inner field too, which practically determined the photoconductive characteristics of NWs.

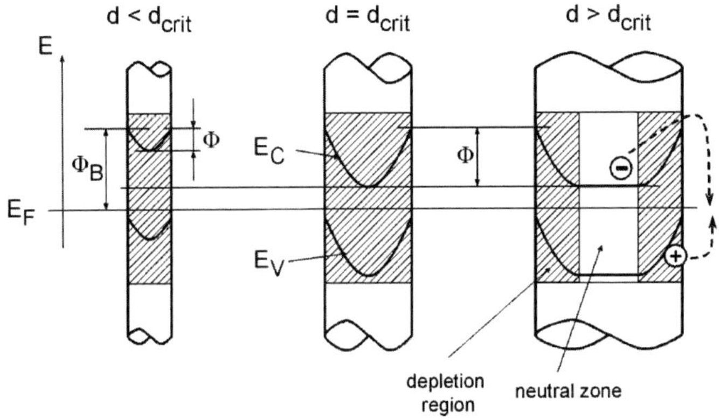

Fig. 14. Dependence of depletion region (shaded), shape of conduction (E_C) and valence band edge (E_V), and recombination barrier Φ on the nanowire diameter d. The detail on the right shows the surface recombination mechanism of the photoexcited carriers. [Adapted from Fig. 6 of Ref. 154. Reprinted with permission from American Chemical Society Publications for Nano Letters. Copyright © 2005, American Chemical Society.]

Responsivity of GaN NWs photodetectors depended on the diameter. There existed a critical diameter, as shown in Fig. 14, below which the device performed like a insulator, otherwise, like a photoconductor.[154] Meanwhile, the response speed of GaN NWs photodetectors were at the magnitude of seconds or even slower[155] and obvious persistent current existed.[154] However, when the surface was passivated, the dark current decreased and responsivity increased.[156] When researchers studied the relationship between illumination power and gain, they found that gain exponentially relied on illumination power with an index of approximate unit.[150] All these results indicated that the surface states brought the GaN NWs photodetectors a high gain.

Limited by the imperfection of GaN NWs growth technologies, the majority of previous NWs photodetectors were photoconductive ones maded by single nanowire that connected two metal electrodes.[143,150,154] With the development of growth technologies, it became available for homogeneous p-n,[157] p-i-n[158] and n-i-n[151] junctions. Besides, heterojunction-based photodetectors made by GaN NWs on Si substrate were occasionally reported,[156,158] and ZnO/GaN heterojunction NWs photodetectors reached a pretty good detection effect.[155,159]

In addition to the previous single-nanowire-based photodetectors, multi-nanowires-based ones have been reported recently,[155,160,161] which simplified the device processes and decreased the costs, but the performances remained to be enhanced. Some researchers combined the graphene with GaN NWs to manufacture a photodetector, which made a good difference on responsivity[132] and sensitivity.[162] Moreover, it was also reported that taking Ag NWs as the electrode material could improve the GaN NWs photodetectors.[161]

5. Conclusions and Outlook

The studies of GaN-based UV photodetectors have realized a lot of exciting achievements. Many types of GaN based UV and deep UV detectors have been developed, including conventional photoconductors, Schottky, MSM, p-n junction and APD *et al.* However, the performances of these photodetectors are still far away from applications, especially for detecting weak signals. In present series of photodetectors, APDs can obtain a gain of 10^5 magnitude with sufficiently low noise, and thus promisingly they can actualize single photon detection. However, the requirement for the quality of GaN based material is very strict.

Up to now, how to grow higher quality GaN based materials and optimize the structure of detector are still the topics that the researches focus on. Fortunately, the surface plasmon or graphene has offered new opportunities to get high performance of GaN based photodetectors and probably accelerate the application of GaN-based photodetectors.

Acknowledgments

This work was supported by the National Natural Science Foundation of China (grant nos 61322406, 61274038, 51472230, 61574142, and 61574021), the Jilin Provincial Science & Technology Department (grant nos 20140520116JH and 20150519001JH), the CAS Interdisciplinary Innovation Team, the Youth Innovation Promotion Association of CAS (grant no. 2012173).

References

1. L. Sun *et al.*, *Appl. Phys. Lett.*, **97**(19), 191103 (2010).

2. Y. Lu *et al.*, *Int. Society for Opt. and Photonics*, 92840P-92840P-7 (2014).
3. H. Amano *et al.*, *Appl. Phys. Lett.*, **48**(5), 353–355 (1986).
4. S. Nakamura *et al.*, *Japanese Journal of Applied Physics*, **30**(10A), L1705 (1991).
5. S. Nakamura *et al.*, *Appl. Phys. Lett.*, **64**(13), 1687–1689 (1994).
6. M. A. Khan *et al.*, *Appl. phys. lett.*, **60**(23), 2917–2919 (1992).
7. S. M. Sze *et al.*, *Physics of semiconductor devices[M].* John wiley & sons, 2006.
8. F. Binet *et al.*, *Journal of applied physics*, **81**(9), 6449–6454 (1997).
9. O. Katz *et al.*, *Appl. Phys. Lett.*, **79**(10), 1417–1419 (2001).
10. S. C. Ling *et al.*, *Appl. Phys. Lett.*, **94**(25)(2009).
11. S. J. Chang *et al.*, *IEEE, Sensors Journal*, **10**(10), 1609–1614 (2010).
12. K. H. Lee *et al.*, *IEEE, Sensors Journal*, **9**(7), 814–819 (2009).
13. K. H. Lee *et al.*, *J. of the Electrochemical Society*, **156**(7), J199-J202 (2009).
14. Z. D. Huang *et al.*, *IEEE, Sensors Journal*, **11**(11), 2895–2901 (2011).
15. S. J. Chang *et al.*, *IEEE Journal of Quantum Electronics*, **44**(10), 916–921 (2008).
16. T. Aoki *et al.*, *Japanese Journal of Applied Physics*, **55**(4S), 04EJ09 (2016).
17. T. Tut *et al.*, *Appl. Phys. Lett.*, **87**(22), 223502–223502 (2005).
18. Z. Huang *et al.*, *Applied Physics Express*, **6**(5), 054101 (2013).
19. C. Rivera *et al.*, *physica status solidi (a)*, **201**(12), 2658–2662 (2004).
20. C. Rivera *et al.*, *Superlattices and Microstructures*, **36**(4), 849–857 (2004).
21. D. Walker *et al.*, *Appl. phys. lett.*, **74**(5), 762–764 (1999).
22. C. H. Chen *et al.*, *IEEE, Photonics Technology Letters*, **13**(8), 848–850 (2001).
23. K. Wang *et al.*, *IEEE Transactions on Electron Devices*, **53**(1), 38–42 (2006).
24. C. K. Wang *et al.*, *Materials Science and Engineering: B*, **112**(1), 25–29 (2004).
25. P. C. Chang *et al.*, *Semiconductor science and technology*, **19**(12), 1354 (2004).
26. J. K. Kim *et al.*, *Appl. Phys. Lett.*, **81**(24), 4655–4657 (2002).
27. X. Sun *et al.*, *Scientific reports*, **5**(2015).
28. S. J. Chang *et al.*, *Electron Device Letters*, IEEE, **24**(4), 212–214 (2003).
29. F. Xie *et al.*, *Solid-State Electronics*, **57**(1), 39–42 (2011).
30. S. W. Chen *et al.*, *Int. Society for Opt. and Photonics*, 97681D-97681D-9 (2016).
31. C. K. Wang *et al.*, *Sensors Journal*, IEEE, **15**(9), 4743–4748 (2015).
32. E. Monroy, F. Calle, E. Munoz *et al.*, *Appl. Phys. Lett.*, **74**(22), 3401–3403 (1999).
33. S. V. Averine *et al.*, *Solid-State Electronics*, **52**(5), 618–624 (2008).
34. I. S. Seo, I. H. Lee, Y. J. Park *et al.*, *Journal of crystal growth*, **252**(1), 51–57 (2003).
35. M. Brendel *et al.*, *Electronics Letters*, **51**(20), 1598–1600 (2015).
36. S. Butun, T. Tut, B. Butun *et al.*, *Appl. Phys. Lett.*, **88**(12), 123503 (2006).
37. H. Y. Liu *et al.*, *IEEE, Sensors Journal*, **15**(9), 5202–5207 (2015).
38. Dabing Li *et al.*, *Applied Physics Letter* **98**, 011108 (2011).
39. Xiaojuan Sun *et al.*, *Applied Physics Letter* **98**, 121117 (2011).
40. Y. K. Su *et al.*, *Japanese Journal of Applied Physics*, **40**(4S), 2996 (2001).
41. P. C. Chang *et al.*, *Japanese journal of applied physics*, **43**(4S), 2008 (2004).
42. P. C. Chang, *et al.*, *IEEE Journal of Quantum Electronics*, **43**(11), 1060–1064 (2007).
43. P. C. Chang *et al.*, *Appl. Phys. Lett.*, **91**(14), 141113 (2007).
44. Z. D. Huang *et al.*, *IEEE,Sensors Journal*, **13**(4), 1187–1191 (2013).
45. L. Sang *et al.*, *Appl. Phys. Lett.*, **98**(10), 3502 (2011).
46. S. H. Abud *et al.*, *Int. J. Electrochem. Sci*, **7**, 10038–10046 (2012).
47. S. H. Abud *et al.*, *Measurement*, **50**, 172–174 (2014).

48. L. Shen *et al.*, *Electron Device Letters*, IEEE, **22**(10), 457–459 (2001).
49. G. Y. Xu *et al.*, *Appl. Phys. Lett.*, **71**(15), 2154–2156 (1997).
50. C. H. Chen *et al.*, *Chem. Phys. Lett.*, **476**(1), 69–72 (2009).
51. Q. Zhou *et al.*, *Appl. Phys. Lett.*, **99**(13), 131110 (2011).
52. J. Simon *et al.*, *Science*, **327**(5961), 60–64 (2010).
53. Q. Chen *et al.*, *Electronics Letters*, **31**(20), 1781–1782 (1995).
54. X. Zhang *et al.*, *Appl. Phys. Lett.*, **67**(14), 2028–2030 (1995).
55. J M. Van Hove *et al.*, *Appl. Phys. Lett.*, **70**(17), 2282–2284 (1997).
56. A. Osinsky *et al.*, *Appl. Phys. Lett.*, **71**(16), 2334–2336 (1997).
57. E. Monroy *et al.*, *Semiconductor science and technology*, **13**(9), 1042 (1998).
58. G. Parish *et al.*, *Appl. Phys. Lett.*, **75**(2), 247–249 (1999).
59. J. C. Carrano *et al.*, *Electronics Letters*, **34**(18), 1779–1781 (1998).
60. C. J. Collins *et al.*, *Appl. Phys. Lett.*, **80**(20), 3754–3756 (2002).
61. T. Tut *et al.*, *Appl. Phys. Lett.*, **92**(10), 103502–103502 (2008).
62. C. A. Hurni *et al.*, *Appl. Phys. Lett.*, **97**(22), 222113 (2010).
63. M. N. Fireman *et al.*, *Appl. Phys. Lett.*, **106**(20), 202106 (2015).
64. M. Qi *et al.*, *Appl. Phys. Lett.*, **107**(23), 232101 (2015).
65. F. Yakuphanoglu *et al.*, *Journal of Alloys and Compounds*, **650**, 671–675 (2015).
66. E. Monroy *et al.*, *Appl. phys. lett.*, **74**(8), 1171–1173 (1999).
67. D. Walker, V. Kumar, K. Mi *et al.*, *Appl. Phys. Lett.*, **76**(4), 403–405 (2000).
68. R. McClintock *et al.*, *Appl. phys. lett.*, **84**(8), 1248–1250 (2004).
69. E. Cicek *et al.*, *Appl. Phys. Lett.*, **103**(19), 191108 (2013).
70. Y. Z. Chiou *et al.*, *Solid-State Electronics*, **46**(12), 2227–2229 (2002).
71. C. Rivera *et al.*, *IEEE Journal of Quantum Electronics*, **42**(1), 51–58 (2006).
72. J. Pereiro *et al.*, *IEEE Journal of Quantum Electronics*, **45**(6), 617–622 (2009).
73. Y. K. Su *et al.*, *physica status solidi (c)*, **6**(S2), S811-S813 (2009).
74. Y. T. Huang *et al.*, *IEEE Photonics Technolgy Letters*, **28**(6), 605–608 (2016).
75. X. Li *et al.*, *IEEE Photonics Journal*, **7**(6), 1–7 (2015).
76. W. Cai *et al.*, *Optics express*, **24**(6), 6004–6010 (2016).
77. J. A. Garrido *et al.*, *Semiconductor science and technology*, **13**(6), 563 (1998).
78. K. S. Stevens *et al.*, *Appl. Phys. Lett.*, **66**(25), 3518–3520 (1995).
79. Z. C. Huang *et al.*, *Journal of electronic materials*, **26**(3), 330–333 (1997).
80. L. B. Flannery *et al.*, *Materials Science and Engineering: B*, **50**(1), 307–310 (1997).
81. B. Shen *et al.*, *Japanese journal of applied physics*, **38**(2R), 767 (1999).
82. D. Walker *et al.*, *Appl. Phys. Lett.*, **68**(15), 2100–2101 (1996).
83. B. W. Lim *et al.*, *Appl. Phys. Lett.*, **68**(26), 3761–3762 (1996).
84. A. Bugallo *et al.*, *Appl. Phys. Lett.*, **98**(23), 3107 (2011).
85. M. A. Khan *et al.*, *Electronics Letters*, **31**(5), 398–400 (1995).
86. W. Yang *et al.*, *Appl. Phys. Lett.*, **73**(7), 978–980 (1998).
87. W. Yang *et al.*, *International Society for Optics and Photonics*, 188–197 (1998).
88. R. Mouillet *et al.*, *Japanese Journal of Applied Physics*, **40**(5B), L498 (2001).
89. M. L. Lee *et al.*, *Appl. Phys. Lett.*, **92**(5), 3506 (2008).
90. J. Kolnik *et al.*, *Journal of applied physics*, **81**(2), 726–733 (1997).
91. A. Reklaitis, L. Reggiani *Journal of applied physics*, **97**(4), 043709 (2005).
92. J. C. Carrano *et al.*, *Appl. Phys. Lett.*, **76**(7), 924–926 (2000).

93. S. Verghese *et al.*, *Device Research Conference Digest*, IEEE, **56**, 54–55 (1998).
94. K. A. McIntosh *et al.*, *Appl. Phys. Lett.*, **75**(22), 3485–3487 (1999).
95. K. A. McIntosh *et al.*, *Appl. Phys. Lett.*, **76**(26), 3938–3940 (2000).
96. S. Verghese *et al.*, *IEEE Transactions on Electron Devices*, **48**(3), 502–511 (2001).
97. P. Kung *et al.*, *Int. Society for Opt. and Photonics*, 64791J-64791J-12 (2007).
98. R. McClintock *et al.*, *Int. Society for Opt. and Photonics*, 69000N-69000N-11 (2008).
99. Z. Vashaei *et al.*, *Appl. Phys. Lett.*, **96**(20), 201908 (2010).
100. R. McClintock *et al.*, *Int. Society for Opt. and Photonics*, 77801B-77801B-13 (2010).
101. R. McClintock *et al.*, *Int. Society for Opt. and Photonics*, 95550B-95550B-10 (2015).
102. X. Wang *et al.*, *2011 3rd Int. Conf. on Computer Research and Development*(2011).
103. X. Wang *et al.*, *Journal of Physics D: Applied Physics*, **44**(40), 405102 (2011).
104. Z. JiYuan *et al.*, *General Assembly and Scientific Symposium*, 2014 URSI, 1–3.
105. Z. JiYuan *et al.*, *ICNS*(2015).
106. R. McClintock *et al.*, *Int. Society for Opt. and Photonics*, 72220U-72220U-12 (2009).
107. X. Wang *et al.*, *Journal of Applied Physics*, **115**(1), 013103 (2014).
108. Z. Ji-Yuan *et al.*, *Chinese Physics Letters*, **29**(9), 097804 (2012).
109. A. V. Sampath *et al.*, *Int. Society for Opt. and Photonics*, 83760O-83760O-90 (2012).
110. R. McClintock, A. Yasan, K. Minder *et al.*, *Applied Physics Letters*, **87**(24)(2005).
111. Y. Huang, D. J. Chen, H. Lu *et al.*, *Applied Physics Letters*, **101**(25), 253516 (2012).
112. A. Osinsky, M. S. Shur, R. Gaska *et al.*, *Electronics Letters*, **34**(7), 691–692 (1998).
113. S. L. Chuang, *Physics of photonic devices[M]*. John Wiley & Sons, 2012.
114. W. Zhang *et al.*, *Appl. Phys. Lett.*, **106**(2), 021112 (2015).
115. G. Bao, D. Li, X. Sun *et al.*, *Optics express*, **22**(20), 24286–24293 (2014).
116. D. Li *et al.*, *Advanced Materials*, **24**(6), 845–849 (2012).
117. S. Butun *et al.*, *Optics express*, **20**(3), 2649–2656 (2012).
118. S. Butun *et al.*, *Nanotechnology*, **23**(44), 444010 (2012).
119. J. K. Sheu *et al.*, *IEEE J. of Selected Topics in Quan. Elec.*,**20**(6), 137–141 (2014).
120. G. H. Chan *et al.*, *The Journal of Physical Chemistry C*, **112**(36), 13958–13963 (2008).
121. C. C. Chen *et al.*, *Nano letters*, **11**(5), 1863–1867 (2011).
122. S. Tongay, T. Schumann, X. Miao *et al.*, *Carbon*, **49**(6), 2033–2038 (2011).
123. X. Li, H. Zhu, K. Wang *et al.*, *Advanced Materials*, **22**(25), 2743–2748 (2010).
124. K. S. Novoselov *et al.*, *Science*, **306**(5696), 666–669 (2004).
125. Z. Chen, I. Santoso, R. Wang *et al.*, *Appl. Phys. Lett.*, **96**(21), 213104 (2010).
126. G. Giovannetti *et al.*, *Phys. Rev. Lett.*, **101**(2), 026803 (2008).
127. H. Zhong *et al.*, *Journal of Applied Physics*, **115**(1), 013701 (2014).
128. H. Zhong, Z. Liu, G. Xu *et al.*, *Appl. Phys. Lett.*, **100**(12), 122108 (2012).
129. S. Tongay *et al.*, *Appl.Phys. Lett.*, **99**(10), 102102 (2011).
130. H. C. Chiamori *et al.*, *Int. Society for Opt. and Photonics*, 949107–949107-7 (2015).
131. C. J. Lee *et al.*, *Applied Mechanics and Materials*, **481**, 146–149 (2014).
132. A. V. Babichev *et al.*, *Appl. Phys. Lett.*, **103**(20), 201103 (2013).
133. C. J. Lee *et al.*, *Japanese Journal of Applied Physics*, **54**(6S1), 06FF08 (2015).
134. S. B. Kang *et al.*, *The 14th International Conference on Quality in Research 2015.*
135. B. J. Kim *et al.*, *Optics express*, **21**(23), 29025–29030 (2013).
136. F. Lin *et al.*, *Appl. Phys. Lett.*, **105**(7), 073103 (2014).

137. M. Musolino *et al.*, *Appl. Phys. Lett.*, **105**(8), 083505 (2014).
138. A. Hurtado *et al.*, *Appl. Phys. Lett.*, **103**(25), 251107 (2013).
139. J. Sim, K. Kim, S. Song *et al.*, *Analyst*, **138**(8), 2432–2437 (2013).
140. H. S. Jung, Y. J. Hong, Y. Li *et al.*, *ACS nano*, **2**(4), 637–642 (2008).
141. H. Kind, H. Yan, B. Messer *et al.*, *Advanced materials*, **14**(2), 158 (2002).
142. C. J. Kim, H. S. Lee, Y. J. Cho *et al.*, *Nano letters*, **10**(6), 2043–2048 (2010).
143. S. Han, W. Jin, D. Zhang *et al.*, *Chem. Phys. Lett.*, **389**(1), 176–180 (2004).
144. M. Yoshizawa *et al.*, *Japanese journal of applied physics*, **36**(4B), L459 (1997).
145. H. Sekiguchi *et al.*, *Journal of crystal growth*, **300**(1), 259–262 (2007).
146. E. Calleja *et al.*, *Journal of crystal growth*, **201**, 296–317 (1999).
147. M. Tchernycheva, C. Sartel, G. Cirlin *et al.*, *Nanotechnology*, **18**(38), 385306 (2007).
148. L. Cerutti *et al.*, *Appl. phys. lett.*, **88**(21), 213114–213114 (2006).
149. G. H. Shen and F. C. N. Hong, *Thin Solid Films*, **570**, 363–370 (2014).
150. R. S. Chen *et al.*, *Appl. Phys. Lett.*, **91**(22), 223106–223106 (2007).
151. F. González-Posada *et al.*, *Nano letters*, **12**(1), 172–176 (2011).
152. N. A. Sanford *et al.*, *Journal of Applied Physics*, **107**(3), 034318 (2010).
153. F. Liu, L. Li, T. Guo *et al.*, *Nanoscale research letters*, **7**(1), 1–9 (2012).
154. R. Calarco, M. Marso, T. Richter *et al.*, *Nano letters*, **5**(5), 981–984 (2005).
155. K. Teker *Sensors and Actuators A: Physical*, **216**, 142–146 (2014).
156. A. de Luna Bugallo *et al.*, *Nanotechnology*, **21**(31), 315201 (2010).
157. M. S. Son *et al.*, *Materials Science and Engineering: C*, **26**(5), 886–888 (2006).
158. Y. B. Tang, Z. H. Chen, H. S. Song *et al.*, *Nano letters*, **8**(12), 4191–4195 (2008).
159. C. H. Chen *et al.*, *Chem. Phys. Lett.*, **476**(1), 69–72 (2009).
160. M. Kang *et al.*, *Japanese journal of applied physics*, **43**(10R), 6868 (2004).
161. Y. Liu, P. Sun, X. Meng *IEEE, Photonics Technology Letters*, **28**(1), 23–26 (2016).
162. S. Kang, A. Mandal, J. H. Chu *et al.*, *Scientific reports*, **5**(2015).

Part II

III-Nitride Materials

CHAPTER 5

SINGLE CRYSTAL ALN: GROWTH BY MODIFIED PHYSICAL VAPOR TRANSPORT AND PROPERTIES

Honglei Wu and Ruisheng Zheng

College of Optoelectronic Engineering, Shenzhen University
Nanhai Ave 3688, Shenzhen, P. R. China
hlwu@szu.edu.cn

Physical vapor transport (PVT) is the most successful and widely used approach for bulk Aluminum nitride (AlN) single crystals. Recent developments of PVT growing bulk AlN single crystals are reviewed and discussed with a focus on crucible materials, the growth setup, the growth parameters, and the type of seeding. The structural and optical properties as well as the sizes of the AlN crystals are affected by the type of seeding or nucleation and the crucible material. It is concluded that precisely controlling the growth conditions, finding the best technical approach and process sequence, and preparing native AlN seed with high quality will be addressed in the further work.

1. Introduction

Aluminum nitride (AlN), gallium nitride (GaN), silicon carbide (SiC), diamond, and other wide-band gap semiconductor materials are often referred to as third-generation semiconductor materials. Compared to first-generation and second-generation semiconductor materials, the third generation is characterized by high breakdown field strength, high thermal conductivity, high electron saturation velocity, and high resistance to radiation; they are suitable for the production of high-frequency, anti-radiation, large power, and high-temperature semiconductor devices. AlN, a typical representative of this group, has a widest direct band gap of 6.2 eV as well as excellent electrical, optical, mechanical, and piezoelectric properties. Table 1 briefly summarizes the technological importance of AlN.

AlN crystals can have either an unstable, zinc-blended, cubic structure or a stable wurtzite (hexagonal) structure under normal conditions.[7] However, zinc-blended AlN rare; therefore, in this study AlN refers to wurtzite AlN, which has a space group of C46v (P63mc) and a point group of C6v (6 mm). Figure 1 shows an AlN unit cell, where $a_0 = b_0 = 0.311$ nm and $c_0 = 0.498$ nm. Each Al (N) atom bonds to the nearest four N (Al) atoms to form a tetrahedral sub-lattice. Pure AlN crystals are colorless and transparent, but in practice, AlN crystals are often yellow or amber due to impurities or native defects such as Al vacancies (V_{Al}) or substitutional impurities of carbon or oxygen,[8] as shown as Fig. 2.

AlN crystals possess excellent chemical and physical properties, which make it one of the best choices for electronic and optoelectronic applications. First, due to its ultra-wide direct bandgap, AlN is a promising material for deep-ultraviolet (DUV) solid-state light-emitting diodes (LEDs) and laser diodes (LDs). AlN has been successfully used to make LEDs with a wavelength of 210 nanometers.[9,10] Second, AlN's excellent thermal conductivity, large breakdown field, high carrier mobility, and high thermal and chemical stabilities make it an ideal material for high-power, high-frequency, and high-temperature electronic applications. Third, the characteristic of high thermal conductivity combined with high volume resistivity, high dielectric strength, and low dielectric loss make AlN a suitable material for substrates and packaging for high-power or high-density assemblies of microelectronic components. High-quality AlN crystals constitute the fundamental material necessary for these applications.

AlN does not occur naturally as a mineral. AlN powder was first synthesized in 1862 from molten aluminum in a nitride atmosphere, and the first single AlN crystal (Φ0.03 mm × 0.3 mm) was isolated in 1956. AlN crystal bullets were grown in 1976.[11,12] Due to its high melting point and a high dissociation pressure at the melting point, AlN crystals are rarely grown from melt at normal pressure conditions. Many crystal-growth methods have been applied to the task of growing AlN crystals in bulk.

The study of AlN crystal-growth technology is a vital research topic for solid state physics in the near future. The main methods for growing

Table 1. briefly summarizes the technological importance of AlN.

Basic	**Value**	**Reference**
Crystal structure	Wurtzite	1
Color	powder: white or gray white crystal: colorless to amber	
Space group	C_{6v}^{4}-P6$_3$mc	1
Lattice constans (nm)	$a = 0.3112$, $c = 0.4982$	1
Band gap (eV)	6.2 (direct)	1
Density (g cm^{-3})	3.28	2
Melting point (°C)	2750 (at 100 ~ 500 atm N_2)	1
Permittivity $\varepsilon(0)$	8.5	1
Permittivity $\varepsilon(\infty)$	4.6	1
Mechanical	**Value**	**Reference**
Bulk modulus (GPa)	210	1
Young modulus (GPa)	329.7	3
Poisson ratio	0.239	3
Tensile strength (MPa)	266	4
Hardness (kg mm^{-2})	800	1
Thermal	**Value**	**Reference**
Thermal conductivity (W cm^{-1} K^{-1})	3.4	5
Thermal expansion coefficient (linear, K^{-1})	$a = 5.72 \times 10^{-6}$, $c = 4.75 \times 10^{-6}$	3
Specific Heat (J g^{-1}°C^{-1})	0.6	1
Thermal diffusion coefficient (cm^2 s^{-1})	1.47	1
Electrical	**Value**	**Reference**
Breakdown strength (V cm^{-1})	$(1.2 \sim 1.8) \times 10^6$	1
Volume resistivity (Ωcm)	$10^7 \sim 10^{13}$	6
Electron mobility (cm^2 V^{-1} s^{-1})	300	1
Hole mobility (cm^2 V^{-1} s^{-1})	14	1
Electron diffusion coefficien (cm^2 s^{-1})	7	1
Hole diffusion coefficien (cm^2 s^{-1})	0.3	1
Electron thermal velocity (m s^{-1})	1.85×10^5	1
Hole thermal velocity (m s^{-1})	0.41×10^5	1

(*Continued*)

Table 1. (*Continued*)

Piezoelectric	**Value**	**Reference**
Piezoelectric Constant e_{15} ($C\ m^{-2}$)	−0.48	1
Piezoelectric Constant e_{31} ($C\ m^{-2}$)	−0.58	1
Piezoelectric Constant e_{33} ($C\ m^{-2}$)	1.55	1
Optical	**Value**	**Reference**
IR refractive idex	2.15	1
Recombination coefficient ($cm^3\ s^{-1}$)	0.4×10^{-10}	1

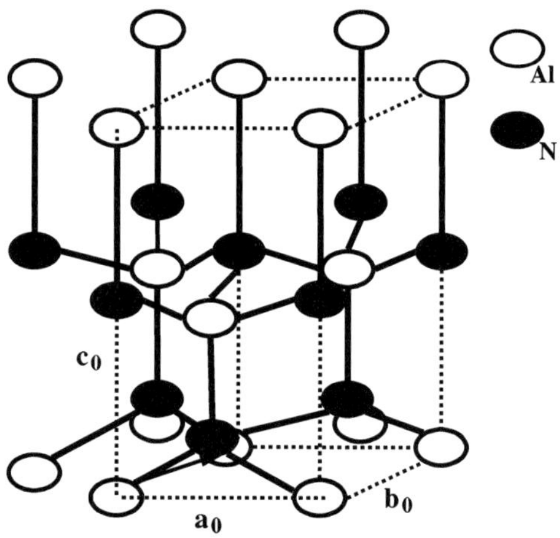

Fig. 1. AlN unit cell.

AlN crystals can be classified into four types: (1) direct nitridation of aluminum, (2) solution growth, (3) chemical vapor-phase growth, like hydride vapor phase epitaxy (HVPE), and (4) physical vapor-phase growth, like physical vapor transport (PVT). Over the past several decades, PVT, also called the sublimation-recondensation method, has proven to be the most successful and widely used approach for bulk AlN crystal growth.[13–15] Different research groups studying AlN crystal growth have developed PVT process technologies. For example, researchers have optimized crucible materials, growth reactor designs, and process parameters, which have a crucial influence on the size, structure, and properties of bulk-grown AlN

Fig. 2. Hexagonal prism AlN single crystal.

crystals. As of the publication date, the maximum size of a single AlN crystal grown by PVT in a laboratory is about 1 inch in diameter.[14,15]

This work proposes a detailed study on the PVT growth of single AlN crystals with sizes ranging from nanometers to centimeters. This study also investigates the structural and optical properties of AlN crystals. Finally, it discusses and contrasts the growth conditions and mechanisms of AlN crystals with different sizes.

2. Growth of Single AlN Crystals

2.1. *Crucible materials and insulation*

PVT crystal-growth processes are generally performed in a graphite or tungsten heating element furnace with a resistive or inductive heating system, as in the case of SiC crystal growth by PVT. Microwave heating techniques can also be employed. Directly heating the source materials with microwaves can successfully grow AlN single crystal platelets up to $2 \times 3\,mm^2$ and needles 1 mm in diameter and 3 mm in length.[16] The induction heating furnace has also been chosen for AlN growth due to its success in SiC crystal growth. Nonetheless, identifying a heating system with the advantages of fast and uniform heating as well as energy savings remains a critical technical problem.

Figure 3 shows a growth unit that operates with a high-nitrogen atmosphere (99.999%) at 60–120 kPa. During AlN growth, the components

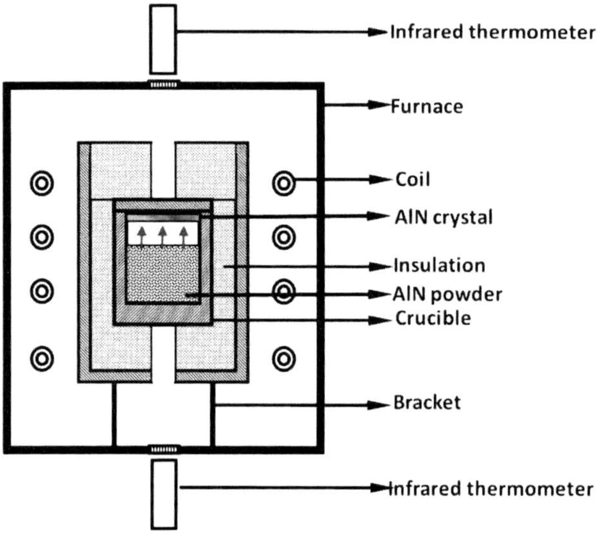

Fig. 3. Scheme of the induction heating unit.

of AlN, Al and N2, sublimate from the source at the hotter bottom of the crucible and re-condense on the crucible lid where the temperature is relatively lower. In this process, the overall reaction AlN(s) $\leftrightarrow$ Al(g) + 1/2N2(g) runs in the forward (sublimation) direction at the source, and in the reverse (recondensation) direction at the top of the crucible. AlN powder sources usually begin to sublimate above 1,800°C. Achieving high-quality AlN crystals in bulk at fast growth rates ($>200\,\mu$m/h) requires high temperatures, above 2,100°C.[17,18] The upper temperature limit for growth is about 2,500°C, at which point the Al vapor pressure reaches 1 atm,[19,20] which often has ruinous results for the crucible (see Fig. 4(a)).

The growth temperature for AlN crystals is so high that crucibles and related heat insulation materials may not remain passive and might react with the source materials, nitrogen and AlN powder. However, incorporating impurities from the crucible or the insulator into the crystal may cause defects and stress. Therefore, the choice of crucibles and related heat insulation materials is critically important. Suitable crucibles and insulator materials must have the following properties[17,18]: (1) the material must have a melting point exceeding 2,500°C, (2) it must not form low melting temperature eutectics with Al, N, or C, (3) it should not react with

Fig. 4. Tungsten crucibles (a) destroyed by the high temperature (>2500°), (b) with graphite insulation after three experimental cycles, (c) with AlN powder insulation after three experimental cycles.

Al vapor or N_2, and (4) its vapor pressure should be much lower than those of Al and N_2.

Through comparative experiments with other materials such as graphite, tantalum, tantalum carbide, and boron nitride, tungsten has been identified as an appropriate material for crucibles in which to grow AlN crystals. Due to their high temperature resistance and good technical properties including thermal shock resistance and low thermal conductivity), graphite felts and foams are commonly used as heat insulation materials. However, they react with tungsten crucibles at high temperature, greatly reducing the lifetime of the crucibles and introducing carbon impurities into AlN crystals. Our study group piloted the use of AlN powder as the thermal insulation material for the PVT growth of AlN crystals.[21] This thermal insulation module includes an AlN ceramic disc, an AlN ceramic circle, AlN powder, a tungsten frame, tungsten flakes, and an alumina crucible. Experimental results show that AlN demonstrates very good thermal durability as an insulator and does not damage the tungsten crucible during the growth process. Figures 4(b) and (c) compare tungsten crucibles with (b) graphite thermal insulators and (c) AlN thermal insulators, respectively, after three experimental cycles. This comparison clearly indicates that using AlN as a thermal insulator can greatly increase the lifetime of the crucibles. However, compared with the graphite insulator, the AlN insulator has higher heat conductivity and may introduce pollution to the furnace due to its sublimation effect at high temperature, a shortcoming that should be addressed in experiments.

2.2. *AlN micro/nanowires and single crystals with millimeter-scale diameter*

Seeds are indispensible for growing AlN single crystals in bulk. Unfortunately, AlN does not naturally occur as a mineral, and AlN seeds were initially unavailable. Researchers have therefore developed different methods for initiating bulk AlN crystal growth, which mainly include (1) self-seeding and subsequent growth of freestanding AlN and (2) growth on SiC substrates.[14,15,20,22] Currently, most AlN single crystals are produced by self-seeded PVT growth.

Figure 3 shows a unit that is employed to grow AlN at high temperatures above 1,800°C in a nitrogen atmosphere of 60–120 kPa. This unit employs AlN polycrystals, which are grown using commercially available AlN powder, as the AlN source; it is resublimed for several hours at 1,850–1,900°C prior to use in order to achieve the highest possible degree of purification.

During the early stages of growth, self-nucleation and the subsequent growth of millimeter-sized freestanding crystals dominate the growth process. The AlN crystal pattern changes from nanometer-sized (in diameter) whiskers to micrometer-sized needles to millimeter-sized, perfect, well-faceted hexagonal prisms as the growth temperature increases from 1,800°C

(a) (b)

(c)

Fig. 5. Optical photographs of AlN whiskers with the nanometer to micrometer-scale diameter (AlN micro/nanowires) at temperature of (a) 1850°C (b) 1950°C (c) 2050°C.

to 2,200°C.[21,23] These AlN crystals are shown as Fig. 5–8. AlN crystal growth demonstrates a strong anisotropy and preferential behavior along the c-axis. However, this habit decreases as the temperature increases. With increasing temperature, polycrystalline AlN boules, shown in Fig. 9, rather than single crystals are produced on the tungsten substrate because of the absence of seeds. This creates the opportunity to develop new techniques to grow bulk AlN single crystals.

2.3. *Bulk AlN single crystals*

2.3.1. *Self-seeding growth*

AlN's sublimation-recondensation process can be described by the simple reaction $AlN(s) \leftrightarrow Al(g) + 1/2N_2(g)$. In fact, the PVT growth of AlN crystals includes the following five sub-processes: (a) sublimation of the

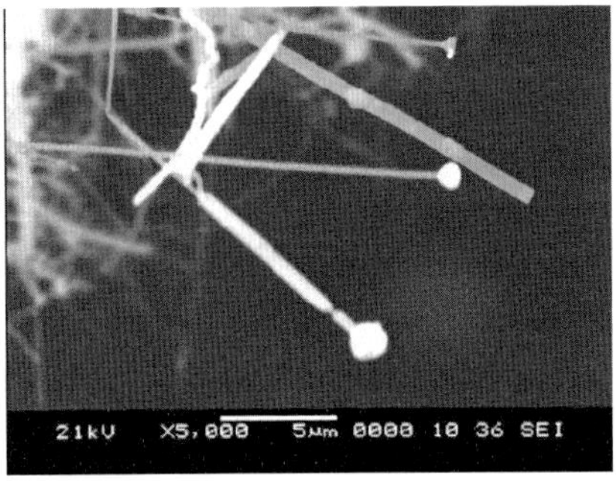

(a)

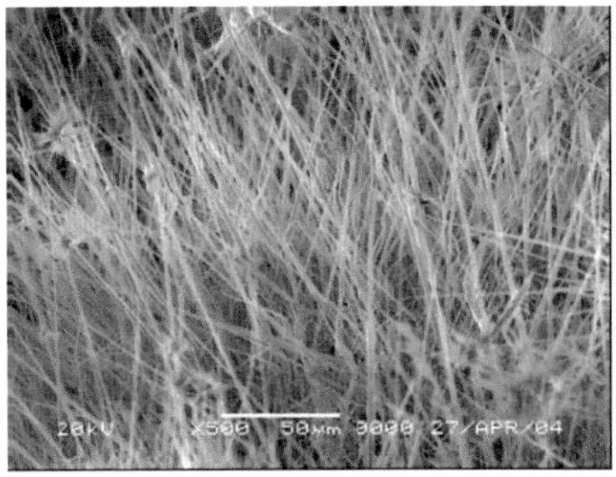

(b)

(c)

Fig. 6. SEM photos of AlN whiskers with the diameter from nano- to micro-meter level.

AlN source, (b) mass transfer of the vapor species, (c) adsorption of the vapor species on the substrate surface, (d) diffusion and nucleation on the surface, and (e) desorption.[25]

The sublimation behavior of AlN, which is described well with experimentally and calculationally relevant curves, mainly relates to growth temperature, usually above 1,850°C, and vapor pressure, 80–120 kPa.[12,25] The gas-phase transportation and adsorption mechanisms of the vapor species have also been discussed extensively.[25–27] Diffusion constitutes the main mass transportation model for the growth species. Gaseous Al and N_2 are the main species during growth, while possible vapor species like metastable Al_xN (g) (x = 2,3,4) can almost be excluded.[28] Growth experiments usually occur in a high-purity nitrogen atmosphere at 80–120 kPa, multiple times above the partial pressure. However, nitrogen molecules have strong binding energy; therefore, we must introduce the

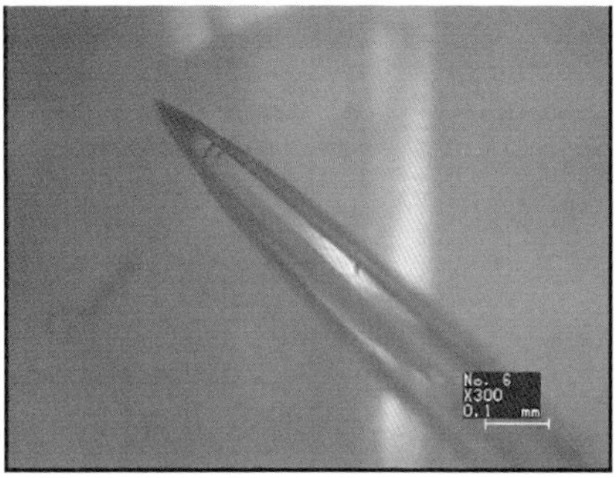

(a)

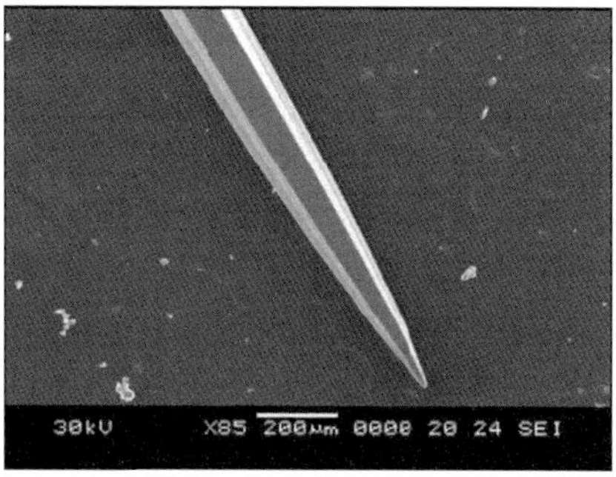

(b)

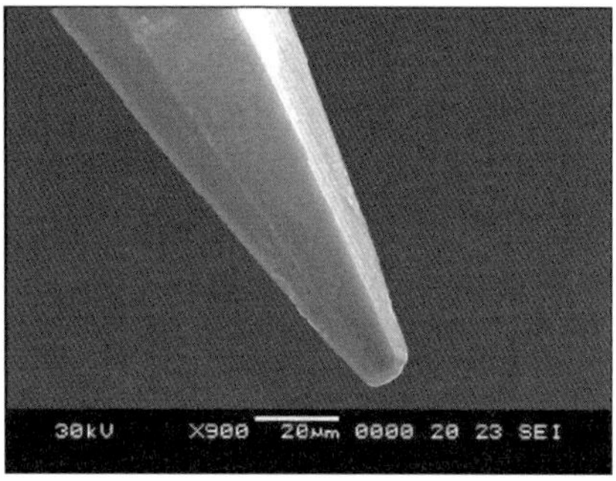

(c)

Fig. 7. AlN needles with micrometer-scale diameter: (a) optical photo, (b)–(c) SEM photos.

(a)

(b)

Fig. 8. Hexagonal prism AlN single crystal with millimeter-scale diameter: (a) optical photo, (b) SEM photo.

(a) (b)

Fig. 9. Polycrystalline AlN boule, 28 mm in diameter, and (a) 3 mm thick and (b) 16 mm thick grown at 2250°C.

sticking coefficients of gaseous N_2 (αN_2), which are much less than one, in order to correct the model while describing adsorption behavior and surface reactions.[29] Higher temperatures lead to enhanced αN_2 values and surface diffusion of adatoms, which improve structural quality in AlN crystals. The maximum temperature of AlN crystals in PVT growth is lower than the decomposition point, about 2,430°C at 100 kPa nitrogen.[30] Considering all of these factors, the growth rate of crystalline AlN, or VG, can be expressed in Eq (2).[26,27,31]

$$V_G = \frac{M_{AlN}}{\rho_{AlN}} \frac{D_{AlO} P_0 \Delta H}{T_0^{1.8} R} \frac{\exp(\frac{\Delta S}{R} - \frac{\Delta H}{RT})}{P_T^{1.5} T^{1.2}} \frac{\Delta T}{\Delta d} = k \frac{\exp(\frac{\Delta S}{R} - \frac{\Delta H}{RT})}{P_T^{1.5} T^{1.2}} \frac{\Delta T}{\Delta d} \quad (2)$$

where k is a constant; ΔS and ΔH represent entropy and enthalpy of sublimation, respectively; R is the universal gas constant; PT is the pressure when the local gas phase temperature is T; δ is the distance from the source to the substrate surface; and ΔT is the temperature difference between the crystallization and sublimation zones. The equation shows that δT constitutes a key factor in the growth rate VG, especially when T and P are determined. Figure 10(a) shows the growth rate curves of AlN crystals. Epelbaum *et al.*[15,24] derived the curve of vapor pressure versus temperature for the AlN-N_2 system, shown in Fig. 10(b), where the left area highlights the common temperature range for the AlN source and the nucleation zone, about 2050–2320°C, which provides sufficient surface mobility and

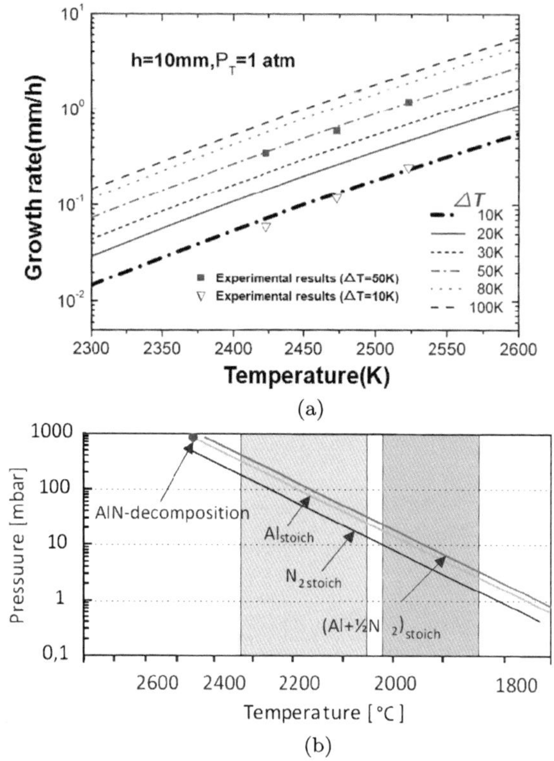

Fig. 10. (a) Theoretical and experimental curves of growth rate of AlN crystals; (b) Vapor pressure versus reciprocal temperature for the AlN-N_2 system, the left areas highlight the temperature windows for the AlN bulk growth.

mass transport for growth rates. The right area in Fig. 10(b) shows the temperature range for the AlN source and the SiC substrate, about 1850–2020°C.

Due to absence of seeds, researchers grow AlN polycrystals spontaneously on planar tungsten crucible lids. The majority of crystalli-es are (0001) oriented, but some are twisted and tilted against each other by several degrees. Polycrystalline AlN boulcs like those in Fig. 9(b) can be produced, but grains above one centimeter rarely occur. At the same time, AlN crystal quality suffers when neighboring grains induce stress due to anisotropic expansion. To address this problem, Slack developed a new kind of tungsten crucible with a conical end, shown in Fig. 11, to select one

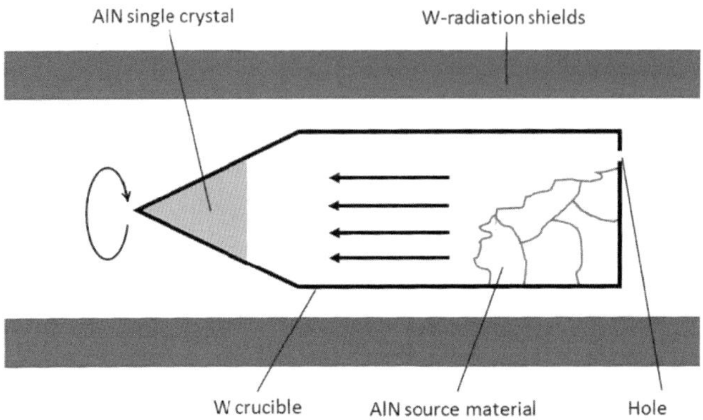

Fig. 11. Schematic cross-section of a tungsten-based growth set-up after Slack.

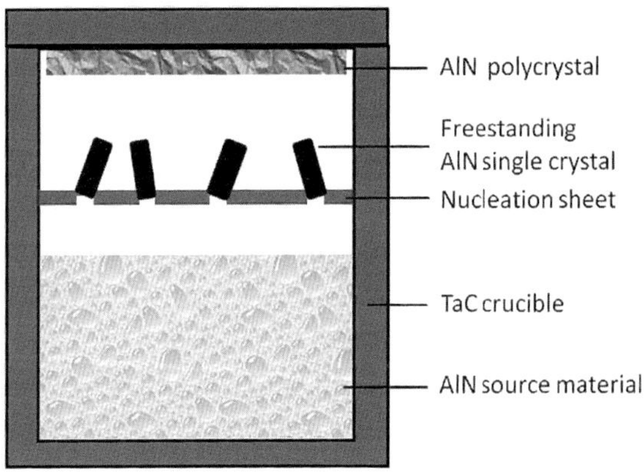

Fig. 12. Crucible design for spontaneous nucleation.

dominating grain and improve the single crystal size.[22] However, no single crystal substrates with a diameter above 10 mm have been grown using this method.

The method of spontaneous nucleation and subsequent freestanding growth has increasingly attracted attention. The growth setup has been improved to promote spontaneous nucleation of large, high-quality AlN seeds. Figure 12 illustrates Hartmann *et al.*'s method of using a perforated sheet inside the crucible above the source material to act as a nucleation area.[15] In this configuration, AlN can be deposited on both the nucleation

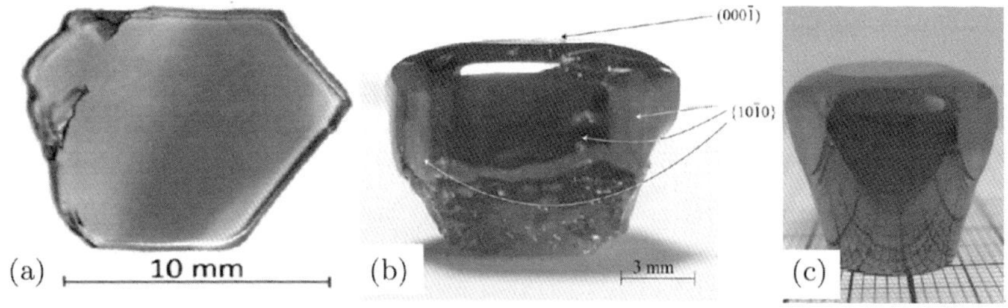

Fig. 13. Homoepitaxially grown bulk AlN crystals; (a) (b) grown by IKZ Berlin, (c) grown by Hexatech Inc.[15] ©IOP Publishing. Reproduced with permission. All rights reserved.

area, in which conditions are close to thermal equilibrium, and on the crucible lid, typically as a polycrystalline layer. This method yields AlN single crystals (see Fig. 13(a)) about 10 mm in diameter. During growth, proper process conditions lead to low nucleation densities and high growth rates. Numerical simulations of the thermal field and supersaturation can determine the optimal parameters. The local supersaturation relative to thermal equilibrium can be defined as[32]

$$S = \frac{p_{Al}^2 \cdot p_{N_2}}{K(T)} - 1 \quad (1)$$

where p_{Al} is the gaseous Al pressure, p_{N2} is the gaseous N_2 pressure, and K(T) is the equilibrium constant. Excessively low supersaturation leads to very low growth rates, and excessively high supersaturation results in defective densities and high-density nucleation points that cause the polycrystallization of AlN. Supersaturation values in the range of $0.25 < S < 0.3$ have proven to be suitable for growing separate freestanding AlN crystals with appreciable size and high structural quality.[25] At a constant ambient pressure in a given growth setup, the variable process parameters are the growth temperature and the temperature difference between crystallization and the sublimation zone. The common temperature for the AlN source and the nucleation zone ranges from 2,050°C to 2,320°C, which provides sufficient surface mobility and mass transport for growth rates greater than or equal to 100 μm/h. Due to the high vapor pressures of Al and N_2 at these temperatures, the temperature gradients must be maintained at 5–20°C/cm.

Further enlargement of the single crystalline diameter and controlled improvement or perpetuation of crystalline perfection is only possible with homoepitaxial bulk growth. The c-plane wafers cut from freestanding crystals are ideally suitable as seeds due to their high structural quality. A strong convex thermal field around the seed can suppress the formation of prismatic {10 −1 0} facets during the first millimeters of growth, as shown in Fig. 13. In this growth stage, the single-crystalline diameter can be enlarged by growth in a conical shape. After formation of the {10 -1 0} facets, diameter enlargement slows down considerably because the growth rate of the {10 -1 0} facets under conditions for homoepitaxial seeding is approximately an order of magnitude lower than growth on the basal plane. At the IKZ Berlin, N-polar bulk AlN crystals were homoepitaxially grown at seed temperatures between 2,050–2,200°C, with growth rates around 200 μm/h.[15,33] The crystals' growth surface shows a well-defined (000 -1) facet with monolayer step-flow features. Similar to wafers from spontaneously-nucleated crystals, wafers cut from these seeded crystals exhibit a zonar structure consisting of zones representing growth on the (000 -1) and {10 -1 0} facets.[14]

In addition to the methods described above, a suitable growth unit can be designed to obtain bulk AlN single crystals. The improved resistively-heated furnace shown in Figure 14 includes a main heater and a top heater that are heated on the sublimation zone and the crystalline zone, respectively. The two heaters establish a vertical thermal gradient. Infrared two-color pyrometers on the body and lid of the crucible separately measure and control the temperatures of the two zones. Growth experiments are carried out in a tungsten crucible, and the thermal field distribution for AlN crystal growth can be controlled accurately. AlN polycrystals function as the source AlN material, and they are resublimated prior to use in order to achieve the highest possible degree of purification. With this two-heater configuration, AlN can be deposited on the nucleation area, where conditions are close to thermal equilibrium, producing bulk single crystals.

Crystal growth is composed of three main sub-processes: (a) heating up, (b) holding, and (c) cooling. The suitable growth temperature and temperature gradient between the crystallization and sublimation zones, that is, the crucible lid and body, were investigated in order to restrain preferential growth of AlN crystals along the c-axis orientation, controlled

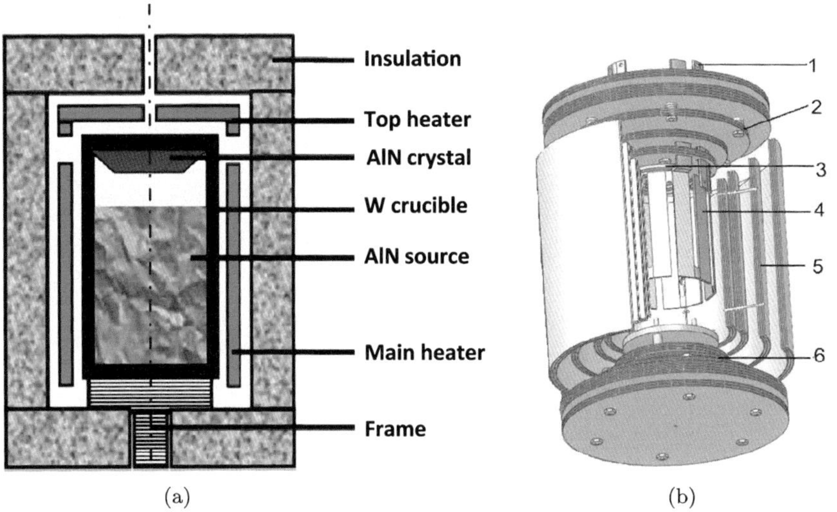

Fig. 14. (a) Schematic diagram of the AlN crystal growth equipment with two resistance heaters; (b) Schematic diagram of the tungsten heater and insulation parts: (1) electrodes, (2) top heat shield, (3) top heater, (4) main heater, (5) surrounding heat shield, (6) bottom heat shield.

by the top heater and the main heater. According to the traditional growth model where the temperature of the crystallization zone is higher than that of the sublimation zone during the heating-up process, or $\Delta T < 0$, the AlN polycrystals form on the tungsten lid, as shown in Fig. 9. Hexagonal AlN single crystals in <0001> orientation form at a growth temperature of 2,250°C, with a temperature difference or ΔT of positive 20–50°C during sub-processes (a) and (c). Maintaining the proper values of ΔT before growth can help avoid too much crystallization on the substrate, where conditions are close to thermal equilibrium and only a few detached AlN single crystals grow, as shown in Fig. 15, and recrystallize in sub-process (c). Transparent yellow or amber crystallites roughly grow in <0001> orientation, with excellent hexagonal morphology. By optimizing experimental parameters, a single AlN crystal has grown to a maximum diameter of 22 mm, as shown in the dash-dotted frame in Fig. 15.

Strong spontaneous and piezoelectric polarization fields affect AlN crystals and other structures grown in the polar c-axis direction. AlN has an anisotropic emission pattern where light is barely emitted from the c-plane and is emitted preferentially from the a-plane. The a-plane emission

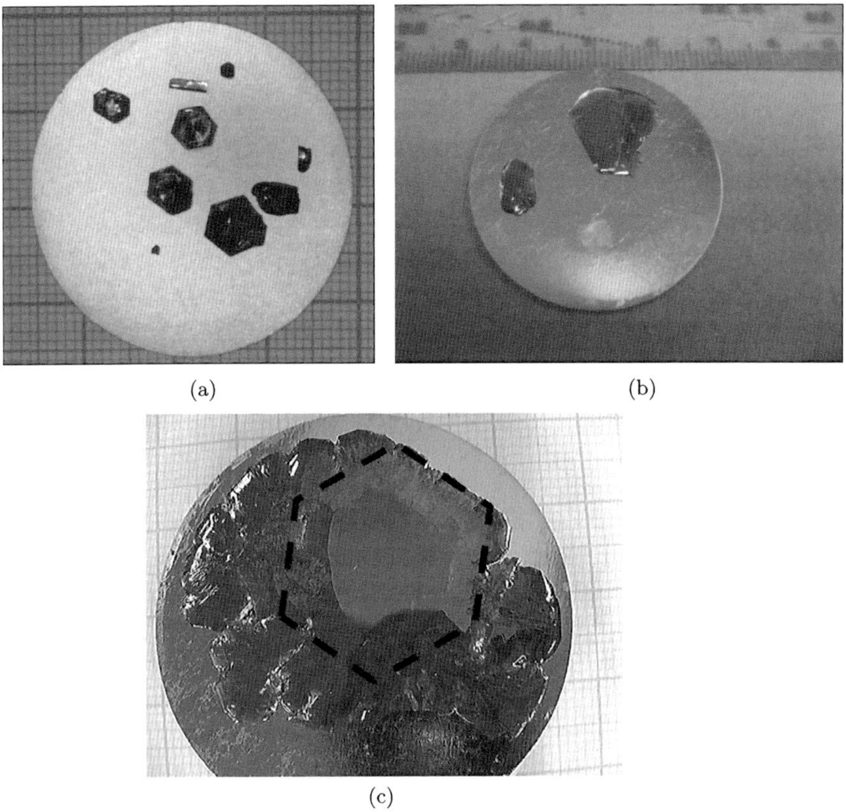

Fig. 15. Optical photograph Growth without seed, several grains (even one grain) on tungsten lids.

intensity is estimated to be 25 times higher than that of the c-plane (see Fig. 16).[34] Therefore, controlling growth direction is the most effective and directive method to grow nonpolar plane AlN crystals.

During sub-process (b) of AlN crystal growth, M-plane AlN single crystals can also form on the tungsten lid when the temperature difference is negative 10–40°C. Reducing the values of ΔT and other parameters allows M-plane AlN crystals to grow easily. Figures 17(a) and (b) show C-plane and M-plane AlN single crystals with ΔT values of 30°C and 15°C, and growth temperatures of 2,200–2,250°C. Figure 18 shows their XRD spectra, and Fig. 18(b) shows the sample's $\omega/2\theta$ XRD data, with only a sharp diffraction peak for wurtzite (100), indicating the growth of M-plane

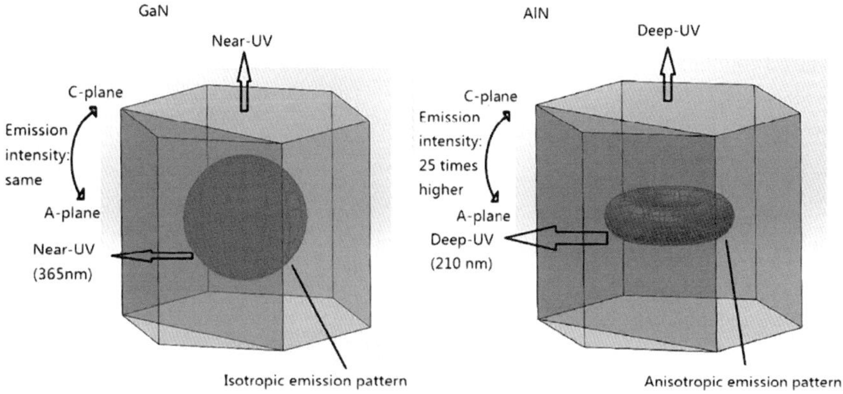

Fig. 16. Emission properties of GaN and AlN.

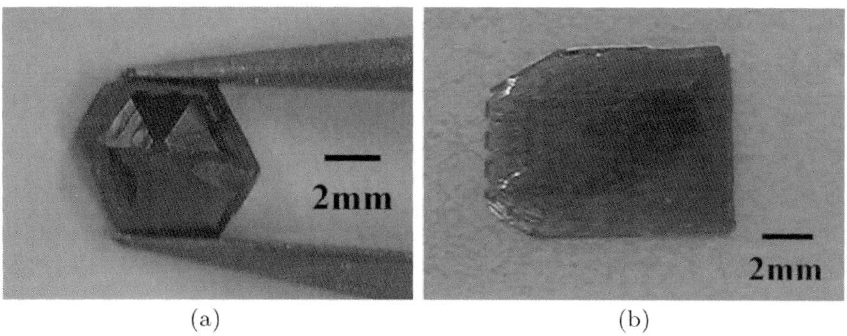

Fig. 17. Optical photographs of (a) polar c-plane AlN single crystal with $\Delta T = 30°C$ and (b) no-polar m-plane AlN single crystal with $\Delta T = 15°C$.

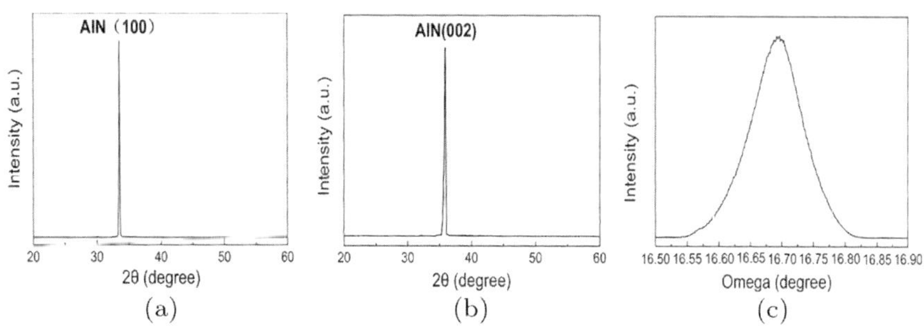

Fig. 18. (a) XRD spectrum of AlN crystals shown as Fig. 17(b), and (b) XRD for Fig. 17(b). Fig. 18(c) is x-ray rocking curve (XRC) of AlN crystals shown as Fig. 17(b).

surface-oriented AlN crystals with a size of about 1 cm. Figure 18(c) shows the symmetric x-ray rocking curve (XRC) on the symmetric (100) plane in a ω-scan. The full width at half maximum (FWHM) is about 300 arcsec.

Strong spontaneous and piezoelectric polarization fields affect AlN crystals and other structures grown in the polar c-axis direction. AlN has an anisotropic emission pattern where light is barely emitted from the c-plane and is emitted preferentially from the a-plane. The a-plane emission intensity is estimated to be 25 times higher than that of the c-plane.[34] Therefore, controlling growth direction is the most effective and directive method to grow nonpolar plane AlN crystals.

Thermal conductivity of tungsten is so high, with the same order as that of AlN, that strong heat exchange occurs between the crystal and its surrounding environment in the radial direction. Therefore, the radial temperature variation near the top is significantly reduced and often ignored. When the value of ΔT is so low that it approaches the radial temperature difference on the crucible cover, axial growth can be compared with radial growth. For the *c*-axis oriented growth habits of AlN crystals, nonpolar M-plane AlN crystals can grow given a proper thermal field around the substrate, for example, where $\Delta T \approx$ the radial temperature difference. With larger temperature differences between the sublimation and crystalline regions, the growth direction of AlN falls along the c-axis. On the contrary, the growth direction is along the a-axis. In summary, when the growth temperature exceeds 2,200° C, AlN crystal habits change from c-oriented growth to a-oriented growth as the temperature difference between the crystallization and sublimation zones decreases. However, excessively low values of ΔT cause low growth rates and are therefore not suitable for growing high-quality AlN crystals with regular shapes.

2.3.2. *Growth with SiC seeds*

Theoretically, seeded growth is the most promising way to produce large-size single crystals, and AlN single-crystalline seeds are the best candidates for the seed-growth of single AlN crystals. However, attempts to enlarge AlN crystals in subsequent PVT growth runs remain unsuccessful. Possible reasons include that (1) large-sized AlN seeds or substrates are not available at present, and (2) polycrystalline AlN materials are almost always found on AlN seeds after the re-growth process. The other choice involves utilizing foreign single-crystalline materials that are available in the form of large-area substrates; these have a small lattice mismatch with AlN, allowing for heteroepitaxial growth, and they are stable in the corresponding growth environment.[35]

SiC has a small a-lattice constant mismatch with AlN of 0.96% for 6H-SiC and 1.2% for 4H-SiC. It may represent a suitable seed because of its large area, up to 4 inches in diameter, and the commercial availability of high-quality single-crystalline SiC wafers. This approach faces two major challenges: (1) The growth temperature is limited to 1,900°C and below in order to slow the decomposition of the SiC seed, resulting in low AlN growth rates of AlN, and (2) Si and C contamination of AlN is inevitable.[36] The left area in Fig. 10(b) shows the temperature range for the AlN source and the SiC substrate, about 1,850–2,020°C. AlN single crystals have been grown on SiC substrates by the PVT method with a diameter of 1 cm and a thickness of up to 4 mm.[37] The FWHM of the XRD rocking curve for the AlN (10-10) reflection is as small as 26 arcsec, and the dislocation density is estimated by the EPD measurement to be less than 10^6 cm^{-2}, in spite of the large lattice mismatch of 1% between AlN and SiC. However, AlN crystals are badly affected by the quality of the SiC substrate. Poor-quality SiC induces defects in AlN crystals, as shown in Fig. 19. By further increasing the diameter and quality of the substrate, it is possible to prepare thick

Fig. 19. AlN single crystal with millimeter-scale diameter on SiC substrate: (a) optical photo, (b) SEM photo.

and crack-free AlN single-crystal layers with diameters exceeding 1 inch. AlN single crystal with a crack-free surface and uniform step-flow growth has been grown on 4° off-oriented 4H–SiC substrate by Radhakrishnan *et al.*[8] It is worth noting that the heterojunction filed effect of AlN layers growing on SiC substrates may be useful for electronic and optoelectronic devices.

3. Optical and Other Properties

3.1. *Raman spectra*

We carried out micro-Raman measurements using a Renishaw Raman spectrometer (inVia Reflex) with a spectral resolution of $1.5\,cm^{-1}$, and a 532 nm laser (Samba CW DPSS) with 3 mW of power for exciting light.[38] The 532 nm laser focused on the m- and c-plane surfaces of the AlN crystal, respectively, via a 50 × objective. Raman scattered light was collected in backscattering geometry and it was then dispersed by a spectrometer with a 250 mm focal length and finely detected by a CCD detector. Figure 20 illustrates the configuration of our experiment. Similar works reported by us in Ref. 38.

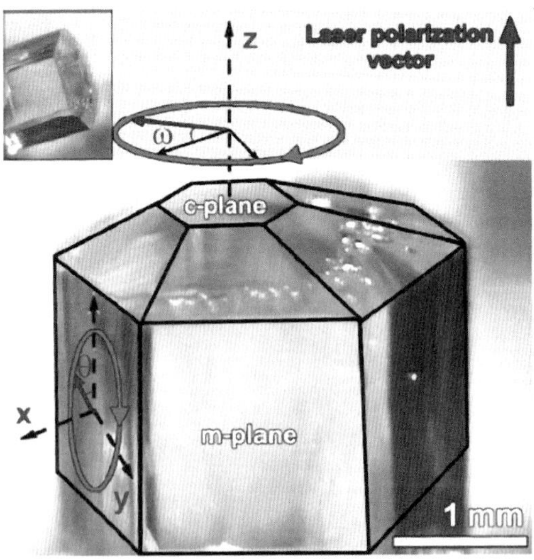

Fig. 20. Schematic illustration of AlN single crystal and Raman scattering coordinate system.

The c-plane and m-plane of AlN single crystal are easy to identify. The c-plane is perpendicular to the z axis (i.e., c axis), while the m-plane is perpendicular to the x axis (i.e., a axis). The arrows denote the polarization vectors of the incident laser. The AlN sample can be rotated around the z axis (x axis) when the wave vector of the incoming laser is along the z axis (x axis), where ω (θ) is the rotation angle with respect to the x axis (z axis).[38]

Figure 21(a) shows the Raman spectra from the m-plane surface of the AlN sample in a range of 560 to 720 cm^{-1} for different rotation angles θ,

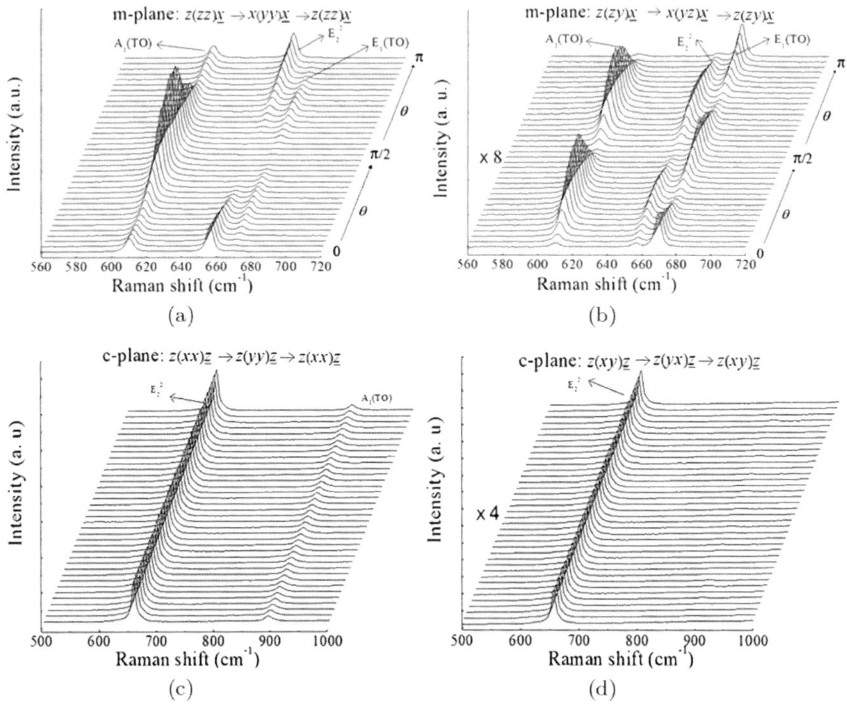

Fig. 21. (a) and (b) show the Raman spectra from the m-plane surface of AlN sample for different rotation angles, shifted by constant amount on x and y axis, for parallel [x(zz)x → x(yy)x → x(zz)x] and perpendicular [x(zy)x → x(yz)x → x(zy)x] polarization vectors of incoming laser and the scattered light, respectively. (c) and (d) record the angle dependent Raman spectra from the c-plane of AlN sample for parallel [z(xx)z → z(yy)z → z(xx)z] and perpendicular [z(xy)z → z(yx)z → z(xy)z] polarization vectors, respectively. For clarity, the ordinates in (b) and (d) have been magnified 8 and 4 times, respectively, comparing to those in (c) and (d).

increasing from 0 to 180° in 5° steps, for parallel polarization vectors of the incoming laser and scattered light [x(zz)x̲ → x(yy)x̲ →x(zz)x̲]. For clarity, the spectra are shifted on both x and y axes by a constant amount. The three modes with FWHM less than 4 cm^{-1} located around 611, 657, and 671 cm^{-1} are ascribed to the allowed Raman-active phonon modes A_1(TO), E_2^2 (e.g. E_2^{high}), and E_1(TO), respectively. Between 0° and 180°, the A_1(TO) mode never disappears completely, and it has a single intensity maximum at $\theta = 90°$ and two minima at 25° and 160°, as shown in Fig. 22(a). The situations of the E_1(TO) and E_2^2 modes differ from that of the A_1(TO) mode. The E_1(TO) mode has two maxima at $\theta = 45°$ and 135° and three minima at $\theta = 0°$, 90°, and 180°, as shown in Fig. 23(c), whereas the E_2^2 mode has two maxima at $\theta = 0°$ and 180° and only one minimum at $\theta = 90°$. Figure 21(b) shows the Raman spectra from the m-plane surface for perpendicular polarization vectors of the incoming laser and the scattered light [x(zy)x̲ → x(yz)x̲ → x(zy)x̲]. Angle dependences of A_1(TO), E_1(TO), and E_2^2 all accord with sinusoidal functions (see Fig. 22(b), (d), and (f)). Both A_1(TO) and E_2^2 have two intensity maxima at $\theta = 45°$ and 135°, and three minima at $\theta = 0°$, 90°, and 180°, while E_1(TO) has three maxima at $\theta = 0°$, 90°, and 180°, and two minima at $\theta = 45°$ and 135°. Compared to the E_1(TO) phonon, the A_1(TO) and E_2^2 mode is phase shifted by 90°, with maximal values where E_1(TO) is minimal, and vice versa.

Angle-dependent Raman spectra can also be obtained from the c-plane surface in a range of 500 to 1000 cm^{-1} for parallel and perpendicular polarization vectors of the incoming laser and the scattered light, as shown in Fig. 21(c) and (d). Figure 21(c) illustrates two allowed modes with FWHM less than 3.8 cm^{-1}, located around 657 and 891 cm^{-1}, which are constant angle dependent and are ascribed to phonon modes E_2^2 and A_1(LO), respectively. The constant angle dependence of these two allowed phonon modes agrees with the calculated selection rules.[38] According to Callender *et al.*, the low intensity of the A_1(LO) signal observable in Fig. 21(c) is due to suppression caused by the deformation potential's destructive interference with the Fröhlich contributions.[30] Constant angle-dependent vibration behaviors of Raman spectra scattering from c-plane surfaces, as shown in Fig. 21(c) and (d), obviously differ from those scattering from m-plane surfaces, as shown in Fig. 21(a) and (b); this is attributed to the isotropic arrangement of atoms in the c-plane. In Fig. 21, only the E_2^2 mode

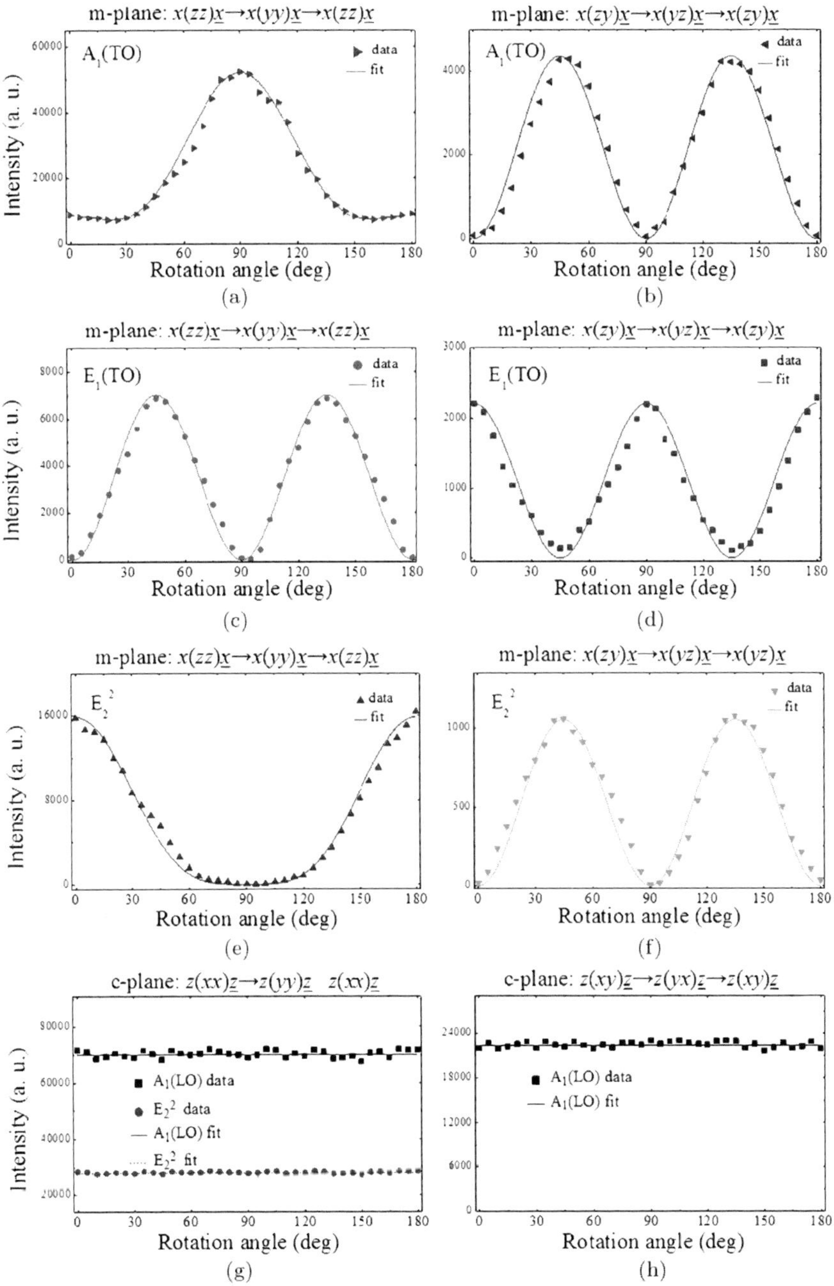

Fig. 22. (a) and (b) show the angle dependent intensity of A_1(TO) signal from the m-plane surface of AlN sample for parallel and perpendicular polarization vectors of incoming laser

Fig. 22 (*Continued*). and the scattered light, respectively. The solid-lines fit for determination of parameters |a| and |b| and φ_{a-b}. (c) and (d) show the intensity of E_1(TO) signal from the m-plane surface vs the rotation angle for parallel and perpendicular polarization vectors, respectively, while (e) and (f) present the intensity of E_2^2 signal vs the rotation angle for different polarization. (a) and (b) record the angle independent intensity of A_1(LO) and E_2^2 signal from the c-plane surface of AlN sample for parallel and perpendicular polarization vs the rotation angle ω.

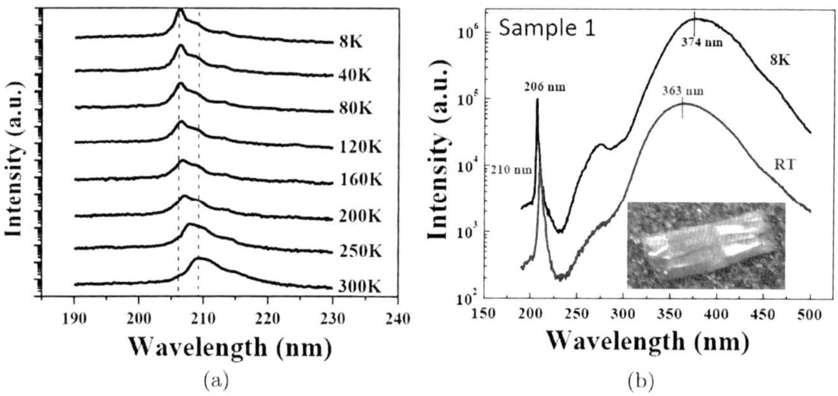

Fig. 23. (a) Near band-gap Cathodeluminescence (CL) of sample recorded at 8–300 K with 15 KV excitation and (b) CL (190–500 nm) at 8 K and room temperature. Inset displays the AlN crystal sample.

with FWHM less than 3.8 cm-1, located around 657 cm^{-1}, is allowed, and it also agrees with the calculated selection rules.[38] In addition, it is worth noting that the E_1(LO) phonon mode does not appear in Figs. 21(c) and (d) because it is forbidden by the backscattering geometry. Sander *et al.* suggested that a weak signal of the E_1(LO) mode will occur when measured geometry deviates from an ideal backscattering geometry.[39] The unallowed E_1(LO) mode reveals that deviations from the measured backscattering geometry are negligible.

Figures 21(a)–(d) show Raman intensities of A_1(TO), E_1(TO), E_2^2, and A_1(LO) phonon modes for parallel and perpendicular polarization geometries, as extracted from Lorentzian fitting. The fitting results, which are plotted against the rotation angle for the two polarization geometries, are shown as squares in Fig. 22(a)–(h). The error bars of the fitting results are too small for the figures to clearly illustrate them. To determine the Raman tensor elements, the angle variations are fitted with functions calculated using least-squares fit procedures. The best fitting curves shown in Fig. 22 agree with experimental data. The ratios of the fit parameters yield the relative values of the Raman tensor elements for AlN, which are summarized in Table 2. As a comparison, Table 2 also lists the relative values of the Raman tensor elements for ZnO crystals, as they appear in Ref. 39.

AlN and ZnO, both of which are advanced wide band-gap semiconductor compounds, can potentially be applied in the optoelectronics field, and they have same wurtzite symmetry as the space group C^4_{6v} ($P6_3mc$). Six modes ($2A_1 + 2E_1 + 2E_2$) for wurtzite AlN and ZnO are active during Raman scattering; among these modes, the A_1 branch is polarized in the z direction. In doped semiconductors, the A_1 branch can influence electronic transport properties due to the resonant coupling of polar phonons and plasmons, or free carrier systems.[40] As shown in Table 1, the ratio of |a|/|d| for the A_1(LO) phonon of AlN is much smaller than that for A_1(TO), which is mainly attributed to A_1(LO) suppression caused by the deformation potential's destructive interference with the Fröhlich contributions.[41] Table 1 also indicates a small ratio of |a|/|d| for A_1(LO) in ZnO; therefore, the Raman tensor elements of |a| and |b| extracted from the A_1(TO) mode are more accurate.

Table 2. Relative values of the Raman tensor elements of A1, E1, and E2 mode in wurtzite AlN are extracted from the theoretical fits to angle-dependent Raman spectra for parallel and perpendicular polarization vectors of incoming laser (e_i) and the scattered light (e_s). Raman tensor elements for all phonon modes are normalized to that of E2 mode.

			e_i // e_s		$e_i \perp e_s$	
Raman tensor elements			AlN	ZnO (Ref. 26)	AlN	ZnO (Ref. 26)
m-plane	A_1(TO)	|a/d|	1.818 ± 0.011	0.611 ± 0.006	1.846 ± 0.005	0.585 ±0 .022
	A_1(TO)	|b/d|	0.730 ± 0.008	0.599 ± 0.005	0.746 ± 0.004	0.413 ± 0.039
	A_1(TO)	φa -b	91.24 deg	93.00 deg	91.18 deg	104.00 deg
	E_1(TO)	|c/d|	0.671 ± 0.007	0.424 ± 0.004	0.721 ± 0.003	0.404 ± 0.001
	E_2^2	|d/d|	1.000 ± 0.009	1.000 ± 0.002	1.000 ± 0.004	1.000 ± .004
c-plane	A_1(LO)	|a/d|	0.408 ± 0.004	0.121 ± 0.001		
	E_2^2	|d/d|	1.000 ± 0.008	1.000 ± 0.002		

For A_1(TO) of AlN, the amplitude of |a| is about twice as strong as that of |d|. This differs from the situation of ZnO, where the amplitude of |a| is about half that of |d|. Determining the ratio of |a|/|d| is necessary because it reflects the Raman tensor anisotropy of the polar A_1 mode. Furthermore, the |a|/|b| ratio of AlN is as much as 2.5, while that of ZnO is only 1.0. The large |a|/|b| ratio in AlN leads to an intensity ratio $I^{xx}_{A1(TO)}/I^{yy}_{A1(TO)}$ [for example, I(90°)/I(0°) in Fig. 21(a)] that exceeds 6.5 as reported previously.[42] In contrast, the $I^{xx}_{A1(TO)}/I^{yy}_{A1(TO)}$ ratio in ZnO is only 1.1,[39] where the superscripts represent the polarization vectors of incoming and scattering light. These results indicate that the Raman tensor of the A1 mode in AlN has strong anisotropy, which could be caused by its large difference in bond polarity.

As shown in Table 1, the phase difference φ_{a-b} between elements |a| and |b| of AlN is 91.24° and 91.18° for parallel and crossed polarized scattering geometries, respectively. The phase difference φ_{a-b} of AlN remains close to that of ZnO, despite the fact that the anisotropy ratios |a|/|b| for AlN and ZnO differ by a factor of about 1.5. This implies that the origin of the anisotropy ratio should differ from that of phase differences for the A1 Raman tensor elements in a wurtzite crystal. While we do not address this problem here, repeating the present Raman scattering measurement on other materials with wurtzite structures could determine the relationship between the anisotropy ratio and the phase difference.

Figures 22(a) and (b) show the angle of independent intensity for A_1(LO) and E_2^2 phonon modes, as recorded from the c-plane surface of the AlN sample for parallel and crossed polarization vectors of the incoming laser and the scattered light, respectively. We can determine the relative values of the Raman tensor elements for c-plane AlN by dividing the square root of the mean values of the fitted areas, which are also summarized in Table 2. The second-ranked Raman tensor R describes the characteristics of a phonon with respect to Raman scattering, which is an entity proportional to the Raman susceptibility[43]

$$\frac{\partial \chi}{\partial \zeta} = V_c^{1/2} \sum_{i}^{unit-cell} \frac{\partial \chi}{\partial u_i} M_i^{-1/2} e_i \tag{2}$$

where χ denotes the polarizability tensor, that is, the dipole moment induced by a unit incident electric field; ξ represents the normal mode coordinate;

Vc is the volume of the primitive cell; *ui* and *Mi* are the static atomic displacement and mass, respectively, of atom *i*; and e_i is the eigenvector of the vibrational mode. In this work, we determined the Raman tensor of AlN using angle-dependent polarized Raman spectra. The Raman tensor can also be calculated using Eq. (2). However, a first-principles calculation is too difficult to achieve in practice. A semi-empirical method based on the concept of bond polarizabilities, which only depends on the band length *l* of two neighboring ions, can estimate $\partial\chi/\partial u$ and interpret the Raman spectra in semiconductors.[44] Wurtzite-type polar crystals, such as AlN, GaN, and ZnO, have two polar bonds with different lengths, in-plane bonds and out-plane bands.[45] The unique Raman scattering behaviors of bulk AlN single crystals are related to its bond polarizabilities.

3.2. *Cathodeluminescence (CL)*

Figure 23 shows the cathodeluminescence (CL) spectra of AlN crystals from a low temperature of 8 K to room temperature or 300 K (a), over both a narrow spectral range of 190–230 nm and (b) a broad spectral range of 190–500 nm. Free exciton (FX) emission peaks in the near band-edge emissions spectra shift from 206 nm to 210 nm as the temperature changes from 8 K to 300 K. However, deep impurity-related emission bands center around 374 nm at a temperature of 8 K, and around 363 nm at room temperature. The peak-position emission line is related to the dopants and defects. Following Van de Walle and Neugebauer,[46,47] we can calculate the formation energy of the defect in charge state q as follows:

$$\begin{aligned} E^f[X^q] &= E_{tot}[X^q] - E_{tot}[AlN, bulk] \\ &\quad - \sum_i n_i\mu_i + q[E_F + E_V + \Delta V(X)], \end{aligned} \tag{1}$$

where $E_{tot}[X^q]$ is the total energy of the defect-containing AlN supercell; E_{tot}[AlN, bulk] is the total energy of the similar AlN supercell without defects; n_i is the number of atoms of type i, both host atoms and impurity atoms, that have been added to ($n_i > 0$) or removed from ($n_i < 0$) the supercell; μ_i is the corresponding chemical potential of the species; E_F is the Fermi level, referenced as the valence-band maximum E_v in the bulk AlN; and ΔV(X) is a correction term to align the reference potential in the defect supercell with that in the bulk. The results confirm the generally accepted finding that the broad emissions peak at ~370 nm (3.4 eV) of the intrinsic AlN crystals occurs due to O as the native defect.[48–50]

4. Conclusion and Future Work

Based on the pioneering work of Slack and McNelly in 1976, the PVT method is recognized as the most promising approach to producing high-quality AlN single crystals. This chapter briefly reviewed and discussed PVT growth and its implications for bulk AlN single crystals. Although great progress has been made in the PVT approach, commercial requirements of bulk AlN single crystals are far from fulfilled, and studying growth technology remains a challenging task. Very high growth temperatures and aggressive gaseous Al growth limit possible crucible materials to tungsten and TaC. The structural and optical properties as well as the sizes of the AlN crystals are affected by the type of seeding or nucleation and the crucible material. Seeding on SiC substrates leads to inch-scale AlN crystal but the AlN crystals reveal a significant mosaicity and dislocation densities inside the undisturbed domains $> 10^5\,cm^{-2}$ even after several subsequent homoepitaxial growth runs. Both spontaneously nucleated AlN crystals and the subsequent homoepitaxially grown crystals have a high structural quality. Now, c-plane AlN bulk single crystals with diameters around 1 inch can be grown, as can no-polar m-plane AlN single crystals with centimeter-scale diameters. However, several problems remain that should be addressed in the near future: (1) precisely controlling growth conditions, especially the temperature distribution in the growth zone, (2) finding the best technical approach and the best process sequence, and (3) preparing native AlN seed with a pre-defined crystallographic orientation. Through continuous efforts to solve these problems, bulk AlN single crystals can play as important a role in the future as silicon single crystals play at present.

References

1. M. E. Levinshtein, S. L. Rumyantsev, and M. S. Shur, *Properties of Advanced Semiconductor Materials*, Wiley-Interscience, p. 42 (2001).
2. L. M. Sheppard, *American Ceramic Society Bulletin*, **69**, 1801 (1990).
3. K. Wang and R. R. Reeber, *Materials research society symposium proceedings*, **537**, 3.18–19 (1999).

4. K. M. Taylor and L. Camille, *Journal of the Electrochemical Society*, **107**, 308 (1960)
5. J. C. Rojo, L. J. Schowalter, K. Morgan, *et al.*, *Materials research society symposium proceedings*, **680**, 1 (2001).
6. D. K. Gaskill, L. B. Rowland, and K. Doverspike, *Properties of group III Nitrides, London*: IEE, p. 101 (1994).
7. J. H. Edgar, *Properties of Group III Nitride, London*: IEE, p. 3 (2004).
8. R. Radhakrishnan Sumathi, *Cryst. Eng. Comm.*, **15**, 2232 (2013).
9. Y. Taniyasu, M. Kasu, and T. Makimoto, *Nature*, **441**, 325 (2006).
10. Y. Taniyasu and M. Kasu, *Diamond and Related Materials*, **17**, 1273 (2008).
11. J. A. Kohn, P. G. Cotter, and R. A. Potter, *American Mineralogist*, **41**, 355 (1956).
12. G. A. Slack and T. F. McNelly, *Journal of Crystal Growth*, **34**, 263 (1976).
13. N. B. Singh, A. Berghmans, H. Zhang, *et al.*, *Journal of Crystal Growth*, **250**, 107 (2003).
14. B. Raghothamachar, R. Dalmau, *et al.*, *Mater. Sci. Forum*, **717**, 1287 (2012).
15. C. Hartmann, A. Dittmar, J. Wollweber, and M. Bickermann, *Semiconductor Science & Technology*, **29**, 3652 (2014).
16. D. Zhuang and J. H. Edgar, *Journal Crystal Growth*, **262**, 168 (2004).
17. WU Hong-lei, ZHENG Rui-sheng, SUN Xiu-ming, *et al.*, *Journal of Synthetic Crystals*, **35**, 1359 (2006).
18. B. Liu, J. H. Edgar, Z. Gu , *et al.*, *MRS Internet Journal of Nitride Semiconductor Research*, **9**, 1 (2004).
19. D. Ehrentraut, Z. Sitar, and MRS Bulletin, **34**, 259 (2009).
20. ZHENG Rui-sheng and WU Hong-lei, *Journal of Shenzhen University Science & Engineering*, **27**, 433 (2010).
21. WU Hong-lei, ZHENG Rui-sheng, and MENG Shu, *Journal of Synthetic Crystals*, **38**, 761 (2009).
22. G. A. Slack 1979 *US Air Force Office of Scientific Research*, (1979) DTIC document ADA085932 www.dtic.mil.
23. B. M. Epelbaum, S. Nagata, M. Bickermann, *et al.*, *Phys. stat. sol. (b)*, **244**, 1780 (2007).
24. B. M. Epelbaum, M. Bickermann, S. Nagata *et al.*, *J. Cryst. Growth*, **305**, 317 (2007).
25. C. Hartmann, A. Dittmar, J. Wollweber, and M. Bickermann, *Semicond. Sci. Technol.*, **244**, 1780 (2007)..
26. B. Wu and H. Zhang, *International Journal of Heat and Mass Transfer*, **47**, 2989 (2004).
27. A. S. Segal, S. Y. Karpov, *et al.*, *J. Cryst. Growth*, **211**, 68 (2000).
28. Y. Li and D. W. Brenner, *J. Appl. Phys.*, **100**, 084901 (2006).
29. S. G. Mueller, R. T. Bondokov, *et al.*, *Phys. Status Solidi a*, **206**, 1153 (2009).
30. B. T. Adekore, K. Rakes, B. Wang, M. J. Callahan, S. Pendurti, and Z. Sitar, *J. Electron. Mater*, **35**, 1104 (2006).
31. L. Liu and J. H. Edgar, *J. Cryst. Growth*, **220**, 243 (2000).
32. C. Hartmann, J. Wollweber, A. Dittmar, K. Irmscher, A. Kwasniewski, F. Langhans, T. Neugut, and M. Bickermann, *Japan. J. Appl. Phys*, **52**, 08JA06 (2013).
33. Hartmann C 2013 Bulk aluminum nitride growth by physical vapour transport http://opus.kobv.de/btu/volltexte/2013/2899.

34. Yoshitaka Taniyasu and Makoto Kasu, NTT Technical Review, **8**, 1 (2010)
35. B. M. Epelbaum, M. Bickermann, and A. Winnacker, *Journal of Crystal Growth*, **275**, e479 (2005).
36. P. Lu, J. H. Edgar, C. Cao, *et al.*, *Journal of Crystal Growth*, **310**, 2464 (2008).
37. M. Miyanaga, N. Mizuhara, S. Fujiwara, *et al.*, *Journal of Crystal Growth*, **300**, 45 (2007).
38. Wei Zheng, Ruisheng Zheng, Feng Huang, Honglei Wu, and Fadi Li, *Photon. Res.*, **3**, 38 (2015).
39. T. Sander, S. Eisermann, B. K. Meyer, and P. J. Kar, *Phys. Rev. B*, **85**, 165208 (2012).
40. D. Olego and M. Cardona, *Phys. Rev. B*, **24**, 7217 (1981).
41. R. H. Callender, S. S. Sussman, M. Selders, and R. K. Chang, *Phys. Rev. B*, **7**, 3788 (1973).
42. W. Zheng, R. S. Zheng, H. L. Wu, and F. D. Li, *J. Alloys Compd.* **584**, 374 (2014).
43. M. Cardona and G. Guntherodt, *Light Scattering in Solid II*, (Springer, 1982), p. 22.
44. S. Go, H. Bilz, and M. Cardona, *Phys. Rev. Lett.*, **34**, 580 (1975).
45. W. Zheng, Z. C. Feng, J. Lee, D. Wuu, and R. S. Zheng, *J. Alloys Compd.*, **582**, 157 (2014).
46. H. L. Wu, R. S. Zheng, W. Liu, S. Meng, and J. Y. Huang, *J. Appl. Phys.*, **108**, 053715.1 (2010).
47. C. G. Van de Walle, and J. Neugebauer, *J. Appl. Phys.*, **95**, 3851 (2004).
48. M. Bickermann, B. M. Epelbaum, and A. Winnacker, *Journal of Crystal Growth*, **269**, (2–4), 433 (2004).
49. Y. Zhang, W. Liu, and H. B. Niu, *Phys. Rev. B*, **77**, 035201.1 (2008).
50. M. Strassburg, J. Senawiratne, N. Dietz, U. Haboeck, A. Hoffmann, V. Noveski, R. Dalmau, R. Schlesser, and Z. Sitar, *J. Appl. Phys.*, **96**, 5870 (2004).

CHAPTER 6

TOWARDS UNDERSTANDING AND CONTROL OF NANOSCALE PHASE SEGREGATION IN INDIUM-GALLIUM-NITRIDE ALLOYS

Yohannes Abate*, Viktoriia E. Babicheva, Vladislav S. Yakovlev and Nikolaus Dietz†

Department of Physics and Astronomy
and
Center for Nano-Optics (CeNO)
Georgia State University, Atlanta, Georgia 30303, USA

**yabate@gsu.edu*
†ndietz@gsu.edu

Phase separations in ternary group III-Nitride semiconductor alloys are a significant challenge that limits the efficiency and performance of optoelectronic devices. Particularly the growth of indium-rich ternary $In_{1-x}Ga_xN$ (InGaN) alloys for light emitting diodes by MOCVD exhibits strong nanoscale phase separations, reflected in compositional and gap energy fluctuations. The suppression of phase separation at small spatial scales is difficult due to the different partial pressures of constituents at a common processing window, as well as our incomplete understanding of the complex thin film growth thermodynamics and lack of detection techniques at all spatial scales relevant to device performance. This chapter introduces a powerful nanoscale spectroscopic imaging technique, scattering-type scanning near-field infrared nanoscopy (s-SNIN), ideally suited for investigation of local nanoscale phase separations in III-Nitride alloys as a function of processing parameters, thickness and composition. To quantify the spectroscopic s-SNIN response, we lithographically patterned InN/sapphire substrate templates in which nanolayers of InGaN as thin as a few atomic layers were deposited. The analysis of those samples enabled us to calibrate the near-field infrared response of the nanolayers as a function of composition and thickness. In addition, a novel theoretical model has been developed that considers the full geometry of the probe tip and its interaction with multiple layers of sample/substrate structures. By analyzing the spectroscopic s-SNIN contracts of the patterned InGaN/InN areas, we identified and quantified phase separations of the various nano-alloys. The scale of phase separation varied widely depending on the target composition and growth processing conditions. We envision wide application of s-SNIN for nanoscale investigation of complex phase phenomena in correlation with growth conditions for plethora of semiconductor alloys.

1. Introduction

Indium-rich group III-Nitride (III-N) alloys [e.g. $In_{1-x}Ga_xN$ (InGaN), $Ga_{1-x}In_xN$ (GaInN), $In_{1-y}Al_yN$ (InAlN), etc.] are at the heart of modern technological application such as light emitting diodes (LEDs), laser diodes (LDs),[1–5] or high-speed optoelectronic emitters and sensors operating from THz to UV region.[6,7] Even though substantial progress has been made in the crystalline quality and the capacity to tailor the bandgaps of these materials, the fabrication of heterostructures and nanostructures that exhibit desired optoelectronic properties is still extremely challenging. This is partly due to a lack of effective methods to establish common processing windows that can realize the formation of homogenous multi-quantum wells (MQWs) and/or quantum dots (QDs) with desirable composition profiles.[8] A further challenge is to link growth processing parameters to growth surface chemistry, surface diffusion and ordering processes that lead to lateral phase segregations. To understand the physical principles related to these segregations requires probing techniques that combine high spatial resolution and chemical identification simultaneously.[3,9–11]

As described in Sec. 2.3, the development of nondestructive probe techniques that inherently offer nanoscale spatial resolution — such as s-SNIN — is the first critical step for a better understanding of phase purity/separation within QDs and better control of the nanoscale composition profiles and, hence, the materials properties that will allow for the design of these alloys in complex structures for targeted device applications.[9–11] Interpretation of s-SNIN nanospectroscopic experimental results to optical constants of the samples, requires theoretical models and simulation as discussed in Sec. 2.4. Detailed knowledge of the nanoscale composition will, in turn, provide essential feedback to the growth processing parameters (e.g. spatial and temporal group III materials supply, precursor decomposition kinetics, afterglow plasma conditioned nitrogen precursor species and concentrations, etc.), discussed in Sec. 2.1. Nanoscale patterning of semiconductor films, such as photolithography discussed in Section 2.2, allows controlled manipulation and investigation of semiconductor alloys.

Phase separation is a general phenomenon observed in many compound materials having highly off-equilibrium growth processing conditions, vastly different partial pressures of the constituents, or large lattice mismatch.[9–13] In thin film compound semiconductor growth containing constituents with vastly different partial pressure, the complex interplay between the potentials and sizes between the various constituents, may lead to local ordering/clustering and strain field induced miscibility effects between multinary alloys. Correspondingly, atoms may segregate to form nonuniform distributions of mixed alloys. Phase separation results in alloys with statistical distributed compositions and modifies bandgap properties that ultimately determine the optoelectronic performances of semiconductor devices. Although transmission electron microscopy (TEM) offers atomic spatial resolution, the quantitative TEM image contrast analysis is a challenge due to dynamic electron scattering changes for thin samples.[14] Optical spectroscopy techniques such as photoluminescence (PL) and Fourier transform infrared spectroscopy (FTIR) are commonly used to analyze composition profiles in semiconductors and to correlate phase separation with various growth methods.[12,15,16] However, these techniques cannot be applied to study phase separation on individual nanostructures and QDs (5 to 100 nm range) owing to the extremely small IR absorption cross-sections and the diffraction-limited spatial resolution. By contrast, scattering-type scanning near-field infrared nanoscopy (s-SNIN) offers spatial resolution several orders of magnitude below the diffraction limit. Its resolution is only determined by the apex radius of the probe tip and not by the wavelength of light.[17] In s-SNIN the sharp metallic probe tip that serves as an IR antenna channeling the incident radiation into a nanolocalized and enhanced near field at its apex. The local near-field interaction between the probe tip and the sample modifies the tip-scattered radiation depending on the optical properties of the sample. The backscattered signal is detected in the far field via an interferometric detection scheme and provides nanoscale-resolved IR near-field images enabling local chemical composition profiling of the sample.[18,19] Elucidation of the experimental IR near-field images depends on theoretical modeling which still is a challenging undertaking. A realistic simulation must take into account the near-field interaction between the whole conical geometry of the probe tip, the complex sample structure and the substrate for a reliable interpretation of experimental data.

As described in Sec. 3, we present experimental data on InGaN alloys which exhibits local phase separation in $In_{1-x}Ga_xN$ alloys at nanometer spatial scale, using spectroscopic high-resolution s-SNIN in the mid-IR spectral region.[20] A photolithographic SiO_2 patterned InN/Sapphire template is used to deposit various $In_{1-x}Ga_xN$ nanolayers with thickness down to a few atomic layers thick. After the $In_{1-x}Ga_xN$ deposition, the SiO_2 mask was removed, and near-field infrared spectroscopy was performed on InN/$In_{1-x}Ga_xN$ nanostructures, as discussed in Sec. 3. Using lithographically patterned samples s-SNIN optical contrasts as a function of composition and thickness of the semiconductor nanolayers were analyzed. We then used this quantitative leads to identify phase separation in individual QDs. The theoretical model based on the finite-elements method (FEM) — Sec. 2.4 — was deployed to interpret the s-SNIN experiments. Unlike previous models that consider the probe conical tip as approximate point dipoles[21] or spheroids, our model considers the full geometry of the tip, as well as all the sample and substrate layers.[20]

2. Experimental and Theoretical Background

2.1. *Growth of III-N's by MEPA-MOCVD*

As discussed above, the vastly different partial pressures between the binaries in the group III-nitrides requires under presently used CVD/MOCVD deposition conditions growth temperatures for AlN of 1100 — 1200°C, for GaN between 950 to 1100°C, and for InN 500 to 650°C, which is a huge challenge in the formation heterostructures that requires the integration in indium-rich layers that have to be embedded into gallium- or aluminum-rich layers grown at higher temperatures. Developing deposition methods that enable a common processing window is therefore paramount for the growth in indium-rich InGaN alloys in their integration in materials device structures. The most common utilized approach is the kinetic stabilization of the growth surface by MBE and plasma-assisted MBE[22] growth that narrows the processing window between InN and GaN. Another approach is the stabilization of the growth surface using superatmospheric nitrogen pressures,[23] i.e. high-pressure chemical vapor deposition (HPCVD).

In our studies, we explore the kinetic stabilization of indium-rich alloys and replace the commonly used nitrogen precursor ammonia (NH_3) through hollow-cathode ionized nitrogen species, utilized in the plasma afterglow regime above growth surface. This approach was introduced by Butcher *et al.*,[24] as migration-enhanced plasma-assisted (MEPA-CVD), and implemented in our customized MOCVD reactor as schematically depicted in Fig. 1. The epitaxial deposition process for InGaN alloys was carried out with simultaneous supply of MO precursors, varying plasma- and MO exposure times, precursor separation times, reactor pressure, and growth temperature. The MEPA-MOCVD reactor has an attached load-lock system and accommodates up to ø 4" substrates (or 3 - ø2" substrates) that can be heated to above 1100°C. The base pressure is below 10^{-8} mbar with a reactor growth pressures from 0.1 mbar to 10 mbar. The showerhead injection system allows for spatially controlled injection of MO sources (e.g., Trimethylindium (TMI), Trimethylgallium (TMG), Trimethylaluminum (TMA), etc.)).

As depicted in Fig. 1, PE- and UV absorption spectroscopy in the wavelength region of 200 nm to 1000 nm is used to analyze the ionized nitrogen fragments. However, at present a gap in understanding exists in the linkage of these ionized nitrogen fragments to the growth surface chemistry depicted in Fig. 2(a), and in the next step to the growth processing conditions. To reduce potential gas-phase reactions between ionized nitrogen fragments and MO precursor fragments in the gas phase, we utilize temporal modulation of active nitrogen species and MO-precursor species towards the growth surface as depicted in Fig. 2(b). These additional processing parameter allows the study of the surface chemistry processes as well as the formation of adduct formations in the gas phase.

For the InN, GaN and InGaN layers growth described here, the MO precursor pulse width was kept constant at 1 sec with a flow rates of

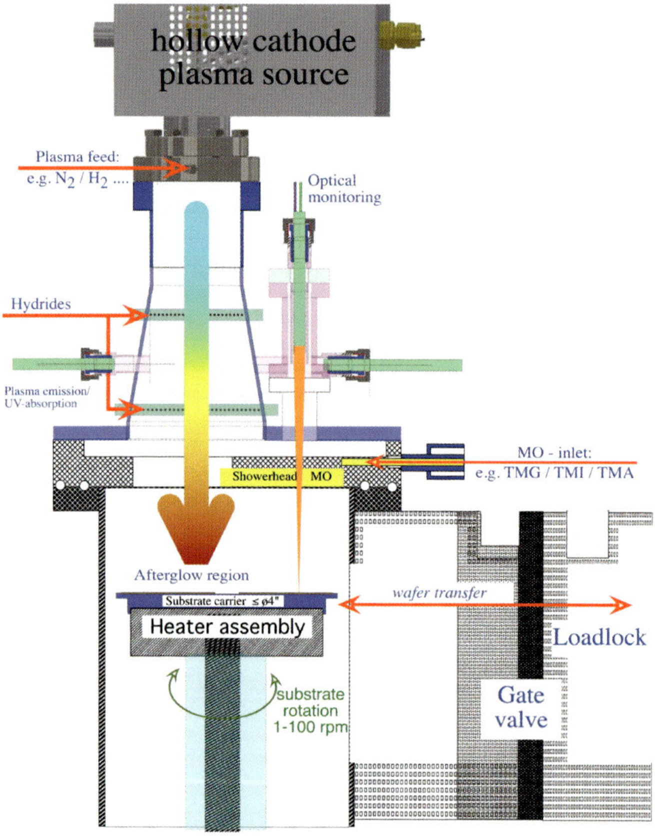

Fig. 1. Schematics of the migration-enhanced, plasmas-assisted MOCVD reactor.

9.6 μmol/min and 11 for TMI and TMG, respectively. The plasma supply was varied indirectly through the rf-power with a constant N_2 flow of 750 sccm through the hollow-cathode tube. For InGaN growth, the TMI and TMG flows were adjusted such that the targeted TMI:TMG ratio and MO surface coverage per cycle sequence are achieved. The total MO flow (μmol/min) for the InGaN alloys was kept constant.

2.2. *Patterned InN/sapphire substrate and InGaN nano-layers on InN/sapphire templates*

Following the InN deposition, the InN/sapphire substrates where subsequent patterned with a two-dimensional (2D) mask of silicon dioxide (SiO_2)

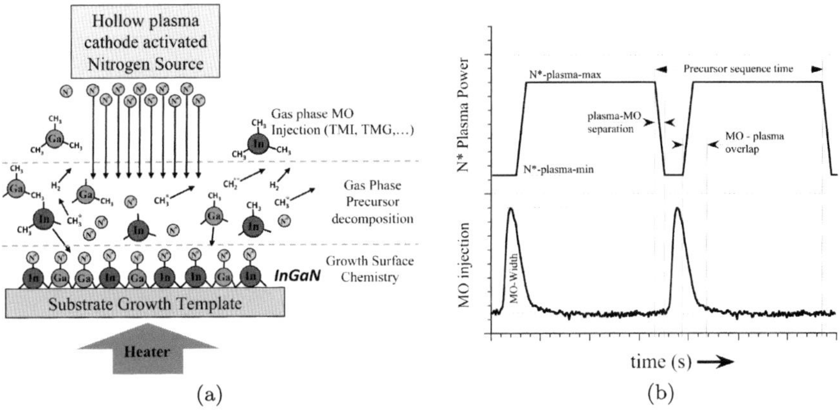

Fig. 2. (a) Schematics of gas phase and growth surface chemistries using plasma-activated nitrogen/hydrogen molecules as nitrogen precursor source for the growth of InGaN epilayers. (b) Temporal modulation of active nitrogen species and MO-precursors supplied to the growth surface.

columns by holographic lithography (Fig. 3).[25] Subsequently, the patterned InN/sapphire templates were overgrown with various $In_{1-x}Ga_xN$ compositions during simultaneous In- and Ga-MO supply. After the overgrowth process, the SiO_2 mask was removed, exposing the underlying InN/sapphire template and acting as a base reference for the deposited GaN and $In_{1-x}Ga_xN$ nanolayers. This patterning procedure allows formation of InGaN nano-layers and exposed reference InN substrate that are ideal for controlled s-SNIN imaging as described below.

2.3. *Near-field nano-imaging and spectroscopy by scattering- type scanning near-field infrared nanoscopy (s-SNIN)*

The demand to fabricate high quality semiconductor films and nanostructures calls for an increased need to detect and map these materials at nanoscale spatial resolution with chemical and spectroscopic sensitivity.[26–28] Optical methods, in addition to being nondestructive and chemically specific, allow investigation of structure–property relationships, effects of size and geometry on light induced resonant properties of nanostructures (such as absorption and scattering), and allow direct measurement of optical constants of complex and phase segregated nanomaterials. Classical light microscopy allows the resolution of sample features down to about half the

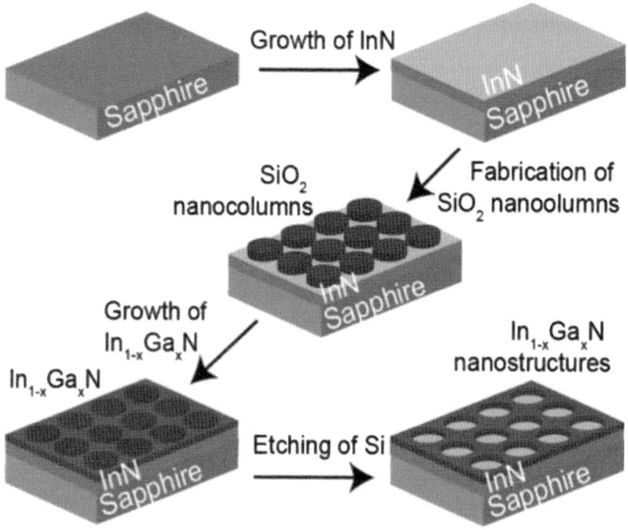

Fig. 3. Schematics showing the lithographic patterning process of InGaN nanostructures.

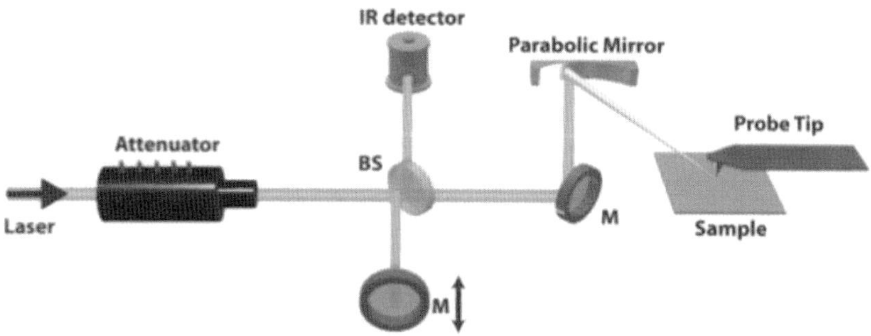

Fig. 4. Schematic illustration of the s-SNIN setup. A focused infrared beam at the probe tip generates near-field scattered light that is collected by the detector via interferometric means.

illumination wavelength (~250 nm in the visible). This limited resolution hinders single particle characterization and spectroscopic investigation of composite nanostructures based on their infrared fingerprints. Scattering-type scanning near-field infrared nanoscopy (s-SNIN) offers ultrahigh optical resolution that depends on the radius of curvature of the probe and not on the wavelength of light used[18,29–39] s-SNIN (shown in Fig. 4) is typically based on an atomic force microscope (AFM), where the sharp metallic probe tip is radiated by a focused laser beam. The tip focuses the incident beam down to a highly localized and enhanced near field. Due to the local near-field interaction between the probe tip and sample, the tip-scattered radiation is modified in both its amplitude and phase, depending on the local dielectric properties of sample. As a result, interferometric detection of the backscattered light provides nanoscale-resolved amplitude and phase images, offering direct nanoscale imaging of the both the real and imaginary part of the dielectric function of the sample. From the near-field images the absorption and scattering properties, local phase composition, nanoscale chemical makeup of materials, and free-carrier concentration, can be extracted.

Successful function of s-SNIN, first of all, requires artifact free topographic mapping, high harmonic demodulation and interferometric detection of tip scattered signal. The capability of providing topographic information simultaneously with optical images is one of the unique advantages of s-SNIN. Quality topographic imaging is essential to interpret simultaneously acquired optical images with s-SNIN. AFM measures the force interaction between a sharp probe tip that is brought in close contact to a sample surface. An internal laser focused on the cantilever of the AFM tip is reflected back to a built-in AFM detector following the vertical deflection of the cantilever allowing topographic mapping of a sample.[40] Current state-of-the-art s-SNIN is based on tapping mode AFM, in which a cantilevered tip vibrating at near-resonant frequency is brought in close proximity to a sample surface (Fig. 4), taps the surface resulting in a modified cantilever oscillation amplitude compared to its free amplitude. A feedback loop that monitors constant tip-sample separation distance allows topographic image formation thorough raster scanning of the sample and adjustments to the vertical displacement of the scanner.

The focusing ability of regular optical lenses is severely limited by diffraction of light. As a result, optical spectroscopic investigation of complex nanostructures is not possible. By focusing light to a sharp AFM metal probe tip s-SNIN imaging way below the diffraction limit is possible. However, when a diffraction-limited focused beam is directed to the AFM tip, the beam illuminates not only the sharp apex of the tip but also the elongated large portion of its shank and a large portion of the sample surface outside the immediate near-field tip-sample interaction region. As a result, the desired near-field signal is in general buried within a large unwanted 'background' scattering. Extraction of the weak localized near-field interaction is only possible by demodulation of the optical signal at higher harmonics of the tip oscillation frequency using a lock in amplifier. Due to the strongly anharmonic response of near-field signal, unlike the nearly harmonic response of the scattered background signal from the tip shank, demodulation at the higher harmonics allows background free near-field imaging of tip-sample interaction. In s-SNIN successful background reduction by harmonic demodulation is checked by taking approach curve measurements, where the probe tip is pulled above the sample surface while collecting scattered signal. A strong near-field signal that decreases exponentially as the tip-sample distance is increased, roughly in the order of the tip apex radius, and completely suppressed beyond the near-field interaction zone indicates background reduction. Amplification of the weak near-field interaction signal modulated at higher harmonics of the tip resonance frequency is achieved using interferometric detection of the scattered signal (Fig. 4). Introducing interferometric detection with an external reference beam allows measurement of background free optical signal especially enabling near-field phase contrast imaging.[18,36,41–43]

2.4. *Theoretical and computational modeling of nano-scale tip-sample interaction*

Few analytical models exist that predict tip-sample interaction when thin films or small particles are probed by s-SNIN. The earliest elegant and most simplified dipole model proposed by Keilmann *et al.*[21] employs a quasi-electrostatic theory where the scattering probe is assumed to be a polarizable sphere radius above the sample and the sample itself is a dielectric or metallic half space described by a complex dielectric constant. If a nanoparticle is probed by s-SNIN, a model based on theoretical work of Gonzegah *et al.* that uses analytical solution of the electrostatic boundary-value problem and takes into account the effect of the substrate for a system of spherical particles can be used.[44] In this approximation, when a nanoparticle is located between the probing tip and a flat sample, the near-field amplitude signal is proportional to the effective polarizability of the system of four dipoles: the probe tip and the nanoparticle are considered as point dipoles, while the sample surface produces their image dipoles. The effective polarizability is then calculated as the sum of the polarizability tensor of the tip and nanoparticle.[44] Alternatively, a finite-dipole model was proposed to calculate the s-SNIN amplitude and phase contrast of particles smaller than the tip-apex.[45,46] In this model, the scattered field is described by the complex-valued scattering coefficient derived from the near-field interaction between the illuminated tip and a flat semi-infinite substrate. The near-field probe is approximated by an isolated metal spheroid located in a uniform electric field. Both the dipole and the finite dipole models are routinely used to understand s-SNIN experimental data.

However, a realistic simulation must take into account the near-field interaction between the whole conical geometry of the probe tip, the complex sample structure and the substrate layer(s) for a reliable interpretation of experimental data.[20] We have developed such a model. In contrast to previous models, where the effective polarizability of the tip in the presence of a sample was evaluated in semi-analytical point-dipole[21] or finite-dipole approximations, our novel model relies on evaluation of the tip-sample polarizability numerically using frequency-domain solver of CST Microwave Studio. Thus, we fully consider the tip-sample near-field interaction, and our model is applicable to any sample (not just thin films) without any fitting parameters.

When a tip is placed close to a sample surface and illuminated by a laser beam, a complex optical interaction between the tip and the sample takes place. The electric field enhanced by the tip induces localized oscillating surface charges in the sample, which, in turn, induce charges at the tip surface. As long as the tip-sample interaction is only efficient over distances much smaller than a light wavelength, its outcomes are well approximated by an effective dipole moment $d_{\mathrm{eff}}(t)$. The field radiated by this oscillating dipole is the scattered light, which depends on tip parameters, sample properties, and the distance between the tip and the sample. This light is measured in the scattering type SNIN interferometrically, by mixing it with a reference beam.

In the s-SNIN technique, it is essential to suppress contributions from other kinds of scattering that do not involve the near-field tip-sample interaction, such as scattering at surface roughness and scattering by the tip shank. This is achieved by homodyne detection. First, the tip vertically oscillates with a frequency of $f_T = 2\pi\omega_T$, which is typically on the order of 0.3 MHz. When the distance between a sample and a tip is small (2–50 nm), the light is highly concentrated within the between them, so that the scattered field strongly depends on the gap size. Second, the reference beam is modulated by making a mirror vibrate with a typical frequency of $f_M = 2\pi\omega_M = 300$ Hz. The fields of the signal and reference beams, $E_S(t)$ and $E_M(t)$, are combined at the beam splitter, and the detector signal is proportional to $|E_S(t) + E_M(t)|^2$ (see Fig. 5). The detector signal is Fourier transformed and analyzed (see below).

To quantify the dielectric constants of various $In_{1-x}Ga_xN$ compositions and thicknesses based on the near-field contrast formation, we developed a theoretical approach that allows for calculations of s-SNIN images. We use the height-dependent reflectivity $r^{(CST)}(h)$ calculated by the CST Microwave Studio solver to evaluate the effective tip-sample polarizability $\alpha_{\text{eff}}(h) = ia_1a_2(r^{(CST)}(h))^*e^{-2ik_0\cos(\theta)(z_{\max}-z_{\text{pos}})}/(2\pi k_0\tan\theta)$. Here, a_1 and a_2 are the dimensions of the simulation box along the x- and y-axes, respectively (periodic boundary conditions are imposed), k_0 is the wave vector of the incident beam, and θ is the angle of incidence, $z_{\max}$ is the upper boundary of the simulation box, and z_{pos} is the coordinate of the substrate surface.

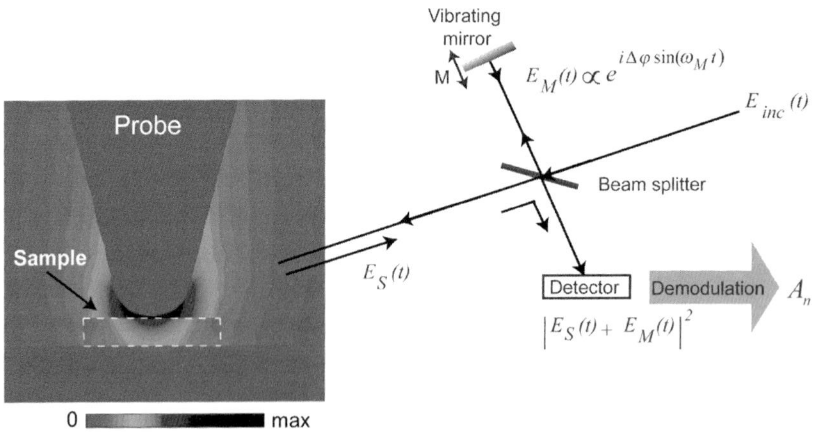

Fig. 5. Schematic illustration of the s-SNIN theoretical model to calculate the near-field interaction of probe and sample based on numerical simulations with finite element method.

The effective polarizability is then used to calculate the far-field radiation given by $E_S(h_0 + \Delta h \sin \omega_T t) = \sum_n s_n e^{in\omega_T t}$, where s_n is the complex s-SNIN signal, n is the demodulation order, and ω_T is the natural frequency of the probe tip. To simulate the experiments, we combine E_S with a reference field E_M that is reflected from an oscillating mirror, implementing a pseudoterodyne interferometer scheme. The main simulation steps and their relation to the experimental setup are schematically shown in Fig. 5. To compare with experiment, we theoretically demodulate the normalized amplitude s-SNIN signal, which is extracted by dividing the signal at the sample by that at the substrate: $(A_{\text{norm}})_n = \frac{|s_n(\text{sample})|}{|s_n(\text{substrate})|}$.

3. Results and Discussions

Figure 6a shows schematic of the probe tip-sample near-field interaction and Figs. 6(b) and (c) show the topography and 2nd-harmonic near-field amplitude optical images (A_2) of a MEPA-MOCVD-grown lithographically patterned InGaN sample on InN substrate (see Sec. 2).

The topographic image and the accompanying line profile (Fig. 6d) show a height of ~18 nm of InGaN. The near-field IR contrast is related

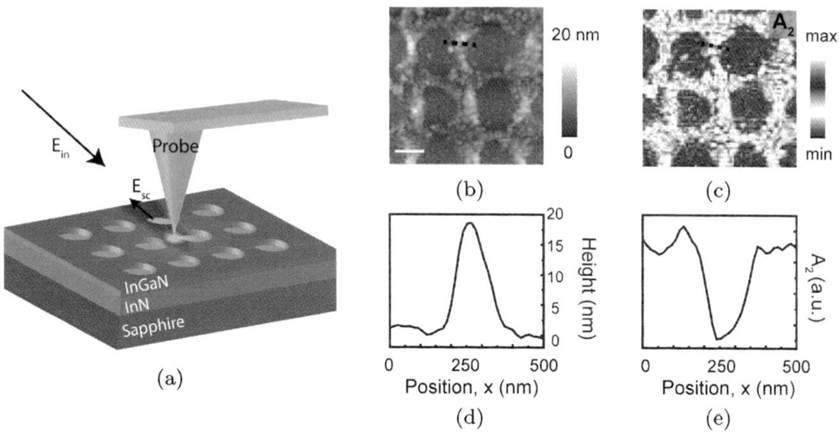

Fig. 6. s-SNIN experimental setup and imaging. (a) Schematics of the s-SNIN setup. An oscillating probe tip illuminated by a focused IR laser images InGaN nanostructures. (b) The topography and (c) the second-harmonic near-field amplitude (A_2) of a lithographically patterned $In_{1-x}Ga_xN$ nanolayer on an InN substrate. (d) Topographic and (e) A_2 line profiles showing higher islands having a lower A_2 contrast.

to the modification of the backscattered IR light ($\lambda = 10.5\,\mu$m) via the near-field interaction of the probing tip and sample. One can see a distinct bright contrast for topographically lower structures and a dark contrast for topographically higher structures as shown in the images and line profiles in Figs. 6(d-e). As well known, the backscattered IR signal offers a map of the dielectric constant (ε) of the sample, where a higher amplitude contrast corresponds to a larger value of Re(ε).[19,47,48] Based on lithographical preparation, topographically higher structures (Figs. 6(b,d)) are made up of InGaN whereas the underlying substrate pattern is InN. A comparison of the topography and s-SNIN IR images suggests that Re(ε) is larger for InN than for InGaN (Figs. 6(c, e)) at the laser wavelength, $\lambda = 10.5\,\mu$m, used for these experiments. In this way, it is possible to quantify the permittivity of $In_{1-x}Ga_xN$ nanolayers, with respect to the InN substrate, depending on their chemical composition and thickness. For this purpose, we prepared several patterned samples with various $In_{1-x}Ga_xN$ compositions and thickness ranging from sub-tens of nanometers down to few-monolayer 2D structures on the InN/sapphire substrate.

Figure 7 shows topography and IR amplitude images for four different compositions and thicknesses: $In_{0.8}Ga_{0.2}N$, thickness 20 nm (Fig. 7(a)), $In_{0.5}Ga_{0.5}N$, thickness 9 nm (Fig. 7(b)), GaN, thickness 1.5 nm (Fig. 7(c)) and Ga, thickness 2.4 nm (Fig. 7(d)). Although AFM topographic mapping indicates the height differences between the two types of materials, it is incapable of chemically specific identification. In contrast, s-SNIN IR mapping clearly identifies the two types of materials by their permittivities and allows further quantification of the role of thickness, composition and substrate on the IR image formation. The impressive capability of our IR s-SNIN imaging is highlighted as the thickness of the film reduces down to a few atomic layers (Figs. 7(c–d)), where the IR amplitude (A_3) contrast between the sample and the substrate is clearly superior to that of the accompanying topographic images. More importantly, these controlled samples allow us to calibrate the experimental near-field contrast and determine its correlation with the dielectric function of complex nanolayers, which depends on their chemical composition and thickness. This calibration further serves as a benchmark for the theoretical models as elaborated below.

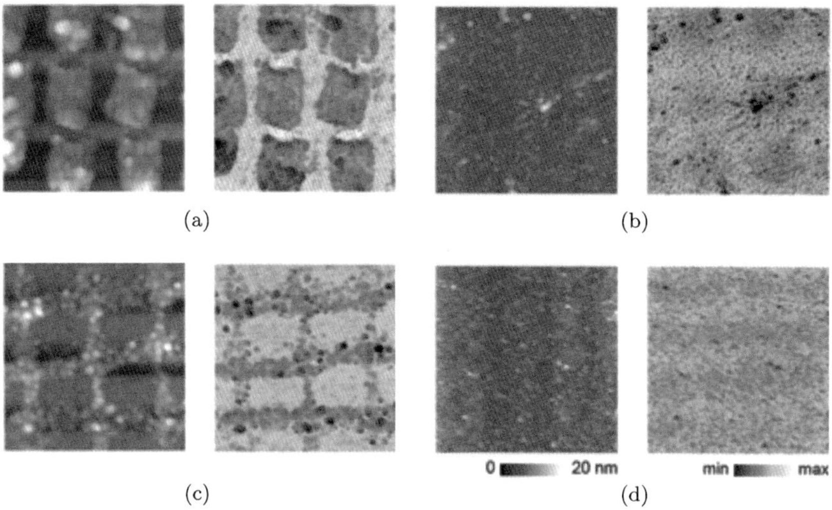

Fig. 7. s-SNIN images of patterned nanolayers. (a) Topography and IR amplitude images of 20-nm-thick $In_{0.8}Ga_{0.2}N$. (b) Topography and IR amplitude images of 9-nm-thick $In_{0.5}Ga_{0.5}N$. (c) Topography and IR amplitude images of 1.5-nm-thick GaN. (d) Topography and IR amplitude images of 2.4-nm-thick Ga.

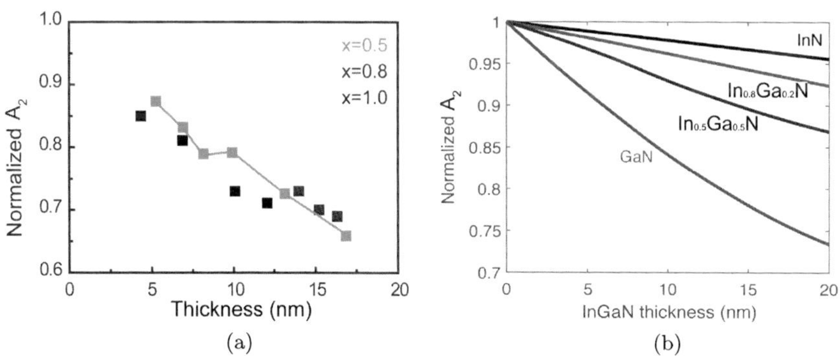

Fig. 8. (a) A plot of the normalized experimental and (b) theoretical s-SNIN amplitude signal as a function of thickness and composition of $In_{1-x}Ga_xN$ alloys.

In Fig. 8(a) we show the dependence of the measured near-field normalized amplitude signal (A_2) on the sample thickness. The red line (drawn as a guide to the eye) connects the experimental dots acquired by normalizing the amplitude (A_2) signal on $In_{0.5}Ga_{0.5}N$ nanolayers to that on the InN/sapphire substrate. The normalized amplitude signal decreases as the thickness increases. Since both composition and thickness contribute to the variation in near-field amplitude signal, a decreasing amplitude (A_2) contrast with decreasing thickness implies the increasing role of the underlying substrate to the optical signal.

To quantify the dielectric constants of various $In_{1-x}Ga_xN$ compositions and thickness based on the near-field contrast formation, we implemented a theoretical approach that allows calculations of s-SNIN response on tip-sample near-field interactions as described above. Using literature reports for the permittivity of Pt tip, $In_{1-x}Ga_xN$ and InN in the mid-IR spectral range, we calculated the normalized amplitude signal as a function of the nanolayer thickness. Figure 8(b) displays the results of these simulations. The theoretical plots reproduce the experimentally observed trend of decreasing amplitude signal with increasing thickness. This is expected since pure GaN ($x = 1$) has a smaller dielectric constant compared to InN. Furthermore, as the Ga content decreases ($x \rightarrow 0$), the dielectric constant of $In_{1-x}Ga_xN$ approaches that of the InN substrate, which decreases the amplitude contrast.

Using the complementary results of our novel theoretical model and the high-resolution experimental results described above on controlled patterned samples, we now address the outstanding issues regarding phase separation and composition profiles on single group III-N semiconductor QDs and complex nano-alloys. In Figure 9 we show the topography and near-field amplitude (A_2) images and line profiles of $In_{0.7}Ga_{0.3}N$ QDs grown on InN/sapphire templates. The topography image (Fig. 9(b)) shows dome-shaped QDs with height ranging 5–14 nm and lateral diameter in the range 60–90 nm. The corresponding amplitude image displayed in Fig. 9(c) shows that some of the QDS exhibit mixed optical contrast.

As an example of such a mixed contrast, we selected a single particle highlighted by a yellow circle on the topography and by a black circle on the amplitude image (Figs. 9(b) and 9(c), respectively) with the corresponding line profiles shown in Figs. 9(d) and 9(e). The selected particle is ~7 nm in height and ~70 nm in width (FWHM). In the amplitude image of this chosen particle we observe dark and bright mixed contrast, which is also experienced by some (but not all) other particles as can be seen in this image. Comparison of the smooth topographic line profile (Fig. 9(d)) and

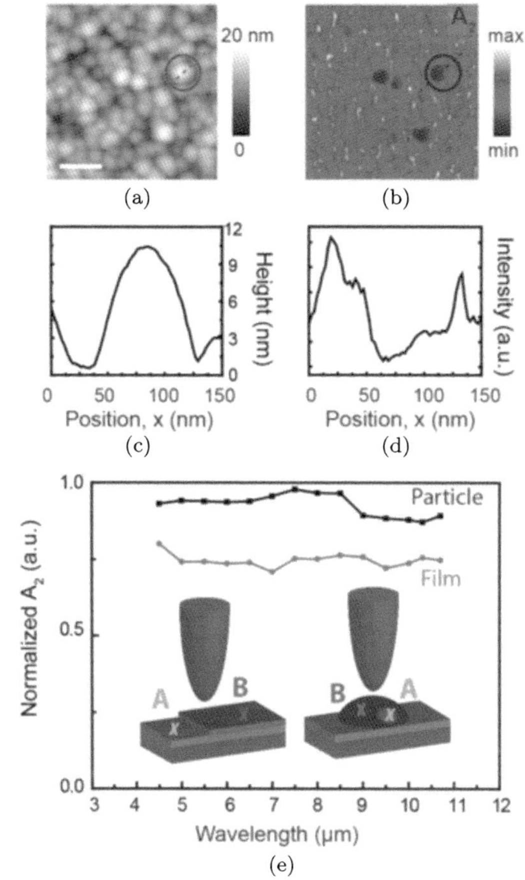

Fig. 9. Experimental s-SNIN images of $In_{1-x}Ga_xN$ QDs. (a) Topography, (b) Near-field amplitude (A_2) of the QDs taken at a laser wavelength of, $\lambda = 10.5\,\mu m$. (c–d) Line profiles of a selected QD of the topography and the corresponding near-field amplitude. (e) Spectroscopic plot of the experimental normalized amplitude signal of a QD (Particle) and a nanolayer (Film).

the amplitude line profile (Fig. 9(e)) displaying varying intensity show that the observed contrast in the amplitude image does not have a corresponding topographic origin.

As discussed above the near-field amplitude image is sensitive to permittivity of the sample; therefore, the existence of mixed contrast on a single particle indicates the presence of two materials that phase segregated on the same QD. To identify the relative permittivity of the two materials on the QD, we performed spectroscopic imaging at several laser wavelengths and compared the spectroscopic responses with that of lithographically patterned samples discussed above with similar composition and height. These experimental spectra are displayed in Fig. 9(e) where the red points are the plotted experimental normalized amplitude values, which were derived by taking the ratio of the signal on $In_{0.7}Ga_{0.3}N$ film to that on InN substrate (point B on the red film in the inset of Fig. 9(e) to point A on the yellow film). The normalized amplitude values for the QD were found by dividing the signal on the dark contrast to that of the brighter contrast (point B in the nanoparticle inset in Fig. 9(e) to that of point A). The two spectra are flat and nearly overlap in the entire wavelength region considered. From the theoretical-experimental calibration of the IR s-SNIN amplitude contrast, which we presented above on controlled patterned samples, we infer that the brighter part of the QD is an In rich region and the darker part is a Ga rich region. The normalized s-SNIN amplitude values are slightly smaller than those of the film throughout the considered spectral region. This small variation indicates that regions A and B, considered for normalization of the signal, are not pure $In_{0.7}Ga_{0.3}N$ and InN regions. Since s-SNIN can only provide a normalized (comparative) measurement, it is not possible to estimate the absolute values of dielectric functions.

To assess general statistical distribution of the phase separation, we took several near-field amplitude images at various locations of the sample. A few of these images are shown in Fig. 10. These results indicate that the scale of the phase separation on a single particle varies over a wide compositional range from particle to particle ranging from statistically Ga-rich to all In-rich QDs. An estimation of the percentage of particles that show any level of Ga-rich phase separation over entire sample region scanned within the resolution and sensitivity of s-SNIN is between $\sim$20–25%. For our $In_{0.7}Ga_{0.3}N$ samples that were grown at 775°C, this estimation is in qualitative agreement with predication based on theoretical phase diagram of $In_{1-x}Ga_xN$ system calculated assuming a constant average value for the solid phase interaction parameter.[49]

In a model used by Stringfellow *et al.*,[49] $In_{0.7}Ga_{0.3}N$ at a growth temperature of 775°C is thermodynamically not stable and may segregate into stable In-rich and Ga-rich InGaN alloy regions. The probability that

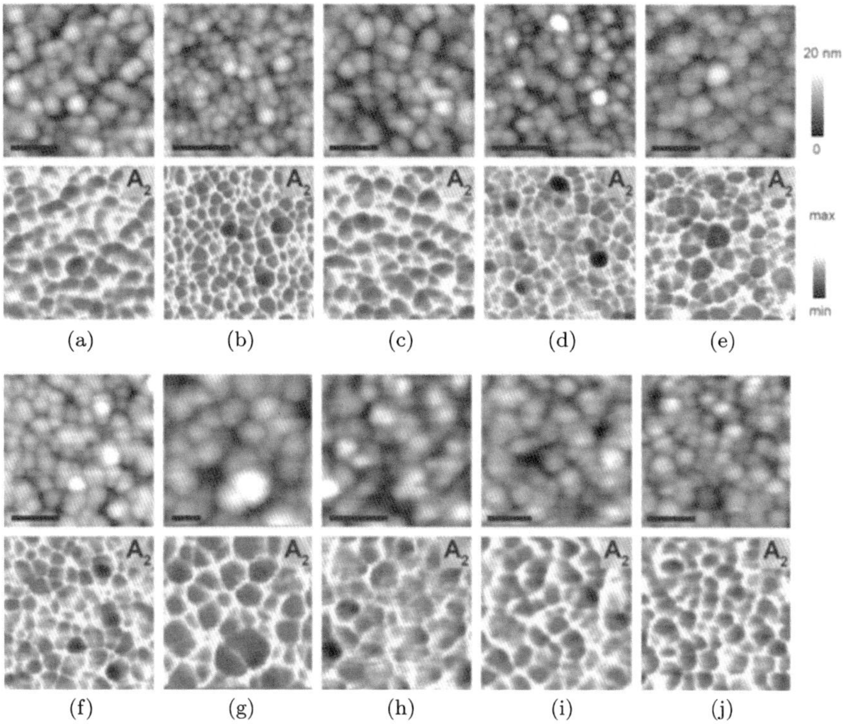

Fig. 10. s-SNIN topography and near-field amplitude (A_2) images taken at several locations on the $In_{0.7}Ga_{0.3}N$ sample at a laser wavelength of $\lambda = 10.5\,\mu m$. (a–j) Topography and accompanying 2nd harmonic near-field amplitude images taken at arbitrary locations on the sample surface. The scale bars on topographic images indicate 200 nm.

it could stabilize to a Ga-rich region depends on the kinetic stabilization of the growth surface, growth temperature, group V/III precursor ratio, simultaneous or sequential precursor supply, and various other processing conditions, which are presently explored by MEPA-MOCVD and verified by s-SNIN.

As described above, our theoretical model allows us to simulate the near-field coupling between the full tip and any size and shape nanoparticle as well as layers of substrate. Since our growth technique covers a wide size distribution, we investigate the dependence of the near-field tip-sample coupling on the geometry and size of the nanoparticles using our theoretical model. The simulated field distributions for two different particle sizes that are both larger than the tip apex are displayed in Figs. 11(a-b).

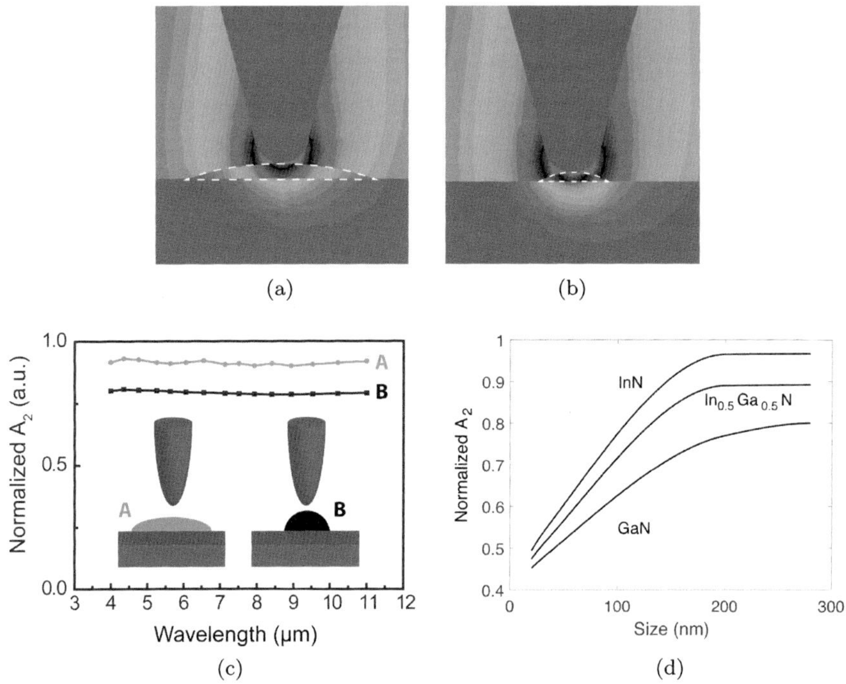

Fig. 11. Numerical simulations of tip-sample interaction and spectra. (a, b) Electric field distribution of tip-sample coupling for large particle (a) and small particle (b). (c) Calculations of the spectra of near-field interaction between large and small particles. A_2 line profiles showing higher islands having a lower A_2 contrast. (d) Dependence on the size of QD for the height = 16 nm.

The lateral size of our particles are comparable or larger than the tip apex (>60 nm diameter), so we do not expect contrast reversal to play a role.[33,34] Spectroscopic simulation using literature data for the permittivity of InGaN on InN substrate in the mid IR wavelength region is shown in Fig. 11(c) for large (red) and small (black) particles. Similar to experimental results, the simulation shows a flat response across the spectral window considered.

The near-field contrast (normalized amplitude signal) is higher for small particles than it is for larger particles (approximating a flat film). This is due to the competing tip-sample and tip-substrate coupling, which is a well-known size effect in s-SNIN.[33,34] For the smaller particle (Fig. 11(b)), the near-field coupling with the substrate is larger compared to the near-field coupling to the particle which will render the signal on the particle weaker. Increasing the particle size (Fig. 11(d)) strengthens the tip near-field coupling to the particle, which results in a stronger signal on the particle as clearly seen in Fig. 11(c). Such important size effects must be considered when analyzing the near-field response of QDs both experimentally and theoretically.

4. Conclusions and Outlook

We directly imaged phase segregation at a nanometer spatial scale in epitaxial $In_{1-x}Ga_xN$ alloys using high-resolution s-SNIN. Using a combination of plasma-assisted CVD and lithographically structured InN/sapphire templates with subsequent deposition of $In_{1-x}Ga_xN$ nanolayers, we calibrated the s-SNIN near-field contrast of $In_{1-x}Ga_xN$ between GaN and InN. In s-SNIN the amplitude signal contrast depends on the local dielectric constant of the sample below the probe tip. The normalized signal contrast is determined by taking the ratio of the signal on the sample to that on the substrate. To guide the experiments in quantifying the near-field contrast and, subsequently, the dielectric constants of $In_{1-x}Ga_xN$ alloys, we have developed an advanced s-SNIN model that fully describes the near-field coupling between the probe tip, the sample, and the substrate. In comparison to previous models, where the effective polarizability of the tip in the presence of a thin film was evaluated semi-analytically in the point-dipole or finite-dipole approximations, our model is more accurate, and it is applicable to samples with a complex geometry. We evaluate the tip-sample polarizability numerically considering the conical shape of the tip and the entire tip-sample near-field interaction accounting for the full structure of the sample layers. The results presented in this work, as well as our experimental and theoretical techniques, present a significant step forward towards clarifying and quantifying the nanoscale phase segregation of indium and gallium in the $In_{1-x}Ga_xN$ lattice, which is a crucial step towards understanding fundamental kinetic processes leading to the nanoscale phase segregation at the growth surface. These allows to tailor a growth surface chemistry to either avoid the segregation process or engineer it to obtain new device functionalities as well as improved optical devices with higher internal device efficiency.

Acknowledgments

YA acknowledge support for this work from the Air Force Office of Scientific Research (AFOSR) grant number FA9559-16-1-0172. VSY was supported by a MURI grant no. N00014-13-1-0649 from the US Office of Naval Research, the US Department of Defense. Support of VB came from a grant no. DE-FG02-01ER15213 from the Office of Basic Energy Sciences, Office of Science, the US Department of Energy. All acknowledge the support of Dorian Alde, Ramon Collazo, Alireza Fali, Sampath Gamage, and Daniel Seidlitz in the characterization and fabrication of samples presented in this contribution.

References

1. I. Akasaki and H. Amano, "Breakthroughs in improving crystal quality of GaN and invention of the p-n junction blue-light-emitting diode," *Jap. J. Appl. Phys. Pt. 1*, **45**(12), 9001 (2006).
2. S. Nagahama, T. Yanamoto, M. Sano, and T. Mukai, "Characteristics of InGaN laser diodes in the pure blue region," *Appl. Phys. Lett.*, **79**(13), 1948 (2001).
3. S. Nakamura, "The roles of structural imperfections in InGaN-Based blue light-emitting diodes and laser diodes," *Science*, **281** (5379), 956 (1998).
4. C. A. Tran, R. F. Karlicek, M. Schurman, A. Osinsky, V. Merai, Y. Li, I. Eliashevich, M. G. Brown, J. Nering, I. Ferguson, and R. Stall, "Phase separation in InGaN/GaN multiple quantum wells and its relation to brightness of blue and green LEDs," *J. Cryst. Growth*, **195**, 397 (1998).
5. D. Zhu, D. J. Wallis, and C. J. Humphreys, "Prospects of III-nitride opto-electronics grown on Si," *Reports on Progress in Physics*, **76**, 106501 (2013).
6. G. Sun, R. L. Chen, Y. J. J. Ding, H. P. Zhao, G. Y. Liu, J. Zhang, and N. Tansu, "Strikingly Different Behaviors of Photoluminescence and Terahertz Generation in InGaN/GaN Quantum Wells," *Ieee J. Selec. Topics in Quant. Elect.*, **19**(1) 8400106 (2013).
7. A. A. Usikova, N. D. Il'inskaya, B. A. Matveev, T. V. Shubina, and P. S. Kop'ev, "Photonic Crystals and Bragg Gratings for the mid-IR and Terahertz Spectral Ranges," *Semiconductors*, **47**(12), 1570 (2013).
8. S. V. Ivanov, T. V. Shubina, T. A. Komissarova, and V. N. Jmerik, "Metastable nature of InN and In-rich InGaN alloys," *J. Cryst. Growth*, **403**, 83 (2014).
9. I. Bormann, K. Brunner, S. Hackenbuchner, G. Zandler, G. Abstreiter, S. Schmult, and W. Wegscheider, "Midinfrared intersubband electroluminescence of Si/SiGe quantum cascade structures," *Appl. Phys. Lett.*, **80**(13), 2260 (2002).
10. N. Grandjean, J. Massies, S. Dalmasso, P. Vennegues, and L. Siozade and L. Hirsch, "GaInN GaN multiple-quantum-well light-emitting diodes grown by molecular beam epitaxy," *Appl. Phys. Lett.*, **74**(24), 3616 (1999).

11. C. D. R. Ludwig, T. Gruhn, C. Felser, T. Schilling, J. Windeln, and P. Kratzer, "Indium-Gallium Segregation in $CuIn_xGa_{1-x}Se_2$: An Ab Initio-Based Monte Carlo Study," *Phys. Rev. Lett.*, **105**(2) 025702 (2010).
12. K. Muraki, S. Fukatsu, Y. Shiraki, and R. Ito, "Surface segregation of In atoms and its influence on the quantized levels in InGaAs/GaAs quantum wells," *J. Cryst. Growth*, **127**(1–4), 546 (1993).
13. K. Muraki, S. Fukatsu, Y. Shiraki, and R. Ito, *Appl. Phys. Lett.*, **61**(5), 557 (1992).
14. J. R. Jinschek and C. Kislelowski, "Time, energy, and spatially resolved TEM investigations of defects in InGaN," *Physica B-Condensed Matter*, **376**, 536 (2006).
15. Z. C. Feng, W. Liu, S. J. Chua, J. W. Yu, C. C. Yang, T. R. Yang, and J. Zhao, "Photoluminescence characteristics of low indium composition InGaN thin films grown on sapphire by metalorganic chemical vapor deposition," *Thin Solid Films*, **498**(1–2), 118 (2006).
16. M. Sato and Y. Horikoshi, "Effect of indium replacement by gallium on InAs/GaAs quantized levels," *Surface Science*, **267**(1–3), 195 (1992).
17. F. Keilmann and R. Hillenbrand, "Near-field microscopy by elastic light scattering from a tip," Philosophical Transactions of the Royal Society of London Series a-Mathematical *Physical and Engineering Sciences*, **362**(1817), 787 (2004).
18. J. M. Stiegler, Y. Abate, A. Cvitkovic, Y. E. Romanyuk, A. J. Huber, S. R. Leone, and R. Hillenbrand, "Nanoscale Infrared Absorption Spectroscopy of Individual Nanoparticles Enabled by Scattering-Type Near-Field Microscopy," *Acs Nano*, **5**(8), 6494 (2011).
19. R. Hillenbrand and F. Keilmann, "Complex optical constants on a subwavelength scale," *Physical Review Letters*, **85**(14), 3029 (2000).
20. Y. Abate, D. Seidlitz, A. Fali, S. Gamage, V. E. Babicheva, V. S. Yakovlev, M. I. Stockman, R. Collazo, D. E. Alden, and N. Dietz, "Nanoscopy of Phase Separation in $In_{1-x}Ga_xN$ Alloys," ACS *Appl. Mater. Interfaces*, **8**(35), 23160 (2016).
21. B. Knoll and F. Keilmann, "Enhanced dielectric contrast in scattering-type scanning near-field optical microscopy," *Optics Communications*, **182** (4–6), 321 (2000).
22. S. V. Ivanov, T. V. Shubina, V. N. Jmerik, V. A. Vekshin, P. S. Kop'ev, and B. Monemar, "Plasma-assisted MBE growth and characterization of InN on sapphire," *J. Cryst. Growth*, **269**(1), 1 (2004).
23. M. Buegler, S. Gamage, R. Atalay, J. Wang, I. Senevirathna, R. Kirste, T. Xu, M. Jamil, I. Ferguson, J. Tweedie, R. Collazo, A. Hoffmann, Z. Sitar, and N. Dietz, "Reactor pressure — growth temperature relation for InN epilayers grown by HPCVD," SPIE Vol. 7784, paper# 77840F-1-7 (2010).
24. K. S. A. Butcher, D. Alexandrov, P. Terziyska, V. Georgiev, D. Georgieva, and P. W. Binsted, "InN grown by migration enhanced afterglow (MEAglow)," *Physica Status Solidi A*, **209**(1), 41 (2012).
25. J. H. Seo, J. H. Park, S. I. Kim, B. J. Park, Z. Q. Ma, J. Choi, and B. K. Ju, "Nanopatterning by Laser Interference Lithography: Applications to Optical Devices," *Journal of Nanosci. and Nanotechn.*, **14**(2), 1521 (2014).
26. A. P. Alivisatos, "Semiconductor Clusters, Nanocrystals, and Quantum Dots," *Science*, **271** (5251), 933 (1996).

27. M. De, P. S. Ghosh, and V. M. Rotello, "Applications of Nanoparticles in Biology," *Advanced Materials*, **20**(22), 4225 (2008).
28. H. Dong, M. Zhu, J. A. Yoon, H. Gao, R. Jin, and K. Matyjaszewski, "One-Pot Synthesis of Robust Core/Shell Gold Nanoparticles," *Journal of the American Chemical Society*, **130**(39), 12852 (2008).
29. F. Keilmann and R. Hillenbrand, "Near-field microscopy by elastic light scattering from a tip," Philosophical Transactions of the Royal Society of London. *Series A: Mathematical, Physical and Engineering Sciences*, **362**(1817), 787 (2004).
30. R. Hillenbrand and F. Keilmann, "Complex optical constants on a subwavelength scale," *Phys Rev Lett*, **85**(14), 3029 (2000).
31. L. Novotny, "Effective Wavelength Scaling for Optical Antennas," *Physical Review Letters*, **98**(26), 266802 (2007).
32. Z. H. Kim and S. R. Leone, "High-Resolution Apertureless Near-Field Optical Imaging Using Gold Nanosphere Probes," *The Journal of Physical Chemistry B*, **110**(40), 19804 (2006).
33. Z. H. Kim, S.-H. Ahn, B. Liu, and S. R. Leone, "Nanometer-scale dielectric imaging of semiconductor nanoparticles: Size-dependent dipolar coupling and contrast reversal," *Nano Letters*, **7**(8), 2258 (2007).
34. Y. Abate, A. Schwartzberg, D. Strasser, and S. R. Leone, "Nanometer-scale size dependent imaging of cetyl trimethyl ammonium bromide (CTAB) capped and uncapped gold nanoparticles by apertureless near-field optical microscopy," *Chemical Physics Letters*, **474**(1–3), 146 (2009).
35. Z. Nuno, B. Hessler, B. Heiberg, R. Damato, T. Dunlap, Y. S. Shon, and Y. Abate, "Nanoscale near-field infrared spectroscopic imaging of silica-shell/gold-core and pure silica nanoparticles," *J. Nanopart. Res.*, (14) 766 (2012).
36. Z. Nuno, B. Hessler, J. Ochoa, Y. S. Shon, C. Bonney, and Y. Abate, "Nanoscale subsurface- and material-specific identification of single nanoparticles," *Optics Express*, **19**(21), 20865 (2011).
37. Y. Ogawa, F. Minami, Y. Abate, and S. R. Leone, "Nanometer-scale dielectric constant of Ge quantum dots using apertureless near-field scanning optical microscopy," *Appl. Phys. Lett.*, **96**(6) 063107 (2010).
38. Y. Abate, S. Gamage, L. Zhen, S. B. Cronin, H. Wang, V. Babicheva, M. H. Javani, and M. I. Stockman, "Nanoscopy Reveals Metallic Black Phosphorus," *Light Science Applications*, **5**, e16162 (2016).
39. J. Zhu, K. Hippalgaonkar, S. Shen, K. V. Wang, Y. Abate, S. Lee, J. Q. Wu, X. B. Yin, A. Majumdar, and X. Zhang, "Temperature-Gated Thermal Rectifier for Active Heat Flow Control," *Nano Letters*, **14**(8), 4867 (2014).
40. F. J. Giessibl, "Advances in atomic force microscopy," *Reviews of Modern Physics*, **75**(3), 949 (2003).
41. N. Ocelic, A. Huber, and R. Hillenbrand, "Pseudoheterodyne detection for background-free near-field spectroscopy," *Appl. Phys. Lett.*, **89**(10), 101124 (2006).
42. S. E. Grefe, D. Leiva, S. Mastel, S. D. Dhuey, S. Cabrini, P. J. Schuck, and Y. Abate, "Near-field spatial mapping of strongly interacting multiple plasmonic infrared antennas," *Physical Chemistry Chemical Physics*, **15**(43), 18944 (2013).

43. S. Mastel, S. E. Grefe, G. B. Cross, A. Taber, S. Dhuey, S. Cabrini, P. J. Schuck, and Y. Abate, "Real-space mapping of nanoplasmonic hotspots via optical antenna-gap loading," *Appl. Phys. Lett.*, **101**(13), 131102 (2012).
44. V. V. Gozhenko, L. G. Grechko, and K. W. Whites, "Electrodynamics of spatial clusters of spheres: Substrate effects," *Physical Review B*, **68**(12), 125422 (2003).
45. A. Cvitkovic, N. Ocelic, and R. Hillenbrand, "Analytical model for quantitative prediction of material contrasts inscattering-type near-field optical microscopy," *Opt. Express*, **15**(14), 8550 (2007).
46. J. M. Stiegler, Y. Abate, A. Cvitkovic, Y. E. Romanyuk, A. J. Huber, S. R. Leone, and R. Hillenbrand, "Nanoscale infrared absorption spectroscopy of individual nanoparticles enabled by scattering-type near-field microscopy," *ACS Nano*, **5**(8), 6494 (2011).
47. Y. Abate, R. E. Marvel, J. I. Ziegler, S. Gamage, M. H. Javani, M. I. Stockman, and R. F. Haglund, "Control of plasmonic nanoantennas by reversible metal-insulator transition," *Scientific Reports*, **5**, 13997 (2015).
48. R. Hillenbrand and F. Keilmann, "Material-specific mapping of metal/semiconductor/dielectric nanosystems at 10 nm resolution by backscattering near-field optical microscopy," *Appl. Phys. Lett.*, **80**(1), 25 (2002).
49. I. H. Ho and G. B. Stringfellow, "Solid phase immiscibility in GaInN," *Appl. Phys. Lett.*, **69**(18), 2701 (1996).

CHAPTER 7

INVESTIGATING STRUCTURAL AND OPTICAL CHARACTERISTICS OF III-NITRIDE SEMICONDUCTOR MATERIALS

Yi Liang*, Xiaodong Jiang*, Devki N. Talwar[†], Liangyu Wan*, Gu Xu*,[‡] and Zhe Chuan Feng*,[§]

**Laboratory of Optoelectronic Materials and Detection Technology, Guangxi Key Laboratory for the Relativistic Astrophysics, School of Physics Science and Technology, Guangxi University, Nanning, 530004, P. R. China*

[†]Department of Physics, Indiana University of Pennsylvania, Indiana, Pennsylvania 15705-1087, USA

[‡]Department of Materials Science and Engineering, McMaster University, Hamilton, L8S 4L7, Canada

A comprehensive review is presented by using multiple spectroscopic methods on binary, ternary and quaternary group III-nitride (GaN, AlN, InN, InGaN and InAlGaN) materials to apprehend their structural, optical, electrical and phonon properties. Ultrathin epifilms exploited in various experimental studies [viz., synchrotron radiation X-ray absorption fine structure (SR-XAFS), X-ray diffraction (XRD), scanning electron microscopy (SEM), transmission electron microscopy (TEM), X-ray photoelectron spectroscopy (XPS), Raman scattering (RS), photoluminescence (PL), Fourier transform infrared (FTIR), spectroscopic ellipsometry (SE), Rutherford backscattering (RBS) and optical transmission (OT)] are prepared under diverse growth conditions on Si substrates by molecular

[§]*Corresponding author: fengzc@gxu.edu.cn*

beam epitaxy (MBE) and on sapphire substrates by metalorganic chemical vapor deposition (MOCVD), respectively. Careful theoretical analyses of the spectroscopic data provided valuable information about the lattice distortions, atomic bond lengths, alloy compositions, dislocation density, phonon-plasmon coupling modes, energy band gaps, and exciton localization effects, etc. A large tensile stress is noticed in samples grown on Si while a small compressive stress appears in films grown on sapphire substrates. Our results provided significant information about the luminescent mechanisms on III-nitrides to help produce high quality samples required for high performance devices in energy and optoelectronics fields.

1. Introduction

Group III-nitride based semiconductor materials (GaN, AlN, InN, InGaN, InAlGaN, etc.) have attracted a great deal of attention in recent years due to their fascinating properties and alluring promises in fabricating the next generation of devices for high temperature, high power, high-frequency electronics, photonics and sensor applications. While the commercial interest in III-nitrides is primarily driven[1–4] to realize improving the performance of blue and ultraviolet light-emitting diodes (LEDs) for automotive and general purpose lighting as well as engineering laser diodes, high frequency power amplifiers for mobile communication wavelength (i.e., $\lambda = 1.55\ \mu$m) and base stations, the scientific interest, however, is focused mostly in the epitaxial growth of III-nitride based multi-quantum wells (MQWs) (e.g., InGaN/GaN) and superlattice structures (SLs), [$(GaN)_m/(AlN)_n$, $(BN)_m/(GaN)_n$, etc.] for designing high-efficiency multi-junction solar-cells, field-effect-transistors and high-performance optoelectronic devices.

Since a sizable nitride substrate has not yet been available for III-N materials' growth — they are frequently prepared on foreign substrates, among which the sapphire and SiC are the two major ones used in the metalorganic chemical vapor deposition (MOCVD) method. Very recently, as an alternative substrate Si has gained considerable interest due to its low-cost, large wafer size, and high crystalline quality. In fact, crack-free GaN epifilms on Si substrates have already been demonstrated by using molecular beam epitaxy (MBE) technique. The knowledge of strain or stress state in epitaxial III-Ns is very important for the understanding of epifilm properties and improving nitride-based opto-electronic and energy devices. As a sensitive, local and non-destructive tool, Raman scattering (RS) spectroscopy has been widely employed to measure stress or strain in semiconductor hetero-structures. In the past several years, Raman scattering measurements for residual strain in III-N epilayers grown on sapphire have been reported.[5–17] There are a few studies devoted to investigating the stress state in films grown on SiC and Si.

The current III-nitride devices have mainly used GaN or InGaN materials with low-In composition. AlN and AlGaN as well as InAlGaN are also employed in some devices. Among the III-nitrides — GaN is a very-well studied material, due primarily to its wider band gap (∼3.4 eV) and use in fabricating optoelectronic devices such as light emitting diodes (LEDs) and semiconductor lasers which can operate in a broader spectral range from ultra-violet to green.[1–3] In contrast, InN has remained the least understood, which extends the emission of nitride-based LEDs from ultraviolet to near infrared region.[1,2] In fact, InN was believed to have a band gap energy of ∼1.89 eV back in 1977. It was not until 2001–2002, when the room temperature (RT) band gap energy of InN, ∼0.65 eV, was finally discovered. Currently, InN has been perceived as one of the most intriguing materials with superior electronic, optical and transport properties. The lowest effective mass, largest electron saturation velocity, and highest mobility made InN quite suitable for engineering high-efficiency solar cells, field-effect transistors and high performance optoelectronic devices, capable of operating at the optical-communication wavelength.[2] The ternary alloy InGaN possesses a tunable direct band gap varying from 0.65 eV (InN) to 3.4 eV (GaN), which covers a wider spectral range from infrared (IR) to visible and to ultraviolet (UV), making InGaN ideal for the whole spectrum solar cells.[1–3] Among all the III-nitride materials, AlN possesses the highest RT band energy gap of 6.2 eV to which much research effort has been devoted.[1–3] AlN is an important material of III-nitride family with many potential applications, including light-emitting devices in the UV and visible region, due to its wide band gap, high dielectric strength, high temperature stability, chemical inertness, high thermal conductivity, and high acoustic velocity. AlN thin films also possess piezoelectric properties, suitable for surface acoustic wave devices.[1,2] Moreover, AlN thin films have been used as buffer layers for the growth of high quality GaN and AlGaN films. However, AlN epitaxial layers grown on sapphire or SiC substrates by metal organic chemical vapor deposition (MOCVD) typically have a large density of threading dislocations (TDs) and point defects, due to the lattice mismatch, low surface mobility of Al species during growth and gas phase adduct formation. AlGaN alloys exhibit large energy band gap variation between 3.4 eV to 6.2 eV at room temperature, and thus have become useful materials for many research topics. Quaternary InAlGaN alloys have also attracted considerable amount of attention in recent years.

To obtain efficient optoelectronic devices, it is vital to accomplish structural and optical analysis, to understand the underlying mechanism of defect structures, and to improve the crystal quality of III-nitride materials. In recent years, a great deal of intensive studies have been carried out on III-nitride based alloys.[5–17] In this chapter, an overview of the structural and optical investigations of III-nitride semiconductor materials has been presented, by employing multiple experimental and theoretical techniques. The experimental data from various methods viz., synchrotron radiation X-ray absorption fine structure spectroscopy (SR-XAFS), X-ray diffraction (XRD), scanning electron microscopy (SEM), transmission electron microscopy (TEM), X-ray photoelectron spectroscopy (XPS), room temperature (RT) and varying temperature (VT) Raman scattering, photoluminescence (PL), Fourier transform infrared (FTIR) spectroscopy, scanning ellipsometry (SE), Rutherford backscattering (RBS), optical transmission are carefully analyzed by using many theoretical and computer modeling techniques. It should be noted that there are so many papers available dealing with the structural and optical properties of III-nitride materials — our discussion here has only been focused on a selection of published results in recent years.

2. Characterization of Group-III Nitrides

2.1. *Structural and optical properties of GaN*

2.1.1. *Synchrotron radiation (SR) X-ray absorption fine structure spectroscopy (XAFS) of GaN*

Synchrotron radiation X-ray absorption fine structure spectroscopy (SR-XAFS) is a pragmatic tool to study the local structure of III-nitrides in the

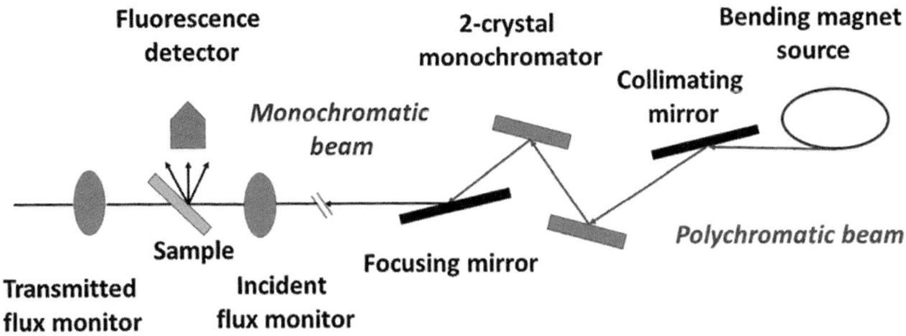

Fig. 1. Schematic experimental set-up for the XAFS measurement. (Reprinted from Ref [18] with permission from Springer.)

first few coordination shells of thin epilayers. In fact, if the fluorescence is used — a monolayer-thin epilayers can be studied if sufficiently intense synchrotron radiation is used. Also, X-ray absorption near edge spectrum (XANES) can provide energy distribution of the electron density in the conduction band corresponding to a particular bond of studied atom with its neighboring atoms. The extended X-ray absorption fine structure (EXAFS) analysis gives information on the local atomic structure, e.g. interatomic distances R, Debye-Waller factors σ and the coordination number N. When these techniques are performed with linearly polarized synchrotron radiation, a direct insight into the shape and anisotropy of the chemical bonds can be obtained. Here, the X-ray absorption fine-structure spectra were employed for measuring four GaN samples at Ga K-edges in X-ray fluorescence yield mode, which is directly related to the distribution of the Ga p-states in the conduction band.

In the case of crystalline solids, XANES spectra extended approximately $\sim$70 eV above the absorption edge is commonly used to probe the conduction band and local electronic structure of crystalline materials. To obtain the anisotropic characteristics of the p-orbital, XANES spectra is measured under normal incidence and grazing incidence. Figure 2 shows the Ga K-edges XANES spectra of four GaN samples measured at 15° and 90° grazing incidence angles. The two undoped samples (NS38 and NS42) of GaN films were grown on (111) Si substrates at 700°C with a less than 10-nm-thick AlN buffer layer by molecular beam epitaxy (MBE) having thicknesses 1.59 and 0.58 μm, respectively. The other two samples (M1044n and M1232n) of GaN films were grown on c-sapphire at 1030°C

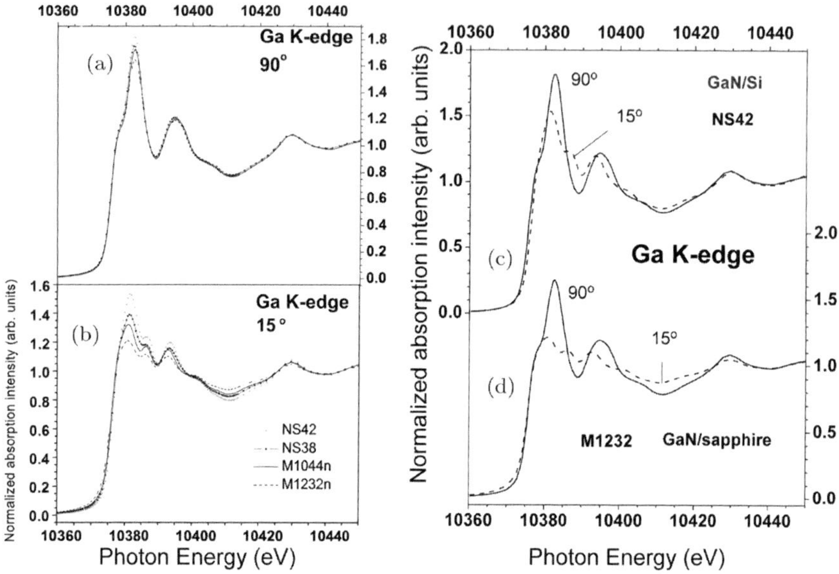

Fig. 2. Ga K-edge X-ray absorption near-edge structure (XANES) spectra under (a) normal incidence and (b) grazing incidence of the samples NS38, NS42, M1044n and M1232n, and comparison of two incidences at a same sample of (c) NS42 and (d) M1232n. (Reprinted from Ref. [20] with permission from Elsevier.)

with a 20-nm-thick GaN buffer layer prepared at 530°C by metalorganic chemical vapor deposition (MOCVD).[20] Clearly, the variations of the peaks in XANES spectra under normal incidence (Fig. 2(a)) and grazing incidence (Fig. 2(b)) angles are obviously different. The comparisons of XANES spectra is shown in Figs. 2(c) and (d) for the MBE grown GaN/Si sample NS42 recorded at two grazing incidence angles of 90° and 15°, respectively. It is noticed that the X-ray information depth into the sample under normal incidence is somehow larger than the depth under grazing incidence and the anisotropic characteristic of p-orbital from GaN/sapphire grown by MOCVD is larger than GaN/Si prepared by MBE. A low anisotropic in wurtzite GaN might imply the inclusions of cubic phase or more portion of polycrystalline hexagonal with random orientation.

The extended X-ray absorption fine structure (EXAFS) spectra of the four samples are exploited to extract the interatomic distances between the absorption atom and its nearest-neighbors, the atomic bond lengths and the lattice constants. The raw EXAFS data is fitted by using

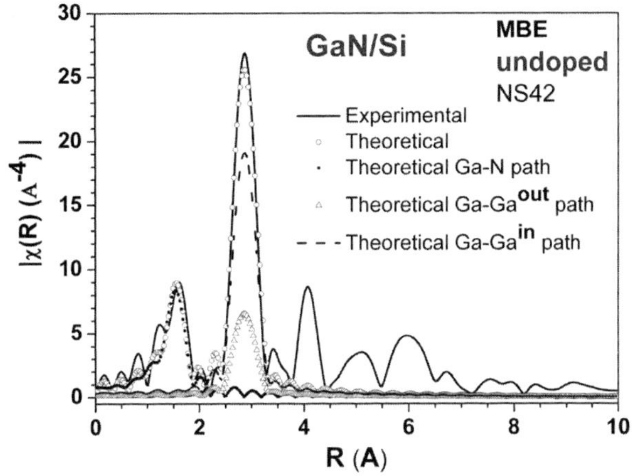

Fig. 3. Fourier transformed EXAFS spectra around Ga K edge of sample NS42. The purple dashed line is the fitting result of the normal incidence of the X-ray beam and green line is the fitting result of the grazing incident of the X-ray. (Reprinted from Ref. [20], with permission from Elsevier.)

Table 1. Results of the quantitative XAFS analysis. (Reprinted from Ref. [20], with permission from Elsevier.)

Sample	R_{GaN}(Å)	σ^2_{GaN}(Å^2)	R^{out}_{GaGa}(Å)	R^{in}_{GaGa}(Å)	σ^2_{GaGa}(Å^2)
M1044n	1.926	0.0025	3.174	3.174	0.0051
M1232n	1.928	0.0017	3.171	3.177	0.0045
NS38	1.931	0.0012	3.172	3.189	0.0056
NS42	1.932	0.0041	3.172	3.189	0.0058

the commercial IFEFFIT program codes (see http://cars9.uchicago.edu/~ravel/software/downloads.html.). Figure 3 is an example of the data fit for assessing the material parameters, such as Ga–N first shell bond length R_{GaN}, Ga–Ga second shell bond length including out of-plane $R^{out}_{GaGa} R^{out}_{GaGa}$ and in-plane R^{in}_{GaGa}, (see Table 1).[20] The perusal of Table 1, reveals the nearest neighbor Ga–N bond length for GaN/sapphire sample is slightly smaller than that of GaN/Si. Moreover, as compared to the bulk reference, the values of in-plane and out-of plane strains caused by the hetero-epitaxial growth can be calculated from the Ga–Ga second shell interatomic distances. More details are reported in Ref. [20].

2.1.2. *Raman scattering spectra of GaN*

Raman scattering spectroscopy is one of the simplest yet very powerful non-destructive technique for probing the vibrational behavior of materials. It provides useful information about the crystal phase, orientation, as well as lattice phonons through a set of well-established selection rules. Earlier, the method has been employed for investigating the vibrational modes of GaN.[19] In an ideal GaN semiconductor, the crystal lattice translation symmetry leads to the plane wave phonon eigen states which indicates that only $q = 0$ phonons at the center of the Brillouin zone (Γ point) participate in the first-order Raman scattering process because of the energy and momentum conservation. However, disorders or finite-size effects may violate the momentum conservation rule, causing the shift and broadening of the Raman peak, as shown in Fig. 4. The perusal of Fig. 4 reveals the high frequency Raman active E_2 mode of the four GaN samples used in SR-XAFS study (cf. Fig. 2, i.e., NS38, NS42, M1044n and M1232n). In order to estimate the average size of the nano-crystals from the Raman

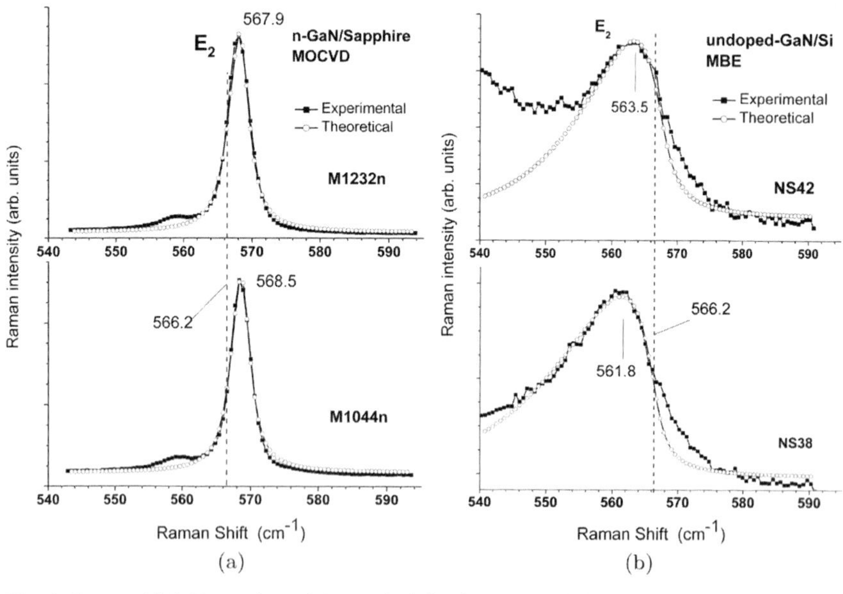

Fig. 4. Raman high E_2 mode and theoretical fits from (a) MOCVD grown n-GaN/sapphire and (b) MBE grown undoped-GaN/Si. (Reprinted from Ref. [20], with permission from Elsevier.)

spectrum, the spatial correlation model[21] is used to fit the Raman active E_2 mode.

In a spatial correlation model, the intensity of the first-order Raman spectrum $I(\omega)$ can be described as

$$I(\omega) \propto \int \exp(-q^2L^2/4)\frac{d^3q}{[\omega - \omega(q)]^2 + (\Gamma_0/2)^2}, \tag{1}$$

where $q = 2\pi/a$ (a is the lattice constant), L is the correlation length in units of a, and Γ_0 is the natural or intrinsic line width. The dispersion relation $\omega(q)$ for optical phonons can be expressed by two analytical forms:

$$\omega^2(q) = A + \left\{A^2 - B\left[1 - \cos(\pi q)\right]\right\}^{1/2}, \tag{2}$$

$$\omega(q) = A - Bq^2, \tag{3}$$

where A and B are adjustable parameters. Table 2 lists the fitting parameters of the high E_2 modes for the four GaN samples using spatial correlation model. The fitting results are displayed in Fig. 4. It is seen that the fits to GaN/Si exhibit larger errors despite trying different Γ_0 values (Table 2). It is simply because the mode frequency of the substrate Si (520 cm^{-1}) has a bigger influence onto the E_2 mode shape of GaN film [Fig. 4(b)].

Also, here, a comprehensive investigation on the phonon anisotropy of wurtzite GaN was reported by employing both angular dependent Raman spectroscopy and theoretical simulations (see Fig. 5). Along the cross section of GaN film, the polarized Raman scattering spectra were measured as a function of the angle between the polarization direction of the incident laser with three different polarization configurations, the perpendicular, parallel, and non-polarized experiment configurations.[22] The theoretical fits were performed via the anisotropic model, which is weighed in the susceptibility contribution and the phase differential of the different

Table 2. Theoretical fitting results of Raman high E_2 mode. (Reprinted from Ref. [20], with permission from Elsevier.)

	A	B	$L(a)$	Γ_0 (cm^{-1})	$\omega(E_2)$ (cm^{-1})
M1044n	568.9	100.0	30.0	3.0	568.5
M1232n	568.8	100.0	27.0	3.0	567.9
NS38	566.8	108.0	7.1	3.0	561.8
NS42	567.9	110.0	8.3	4.0	563.5

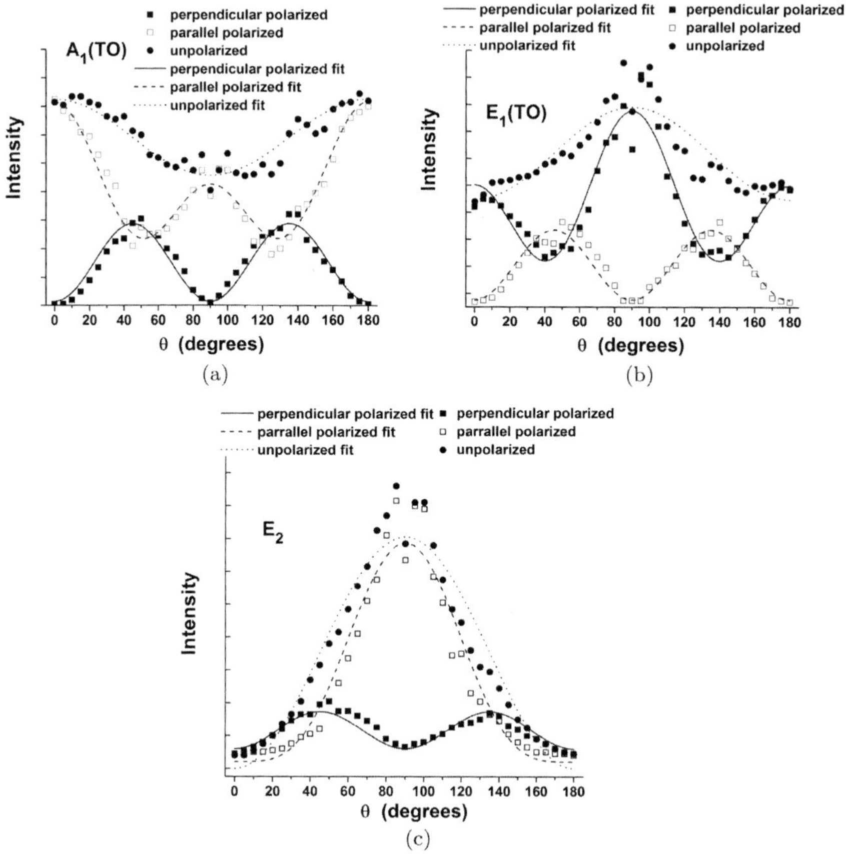

Fig. 5. (a) Raman peak intensities of the A_1(TO) mode as a function of rotation angle in three polarization configurations. (b) Raman peak intensities of the E_1(TO) mode with rotation angle. (c) Raman peak intensities of the E_2 mode with rotation angle. (Reprinted from Ref. [22] with permission from American Institute of Physics.)

vibrating elements. From Fig. 5, it is clearly seen that the Raman intensities of the A_1 (TO) and E_1 (TO) modes showed a sinusoidal dependence on the rotating angle. However, E_2 mode shows a quite different dependence because it has the same vibrating elements whereas the A_1 and E_1 have two types of elements and will have the phase differential contributing to the susceptibility of the polarized angle.[22]

Three m-plane GaN films grown on sapphire by MBE with varying III/V ratios of 70, 80 and 90, and designated as MO267, MO268 and MO269 are

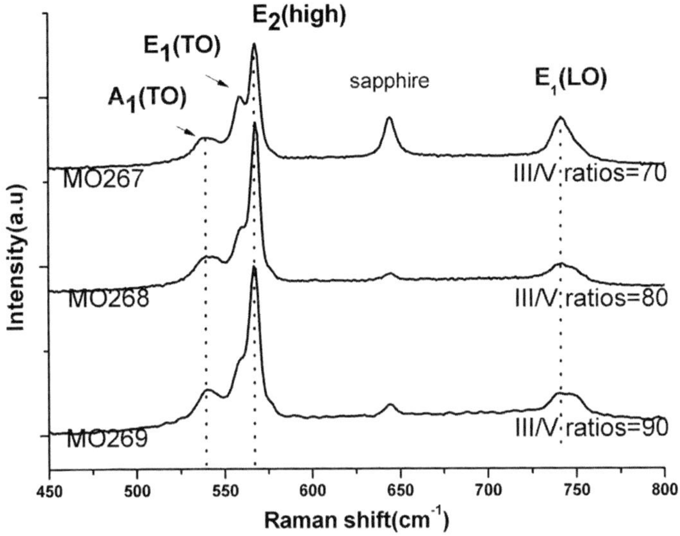

Fig. 6. Raman spectra of m-GaN with different III/V ratios. (Reprinted from Ref. [23] with permission from Trans Tech Publications.)

studied by Raman scattering spectroscopy. The results displayed in Fig. 6 reveal Raman shifts of A_1 (TO), E_1 (TO), E_2(high) and E_1 (LO) modes around 540 cm^{-1}, 558 cm^{-1}, 567 cm^{-1}, 741 cm^{-1}, respectively. The broad peak at 741 cm^{-1} contains E_1 (LO) mode of *m*-plane GaN and E_g mode of sapphire as the phonon peaks of sapphire substrate are noticed around 644 cm^{-1} and 750 cm^{-1}. Raman active E_2 (low) peak is not shown here as its frequency at 142 cm^{-1} is too low. However, it should be noted that the quality of epitaxial wafer can be analyzed from the linewidth of E_2 (high) mode whose peak position indicates that compressive stress comes from the lattice mismatch. Fitted by Lorentzian model, the FWHM of E_2 (high) modes of MO267, MO268 and MO269 is 5.9 cm^{-1}, 5.3 cm^{-1} and 6.4 cm^{-1}, respectively, which reveals that MO269 with Ga-rich condition suffers from a higher compressive stress at room temperature and the film with a III/V ratio 80 exhibits the best crystalline quality.

Figure 7 shows the varying-temperature Raman spectra of GaN epitaxial film grown on *m*-plane sapphire with the III/V ratio = 70. The peaks of the active phonon modes A1 (TO), E1 (TO), E2 (high) and E1 (LO) are broadened and shifted to lower frequencies while the temperature increased,

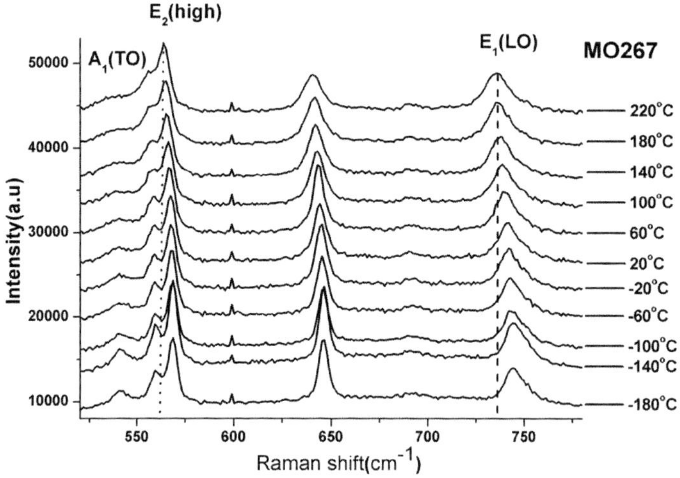

Fig. 7. Varying temperature Raman scattering spectra of GaN. (Reprinted from Ref. [23] with permission from Trans Tech Publications.)

i.e., the Raman shifts can be described as a function of temperature. The redshifts of these Raman modes are the results of harmonic coupling to phonons of other branches, and/or the heating caused by volume expansion or crystal lattice dilation. Actually, it is found that the three-phonon processes are dominant in the redshifts of E1 (LO) and E2 (high).[23]

2.2. *Structural and optical properties of InN*

2.2.1. *XRD spectra of InN grown on $LiAlO_2$ substrates*

X-ray diffraction (XRD) is a non-destructive and rapid characterization method employed for assessing the quality of semiconductor thin films and devices. Figure 8(a) shows the $\theta/2\theta$ scan X-ray diffraction patterns of InN films grown on γ-$LiAlO_2$ (LAO) substrate by plasma-assisted MBE (PA-MBE) with different growth conditions (see: Table 3). The XRD peaks of InN($10\underline{1}0$), InN(0002), In(101), and LAO (200) planes are located at around 29.0°, 31.3°, 32.9°, and 34.7°, respectively. Apparently, (cf. Fig. 8) the growth orientation of InN films on LAO substrate is very sensitive to the growth condition. High-content c-plane InN film can be grown with higher plasma power and higher growth temperature (sample A). Moreover,

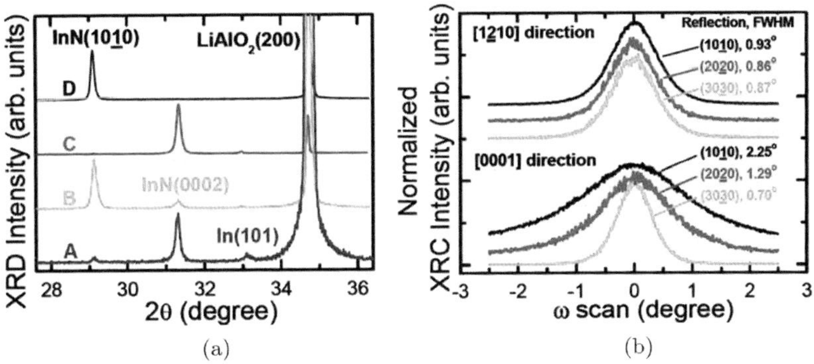

Fig. 8. (a) $\theta/2\theta$ scan XRD patterns of InN thin films grown on LAO substrate with different growth conditions (b) XRCs of pure m-plane InN taken along a-axis [1210] and c-axis [0001] directions. The FWHM measured along c-axis direction shows strong index dependence. (Reprinted from Ref. [24] with permission from American Institute of Physics.)

Table 3. Growth conditions of the InN films. (Reprinted from Ref. [24] with permission from American Institute of Physics.)

Sample	*Outgassing temperature* (°C)	*Nitridation*	*rf power* (W)	Growth temperature (°C)
A	600	No	200	460
B	600	No	100	390
C	800	Yes	100	390
D	800	No	100	390

when InN films were grown on nitridated surface or without proper high-temperature outgassing, the films were mixed with m-plane and c-plane InN (samples B and C), which indicates that substrate surface condition affects the InN growth as well. Actually, almost pure c-plane InN can be obtained via the nitridation on cleaned substrate surface (sample C) and pure m-plane InN can be grown on LAO substrate with proper treatment under lower plasma power of 100 W and growth temperature of 390°C (sample D).[24]

Symmetric X-ray rocking curves (XRC) have been broadly performed to study mosaic and thus dislocations in thin films. To investigate the crystal structure of the above pure m-plane InN, the ω-scans of InN(10$\underline{1}$0), InN(20$\underline{2}$0), and InN(30$\underline{3}$0) reflections measured along c-axis (0001) and

a-axis $(1\bar{2}10)$ directions were further investigated through X-ray rocking curve, as shown in Fig. 8(b). Obviously, when measured along a-axis, $(30\bar{3}0)$, $(20\bar{2}0)$, and (1010) crystal plane reflections have almost identical full widths at half maximum (FWHM). However, if measured along c-axis, $(30\bar{3}0)$ and $(20\bar{2}0)$ reflections present narrower FWHM than $(10\bar{1}0)$ reflection, i.e., almost one-third and on-half, respectively. High density basal-plane stacking faults (BPSFs) in the m-plane nitride crystals lead to the broadening of $(10\bar{1}0)$ reflection measured along c-axis, rather than crystal mosaic tilt only,[25,26] which can be proved by applying modified Williamson–Hall (WH) analysis.[24,26,27]

In order to further analyze the effects of growth conditions on InN samples — the field-emission scanning electron microscopy (FESEM) images, electron back scatter diffraction (EBSD) patterns, and corresponding atomic models of samples C and D were also studied and the results are displayed in Figs. 9(a–f). Obviously, before InN growth, applying substrate nitridation not only changes growth orientation [Fig. 8(a)] but also the morphology [Figs. 9(a) and (b)]. The m-plane InN film is much flatter than c-plane InN columnar film grown on LAO substrate, even though it shows stripe-like feature. According to the generated EBSD pattern shown in Fig. 9(c), the EBSD INCA software can automatically examine the crystal orientations and the atomic models. It is demonstrated that the columnar film exhibits six-fold symmetric bright belts intersected with a pole along c-axis direction. Three belts are associated to m-planes $\{10\bar{1}0\}$ and the other three are a-planes $\{1\bar{2}10\}$. Based on this pattern, a corresponding atomic model was plotted and a c-face hexagon model is obtained in Fig. 9(e). Counter to that, m-plane film shows a two-fold symmetric pattern with a pole along the m-axis $[10\bar{1}0]$ and a m-face hexagon model [Fig. 9(f)]. Combined with SEM images, it is found that the a-plane $(1\bar{2}10)$ orientation is along the stripe-like long-axis direction while the c-plane is perpendicular to the stripe-like direction, which is caused by the formation of high density BPSFs lying on the c-plane as implied by the modified WH analysis.[24] Actually, m-plane InN grown on free-standing m-plane GaN substrate and m-plane GaN grown on LAO substrate show a similar morphology.

Figure 10(a) shows a high-resolution TEM image of a 230-nm-thick m-plane InN film (sample D), which is employed to study the

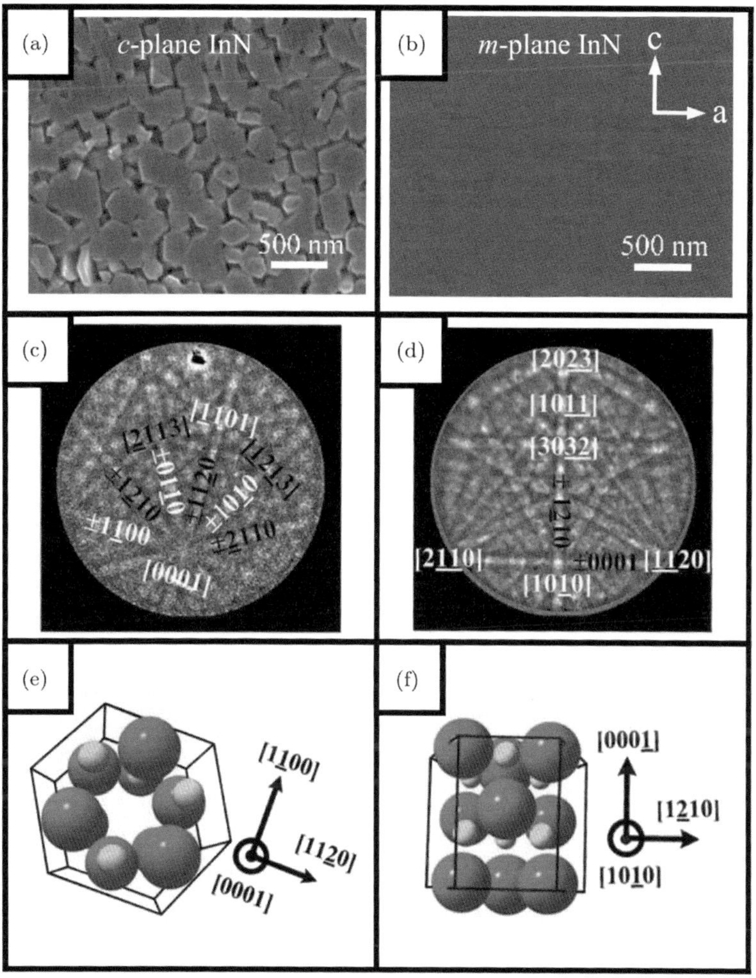

Fig. 9. Field-emission electron microscope images, electron back scatter diffraction (EBSD) patterns, and atomic models of (a), (b), and (e), *c*-plane InN columnar film (sample C) and (b), (d), (f) *m*-plane InN epifilm (sample D). The atomic models are generated from corresponding EBSD patterns by INCA software. (Reprinted from Ref. [24] with permission from American Institute of Physics.)

microstructural properties. Apparently, the as-grown InN film still show flat surface even though the surface of LAO substrate is rough (owning to the mechanical polishing process). Due to the vaporization of the LAO substrate under high-energy electron beam irradiation, some voids can be observed at

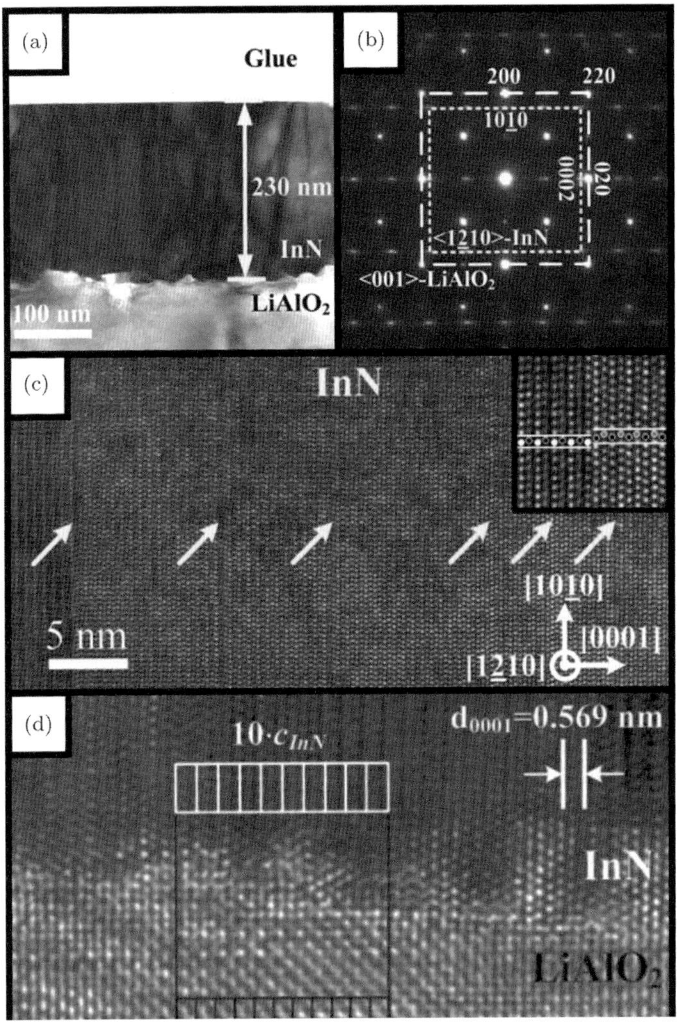

Fig. 10. (a) TEM images of the pure m-plane InN epifilm. (b) The corresponding SAED pattern. (c) and (d) High resolution lattice images taken at film and interface region, respectively. (Reprinted from Ref. [24] with permission from American Institute of Physics.)

interface region, which cannot be found in the m-plane InN film. So the film exhibits highly structural thermal stability. Moreover, the crossing of most visible threading dislocations (TDs) through the entire film starts at interface rather than the film, indicating a thicker film has lower TDs density. When a low-temperature (LT) buffer is used to eliminate TDs extension, sparse TDs closed loop is formed near the interface as commonly observed in c-plane films grown on sapphire substrate. In other words, m-plane InN can be directly deposited on LAO substrate without LT-buffer layer assistance. Figure 10(b) presents a corresponding selected area electron diffraction (SAED) pattern taken at the interface, i.e., in-zone condition for both film and substrate, simultaneously. It indicates that almost no tilting between film and substrate. Figures 10(c) and (d) show high-resolution lattice images taken at the film and interface region.

Furthermore, the inter-planar spacings of LAO(010) and InN(0001) are extracted as $d_{(010)-\mathrm{LAO}} = 0.516\,\mathrm{nm}$ and $d_{(0001)-\mathrm{InN}} = 0.569\,\mathrm{nm}$, respectively. The measured lattice parameters of the InN film and LAO agree well with the bulk lattice constants of InN ($a_{\mathrm{InN}} = 0.354\,\mathrm{nm}$ and $c_{\mathrm{InN}} = 0.570\,\mathrm{nm}$) and LAO ($a_{\mathrm{LAO}} = b_{\mathrm{LAO}} = 0.517\,\mathrm{nm}$ and $\mathrm{c}_{\mathrm{LAO}} = 0.627\,\mathrm{nm}$), which implies that the InN film reveals single crystal wurtzite structure except for some BPSFs parallel to the c-plane face as pointed by arrows. Enlarging BPSFs' area as shown in the inset of Fig. 10(c) can observe an ...ABABCBCB... stacking sequence assigned as the I_1 type BPSFs.[28,29] Here, to clarify the stacking sequence, A, B, and C, represent In-N atomic bilayers and are marked as white, black, and gray circles, respectively. The type-I_1 stacking fault has the lowest formation energy in wurtzite semiconductors. The density of BPSFs is estimated around $\sim 1.5 \times 10^6\,\mathrm{cm}^{-1}$, which is quite close to the density estimated from modified WH analysis. Short lateral coherence length (5–10 nm) of the BPSFs also elongate reciprocal lattice along c-direction shown in Fig. 10(b), leading to the broadening of the (1010) reflection measured by XRC. Figure 10(d) shows epitaxied InN film on LAO at interface region. Interestingly, the InN lattice with 10 c-lattice constants marked with a white grid matches with 11 a-lattice constants of LAO marked with a black grid. Usually, epitaxially growing m-plane InN on LAO would be difficult because of the large lattice mismatches between m-plane InN and LAO along $[1210]_{\mathrm{InN}}//[001]_{\mathrm{LAO}}$ and $[0001]_{\mathrm{InN}}//[010]_{\mathrm{LAO}}$ orientations (13% and 10%). However, This heteroepitaxy for the m-plane GaN deposited on LAO has only ~1.4% and ~0.1% along $[1210]_{\mathrm{GaN}}//[001]_{\mathrm{LAO}}$ and $[0001]_{\mathrm{GaN}}//[010]_{\mathrm{LAO}}$ orientations, respectively. Hence, the m-plane InN can be epitaxied on LAO under such large lattice-mismatched condition when commensurately lattice-matched condition with a small misfit strain of ∼0.2% is satisfied.[24]

2.2.2. *Rutherford backscattering characterization of InN grown on sapphire*

Rutherford backscattering (RBS) is a useful technique for the analysis of sample composition and thickness. The RBS was used to appraise the microstructure of InN/sapphire samples with N_2-plasma gun power at 350 W and 400 W, respectively. Figures 11(a) and (b) show the experimental and simulated RBS spectra, where the labeled vertical black colored arrows indicate the energy for backscattering from various elements (In, N, Al and O) at the surface, respectively. After fitting with SIMNRA (a Microsoft Windows program used for the simulation of RBS) codes, the indium concentration and InN thickness of these two samples were obtained. The estimated thickness of InN layer is quantified around 300 nm for In: $N = 0.50{:}0.50$ ratios. One must note that based upon a simple single layer/substrate model, the simulations of the RBS data get bad fits for the sample grown at higher plasma power (e.g., see, Fig. 11(b) — the circle regions on the sample at 400 W). On the other hand, when using two or more layered structure with a non-stoichiometric ratio, good fits can be obtained. Figure 12 shows the results of better simulation for InN films at 400 W, which uses a surface layer and an intermediate layer between the substrate and InN film to fit. Clearly, the InN film samples with higher power lead to

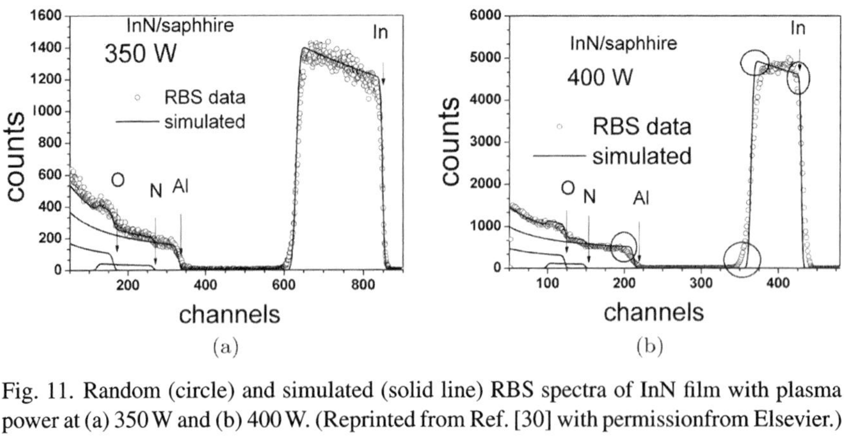

Fig. 11. Random (circle) and simulated (solid line) RBS spectra of InN film with plasma power at (a) 350 W and (b) 400 W. (Reprinted from Ref. [30] with permissionfrom Elsevier.)

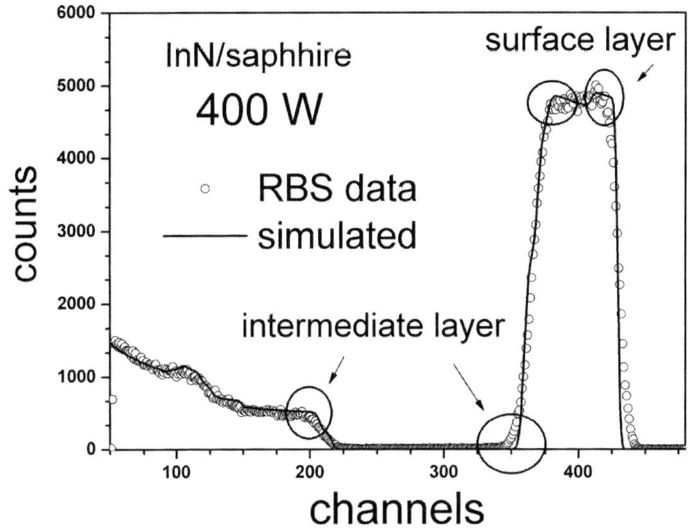

Fig. 12. Random (circle) and simulated (solid line) RBS spectra, including the surface effect and inter-diffusion between InN layer and substrate, for an InN/sapphire with plasma power of 400 W. (Reprinted from Ref. [30] with permission from Elsevier.)

an inter-diffusion between InN and sapphire.[30] In fact the PA-MBE InN film samples using plasma power of 350 W with carrier concentration of about $\sim 1 \times 10^{20}\,\text{cm}^{-3}$ possess high crystalline perfection. As the plasma power reaches 400 W — the carrier concentration increases to a higher value of $\sim 1.5 \times 10^{20}\,\text{cm}^{-3}$ — the InN film shows a slightly inferior crystallinity.[30]

2.2.3. *Temperature-dependent photoluminescence of InN/sapphire*

Figure 13 shows the temperature dependent photoluminescence spectra (recorded at temperatures from 20 to 300 K) for a typical InN sample grown on sapphire by MBE with the plasma power of 200 W. It is noticed that because of lattice expansion and electron–phonon interaction, when the temperature increases, the intensity of the photoluminescence (PL) peak decreases and PL peak exhibits broadening and redshift. While the temperature dependence of PL shift or band gap for epitaxially grown InN can be described by Varshni's[31] empirical formula, however, it is found Pässler's[32] equation incorporating electron-phonon interaction to be slightly better.

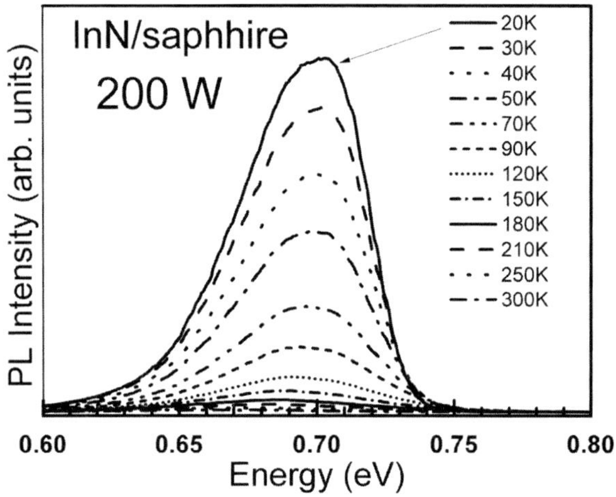

Fig. 13. The temperature-dependent photoluminescence of an InN grown on sapphire with plasma power at 200 W. (Reprinted from Ref. [30] with permission from Elsevier.)

2.2.4. *Raman scattering spectra of InN/sapphire grown with different plasma power*

Three InN samples labeled as CC-21, CC-22 and CC-24 were grown on sapphire by MBE with 230, 300 and 350 W plasma power, respectively. Room temperature (RT) first-order Raman scattering (RS) spectra (cf. Fig. 14) in the backscattering geometry $[z(x, y)\bar{z}]$ is used to characterize these samples. The perusal of Fig. 14 clearly reveals three strong phonon features near $\sim$590 cm^{-1}, $\sim$490 cm^{-1}, $\sim$88 cm^{-1} corresponding to the A_1(LO), E_2^{high}, E_2^{low} modes, respectively. Several other phonon features are uncovered near $\sim$160, $\sim$202, $\sim$410, $\sim$446, $\sim$475, and $\sim$562 cm^{-1} — shown by vertical arrows in Fig. 14.

Furthermore, it is found the main A_1 (LO) mode exhibiting a relatively large FWHM $\sim$30 cm^{-1} (i.e., 575–610 cm^{-1}), which is combined by the A_1 (LO) phonon centered at around 590–591 cm^{-1} and E_1 (LO) mode located at about 593–595 cm^{-1}. Meanwhile, there is a phonon band near $\sim$562 cm^{-1} (possibly related to the surface optical mode) at the low frequency-edge of A_1 (LO), which shows an increased intensity with the increase of plasma power. Also, on the left hand side of the strong E_2^{high}, E_2^{high} peak, three shoulders (near $\sim$410, $\sim$446 and $\sim$475 cm^{-1}, the third shoulder represents E_1 (TO) mode) are observed in all InN samples and these features become

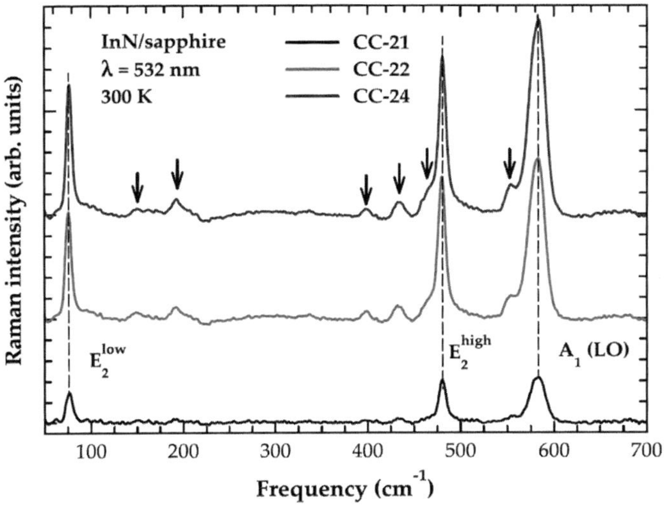

Fig. 14. Room-temperature RS spectra for three InN/sapphire (CC-21, CC-22 and CC-24) samples by PA-MBE with different plasma power. (Reprinted from Ref. [33] with permission from Elsevier.)

stronger in films prepared with higher plasma power. This characteristic is assigned as a low LO-plasmon coupled mode ω_- arising from the interaction of LO phonons through macroscopic field with collective excitations of free charge carriers. It is emphasized that the Raman active mode near $\sim$410 cm^{-1} falls possibly within the forbidden gap of acoustical and optical phonon branches ($\sim$225–445 cm^{-1}) of InN. In the absence of realistic calculations of defect vibrational modes (e.g., an average–t–matrix Green's function theory[34]), one cannot exclude the observed phonon feature near $\sim$410 cm^{-1} as a gap-mode of N atoms surrounded by In vacancies in InN. Again, two other phonon bands emerged at $\sim$160 and $\sim$202 cm^{-1} are dominated possibly by the overtones of acoustical phonons. It is demonstrated that an acoustic phonon near a critical-point A generate the band around $\sim$160 cm^{-1} and overtones of TA modes either near K or M critical point generate a band close to $\sim$202 /cm.[35]

It is found that when the electron concentration is high (i.e., $\eta >$ $10-^{18}\,\mathrm{cm}^{-3}$) — the as-grown InN/sapphire films exhibit n-type conductivity, indicating that the plasmon frequency ω_p to be higher at large η. As a result of the polar nature of InN films, the strong interaction between ω_p

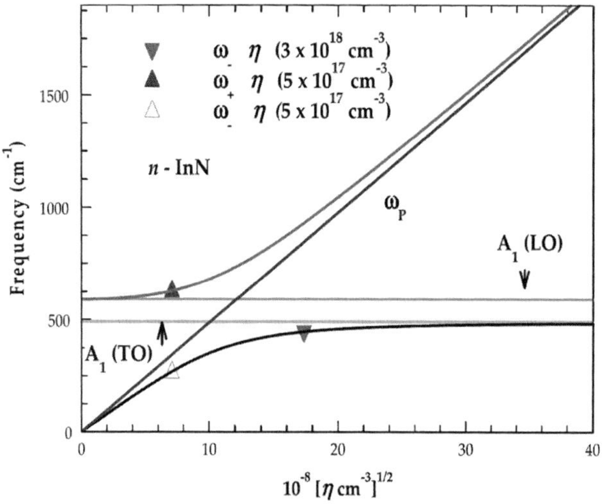

Fig. 15. Calculated LO phonon–plasmon coupled modes ω_+/ω_- as a function of the free carrier concentration in InN. (Reprinted from Ref. [33] with permission from Elsevier.)

oscillations and LO-phonons will instigate (see Fig. 15) two coupled modes ($\omega_{+}/_{-}$). When η decreases, the lower mode ω_- drops from A_1 (TO) to lower value while the upper mode ω_+ becomes closer to A_1 (LO) frequency. For low $\eta = 5.0 \times 10^{17}\,\text{cm}^{-3}$, the calculated ω_- ($\sim$269 cm^{-1}) and ω_p ($\sim$345 cm^{-1}) are smaller and $\omega+$ ($\sim$629 cm^{-1}) differs from A_1 (LO) by a few wave numbers (cf. Fig. 15). As η ($\sim 3.0 \times 10^{18}\,\text{cm}^{-3}$) approaches close to the range of charge carriers in as-grown InN, our simulations provide two LO-plasmon coupled modes near ω_- ($\sim$446 cm^{-1}) and ω_+ ($\sim$930 cm^{-1}). Due to a significantly large phonon broadening at higher η, the low frequency coupled mode ω_- can be seen in Fig. 15 while the higher energy mode ω_+ has not been detected in the Raman scattering studies.[33]

2.2.5. *Infrared (IR) reflectivity of InN/sapphire*

Fourier transform infrared (FTIR) spectroscopy is a useful tool for examining the optical properties of thin film and obtaining their structural information such as thickness and free carrier concentrations. Here, a Fourier transform infrared (FTIR) spectrometer was used to examine the phonon characteristics of two PA-MBE grown InN/sapphire samples. Figure 16 displays the measuring room temperature reflectivity spectra

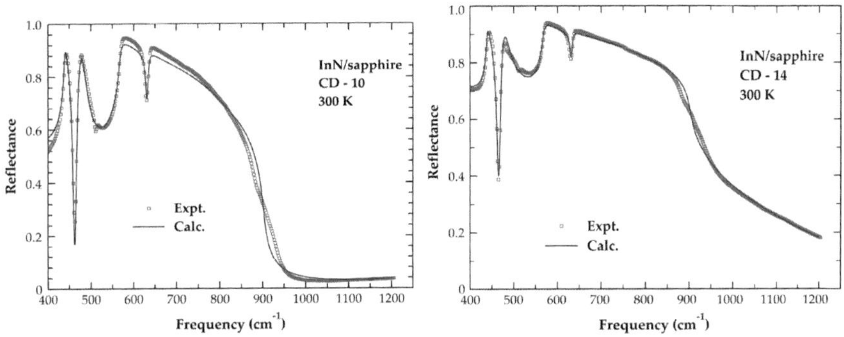

(a) (b)

Fig. 16. Comparison of the experimental (open squares) room temperature infrared reflectivity spectra of InN/sapphire samples grown by PA-MBE with the theoretical results (full lines) using the parameter values of Table 2: (a) for sample CD-10 and (b) for sample CD-14. (Reprinted from Ref. [33] with permission from Elsevier.)

and the corresponding numerical simulations of reflectivity performed by exploiting a standard methodology of multilayer optics within the transfer-matrix approach. Clearly, the typical FTIR reflectance spectra (open squares) of InN/sapphire samples (CD-10 and CD-14) are found to be in very good agreement with the simulated curves (solid lines). From Fig. 16, the thickness of these two InN/sapphire samples (CD-10 and CD-14) are obtained: $d_{\text{CD}-10} = 0.24\,\mu\text{m}$ and $d_{\text{CD}-14} = 0.28\,\mu\text{m}$. Moreover, from Figs. 16(a) and (b), the free carrier concentrations are estimated ($\eta \sim 4.73 \times 10^{18}\,\text{cm}^{-3}$ and $\eta \sim 8.47 \times 10^{18}\,\text{cm}^{-3}$) by using the classical harmonic Lorentz oscillator approach[33,36] to fit the IR reflectivity of the two samples.

2.3. *Investigation of optical parameters for boron doped AlN*

2.3.1. *Spectroscopic ellipsometry measurements*

Spectroscopic ellipsometry (SE) is considered as one of the most powerful characterization techniques to obtain information of the optical properties for III-N semiconductor materials. Specially, the SE technique is utilized to evaluate the materials dielectric functions $\varepsilon(E)$ including refractive indices and absorption coefficients, which are essential for the designs and analyses of hetero-structure based lasers and other wave-guiding devices. Here, the

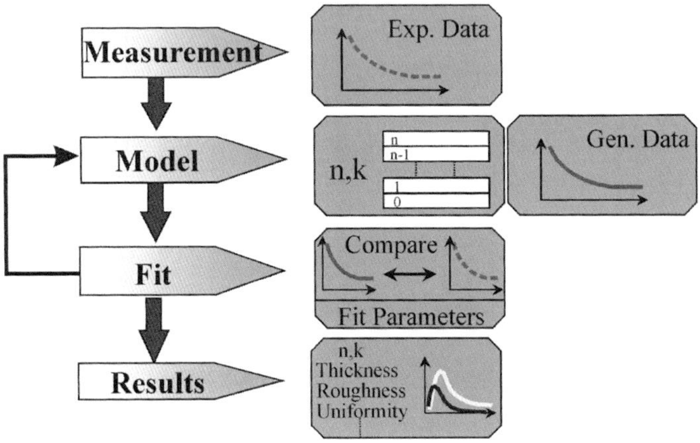

Fig. 17. The flow chart used in the spectroscopic ellipsometry fitting procedure (see online version for color). (Reprinted from Ref [37] with permission from Inderscience Enterprises Ltd.)

dielectric function of diamond substrate and AlN films doped with boron (B) contents of 0%, 3%, 5% are extracted by using the SE. Figure 17 shows the detailed methodology used for the analysis of SE data and the corresponding fitting results by using WVASE32 software.

Figure 18 displays the experimental and fitting results for four B-*doped* AlN samples. Clearly, the simulations agree well with the SE data. So, the accurate thicknesses of the four samples can be obtained, i.e., the thickness of polycrystalline diamond film is around 10 μm, and that of B-*doped* AlN films are in the range from 750 nm to 850 nm. Furthermore, it is found that three B-doped AlN samples exhibiting similar dispersions, except that the index of refraction is 2.30 for AlN, 2.12 for BAlN100 and 2.35 for BAlN150 at around 600 nm wavelength, respectively (see Fig. 19). Absorption band for AlN is at around 282 nm while that for BAlN100 is around 313 nm (see Fig. 20). So, the dielectric functions of the B-doped AlN samples are found strongly dependent on the film preparation process, i.e., the B concentration.

2.3.2. *X-ray diffraction of boron doped AlN*

FWHMs of X-ray diffraction analysis of the above B-doped AlN films with various boron concentration at (002) reflection are 0.61°, 1.01° and 1.67° for AlN, BAlN100 and BAlN150, respectively.[38] Moreover, variation of *c*- and

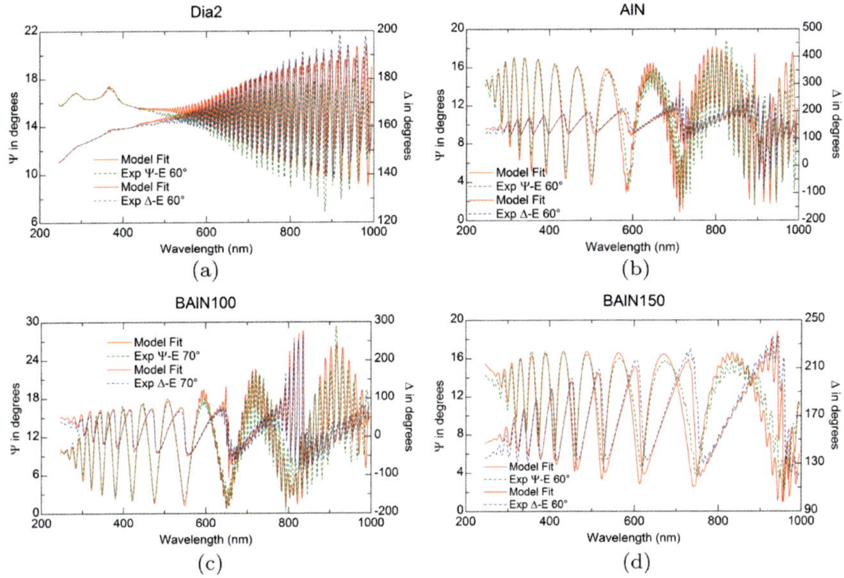

Fig. 18. SE experimental and fitting results of (a) Dia2, (b) AlN, (c) BAlN100, and (d) BAlN150. (Reprinted from Ref [37] with permission from Inderscience Enterprises Ltd.)

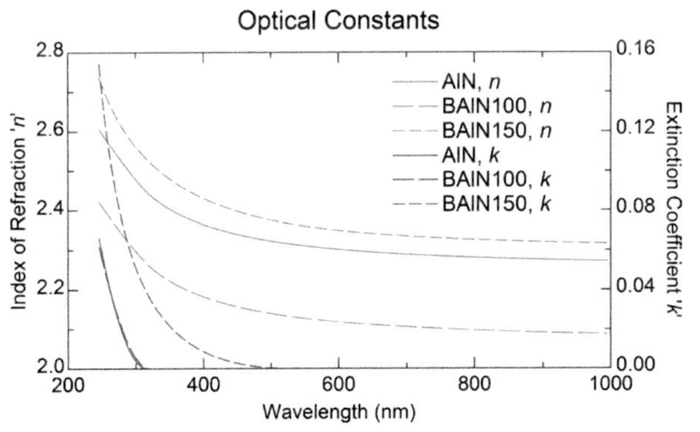

Fig. 19. Dielectric functions of AlN, BAlN100 (with boron content of 3%) and BAlN150 (with boron content of 5%). Red curves arc index of refraction and blue curves are extinction coefficient (see online version for color). (Reprinted from Ref [37] with permission from Inderscience Enterprises Ltd.)

a-lattice constants of these deposited B-doped AlN film including bulk AlN is obtained by x-ray diffraction analysis, as shown in Table 4. Clearly, with the increase of boron concentration the *c*- and *a*-lattice constants of the deposited films decrease. It should be noted that the lattice constants

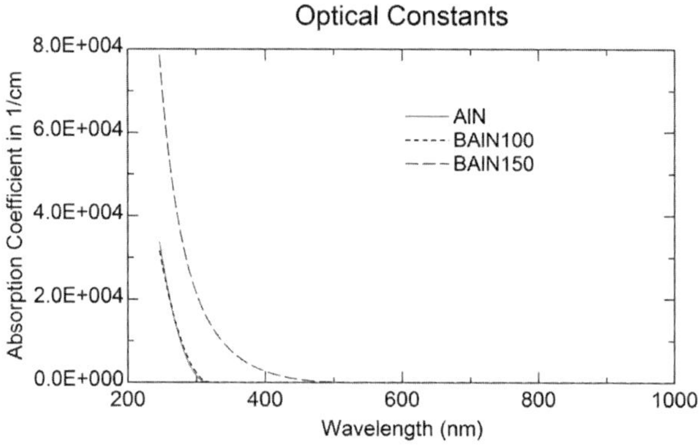

Fig. 20. Red curve, black curve and blue curve represents absorption coefficient of AlN, BAlN100 (with boron content of 3%) and BAlN150 (with boron content of 5%), respectively (see online version for color). (Reprinted from Ref [37] with permission from Inderscience Enterprises Ltd.)

Table 4. Variation of c- and a-lattice constant of deposited (B)AlN film with different boron contents as well as bulk AlN. (Reprinted from Ref [37] with permission from Inderscience Enterprises Ltd.)

Sample	*Boron content* (%)	*c-lattice constant* (Å)	*a-lattice constant* (Å)
AlN	0	4.941	3.133
BAlN100	3	4.933	3.106
BAlN150	5	4.922	3.072
Bulk AlN	0	4.980	3.110

of bulk AlN are different from those of the deposited AlN film. The a-lattice constant of AlN film is found larger than the bulk one, while the c-lattice constant of AlN film is smaller. Especially, the a-lattice constant of BAlN100 (with a boron content of 3%) is 3.106 Å — very similar to that of the bulk AlN (3.110 Å), which indicates good lattice match and BAlN100 could be an excellent buffer layer for the AlN layer growth. In accordance with SE results, the XRD results also reveal that the BAlN150 film (with boron atom fraction of 5%) possesses a larger band tail, which is caused by the structural disorders in B-doped AlN that instigated deterioration in the band gap as well as the exaggeration of the band tail. Meanwhile, the induction of boron atoms caused a downward shift in the index of refraction (n) of BAlN100 (with boron atomic fraction of 3%), the red shift of the extinction coefficient results in the upward shift for n in BAlN150, in corroboration with the results of Kramers-Krönig analysis. Therefore, higher B contents may cause red shift in the band gap and deteriorate the materials' crystallinity — the crystal properties of AlN layers are better when the B-doped AlN films with B concentration is lower than 5%.[37]

3. Group-III Nitride Ternary Alloys

3.1. *Structural and optical properties of $Al_xGa_{1-x}N$*

3.1.1. *High resolution X-ray diffraction*

Figure 21 shows the high-resolution XRD spectra of the (002) plane diffraction for four $Al_xGa_{1-x}N$ epilayers samples with diverse Al compositions (0.19, 0.32, 0.45, and 0.58, evaluated by fitting the XRD measurement data by using a commercial software, "X'Pert Epitaxy"[18]). The perusal of Fig. 21 reveals a diffraction peak around 36.02° originating from the AlN template while the other diffraction peaks coming from the $Al_xGa_{1-x}N$ epilayers. Based on the XRD measurements, one can evaluate the lattice constants $a, c, c/a$ ratios, and the internal parameter $u = a^2/(3c^2) + 0.25$ for the $Al_xGa_{1-x}N$ epilayers as summarized in Table 5.[18]

The threading dislocation density for the $Al_xGa_{1-x}N$ epilayers with different Al compositions is also investigated (see Fig. 22) by high resolution

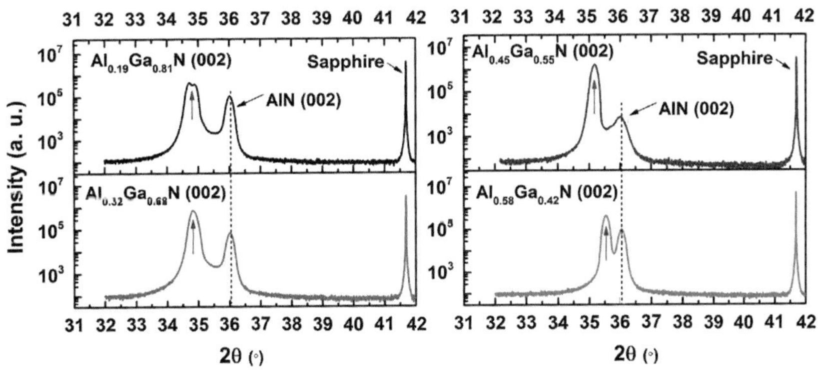

Fig. 21. High resolution XRD 2θ-ω scan curves for the (002) plane diffraction of $Al_xGa_{1-x}N$ epi-layers. (Reprinted from Ref. [18] with permission from Springer.)

Table 5. Summary of the lattice constants a, c, c/a ratios and the internal parameters u for $Al_xGa_{1-x}N$ epi-layers (from [18], reproduced with permission by Springer).

Lattice parameters	$Al_xGa_{1-x}Nepi - layers$			
	$x = 0.19$	$x = 0.32$	$x = 0.45$	$x = 0.58$
a (Å)	3.17193	3.16245	3.15297	3.14272
a (Å)	5.14073	5.11549	5.09026	5.06291
c/a	1.62069	1.61757	1.61443	1.61101
u	0.37691	0.37739	0.37789	0.37843

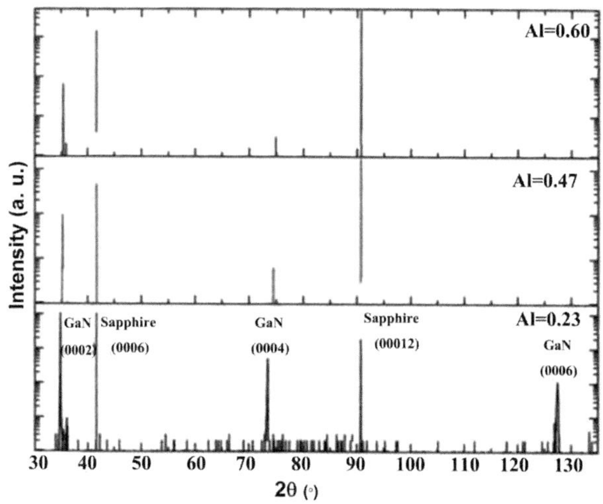

Fig. 22. High resolution 2θ-ω scans from 30 to 130° for a set of Al-content (0.20–0.60) $Al_xGa_{1-x}N$ thin films. (Reprinted from Ref. [41] with permission from The Japan Society of Applied Physics.)

X-ray diffraction (HRXRD) performed for (0002), (0004), and (0006) reflections. Generally, composition and strain differences will induce the variation in unit cell parameters.[39] Assuming the validity of Vegard's law,[40] the composition (0.23, 0.47, 0.60) of $Al_xGa_{1-x}N$ layers can be extracted from 2θ-ω scans of the (0002) reflection based on a relationship between the lattice parameter and the chemical composition. Combined with Williamson–Hall (WH) plots which extract the values of lateral coherence length ($L_{\parallel}$), vertical coherence length ($L_{\perp}$), tilt angle (α_{tilt}),

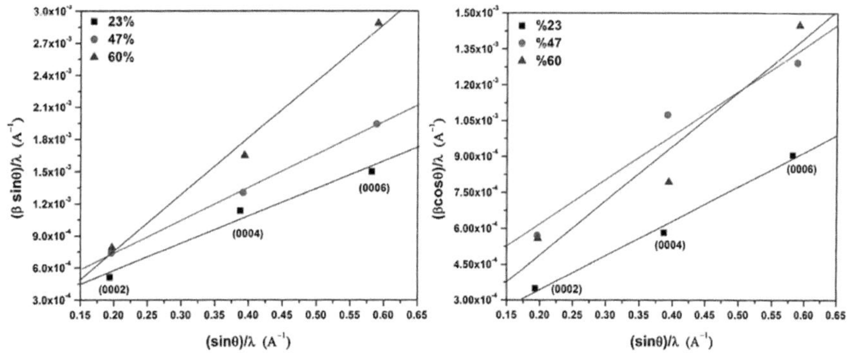

Fig. 23. Williamson–Hall plots for $Al_xGa_{1-x}N$ thin films. (Reprinted from Ref. [41] with permission from The Japan Society of Applied Physics.)

Table 6. Analysis summary of HRXRD measurements of $Al_xGa_{1-x}N$ thin films (Reprinted from Ref. [41] with permission from The Japan Society of Applied Physics.)

	$Al_{0.23}Ga_{0.77}N$	$Al_{0.47}Ga_{0.53}N$	$Al_{0.60}Ga_{0.40}N$
Lateral coherence length $L_{\|}$ (μm)	0.74	0.36	NA
vertical coherence length $L_{\perp}$ (μm)	0.79	0.17	1.02
Tilt angle α_{tilt} (arcsec)	526	633	1206
Screw dislocation (N_{screw}) (cm^{-2})	0.55×10^9	0.80×10^9	2.92×10^9
Strain parallel to c lattice ($\varepsilon_{\perp}$)	3.6×10^{-4}	4.6×10^{-4}	5.6×10^{-4}

screw dislocation (N_{screw}), and the heterogeneous strain ($\varepsilon_{\perp}$), the effect of different Al compositions on the structural properties of AlGaN layers are investigated.

Figure 23(a) shows the WH plots used to obtain the $L_{\|}$ and α_{tilt} (combining the Burgers vectors of c-axis for AlGaN thin films ($|b_c| = 5.185$ Å) can calculate screw dislocation density) and Fig. 23(b) determines the physical sizes of the columnar structure, $L_{\perp}$ and strain component, $\varepsilon_{\perp}$. All of the mosaic structure factors of AlGaN samples measured from HRXRD are summarized in Table 6.

3.1.2. *SR-XAFS spectra of $Al_xGa_{1-x}N$*

Here, the local electronic structure of $Al_xGa_{1-x}N$ materials is studied by SR-XAFS having various Al compositions at the Ga K-edge with an incident photon direction perpendicular to the surface of the samples

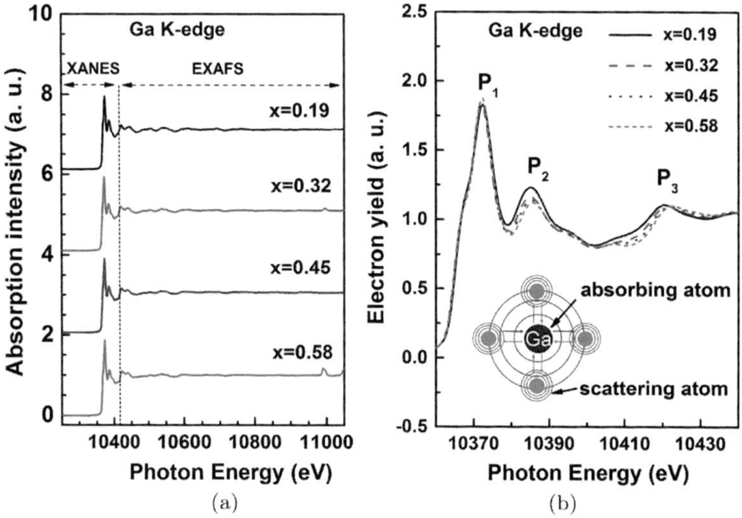

Fig. 24. (a) Ga K-edge absorption intensity versus the photon energy of X-ray for the $Al_xGa_{1-x}N$ epi-layers. (b) Measured Ga K-edge XANES spectra for $Al_x Ga_{1-x}N$ epilayers. The inset shows the schematic diagram of Ga atom which absorbs X-ray and is surrounded by four N atoms. (Reprinted from Ref. [18] with permission from Springer.)

(see Fig. 24(a)). We noticed that the Ga absorption edges for all the samples are nearly 10,365 eV. Generally, due to the electron dipole transitions from initial atomic orbital states with l character (where l is the orbital angular momentum of electron) to final states in the conduction band (or to states of anti-bonding molecular orbitals in the case of molecules) with $l \pm 1$ character,[42] the part of the XAFS spectra extended approximately 70 eV above the absorption edge is the X-ray absorption near-edge structure (XANES) spectra. XANES spectra can be used to probe the conduction band and local electronic structure of the crystalline materials, as a result of the dependence of the XANES spectra on the material and the local environment around the central atom which absorbs X-ray. Figure 24(b) displays the measured Ga K-edge XANES spectra for $Al_x Ga_{1-x}N$ epilayers. According to the dipole transition selection rule, the Ga K-edge XANES spectra of $Al_xGa_{1-x}N$ epilayers indicates the electron transitions from Ga *1s* state to empty *4p* states. The inset in Fig. 24(b) shows the schematic diagram of Ga atom which absorbs X-ray and is surrounded by four N atoms. Obviously, local electronic structure and the chemical state of the central Ga atom absorbing X-ray make multiple-scattering of photon-electron happen, which generate some feature peaks in the XANES spectra (see features P_1, P_2 and P_3 shown in Fig. 24(b)). It should be emphasized that, due to the lattice differences between AlN and GaN, the feature P_3 exhibits an obvious energy shift.

Removing the background by using ATHENA program, the K-space oscillation signals from the EXAFS spectra are extracted, as shown in Fig. 25(a).[43] After Fourier transform in a suitable K range (from 2.635 to 11.519 Å^{-1}), the R-space signals shown in Fig. 25(b) can be obtained. The best fitted results indicated by blue circles are also shown in Fig. 25(b). Clearly, the peaks amplitudes or positions of the Ga–N-related peaks for all the $Al_xGa_{1-x}N$ epilayer samples are almost similar. In contrast,

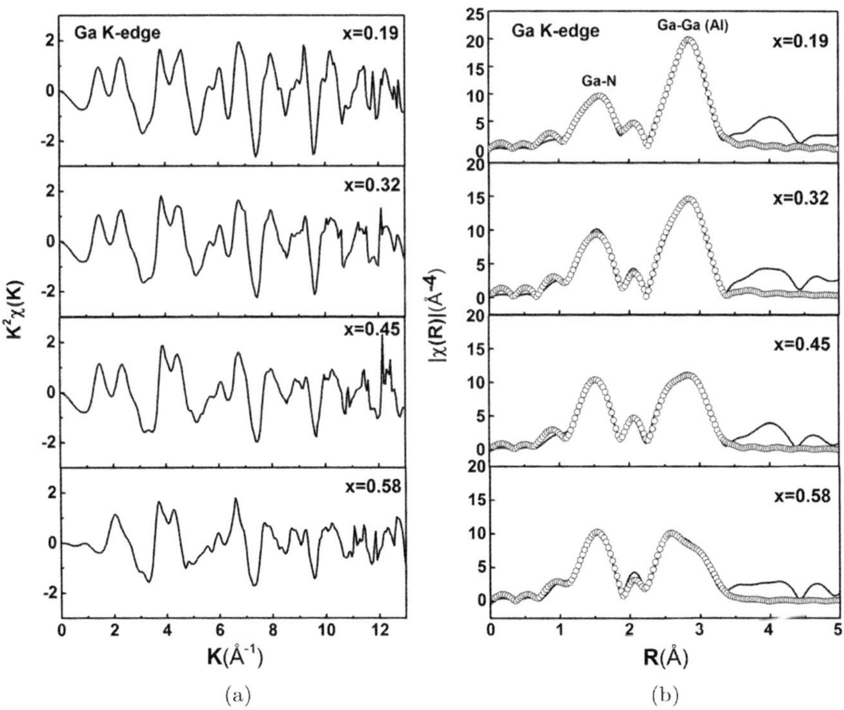

Fig. 25. (a) The Ga K-edge oscillation $k^2\chi(k)$ of $Al_xGa_{1-x}N$ epi-layers; (b) magnitude of the Fourier transforms of $k^2\chi(k)$ of (a) (indicated with solid line) as a function of spacing R together with the best fitted results (Reprinted from Ref. [18] with permission from Springer.)

both peak amplitude and position of the large peak (Ga–Ga-related peak which is attributed primarily to the second shell backscattering) vary with increasing Al composition. That is because the Al atom has a smaller radius than the Ga atom, which indicates that it has smaller backscattering amplitude. In order to analyze this case quantitatively, in the fitting, a Debye–Waller factor (the relative mean square deviation in bond length, σ^2) is adopted for the Ga–Ga (Al) bond as a function of Al composition. Moreover, to extract the interatomic distance between the Ga atom and its neighbors, a structural model based on the wurtzite GaN structure was built to fit the EXAFS data in the R range of 1.005–3.324 Å using the IFEFFIT package. Detailed fitting procedure of the spectra can be found in Ref. [18].

Figure 25(b) shows the fitting curves (the blue circles), which are in good agreement with the experimental data. It should be noted that the intensities of Ga–Ga- and Ga–Al-related peaks mainly come from the three nearest N atoms sitting on three equivalent tetrahedral positions and the twelve next-nearest Ga and Al atoms. Moreover, the EXAFS oscillation is mainly caused by the backscattering of the electrons from non-coaxial atoms in principle because the incident direction of X-ray was along the *c*-axis of the samples in the EXAFS measurements. It is known that Ga-N bond length and Ga–Ga (Al) bond length are the distances between the central Ga atom and its three nearest N atoms or its twelve nearest in-plane Ga or Al atoms sitting on three equivalent tetrahedral positions. From Fig. 25(b), the best fit results for the interatomic distances are obtained and summarized in Table 7.

Table 7. Fitting parameters of EXAFS data for all the $Al_xGa_{1-x}N$ epi-layers. (Reprinted from Ref. ..[18] with permission from Springer.)

Calculated Results	$Al_xGa_{1-x}N$ epi-layers			
	x = 0.19	$x = 0.32$	$x = 0.45$	$x = 0.58$
Ga-N (Å)	1.9579(5)	1.9504(1)	1.9528(6)	1.9542(1)
Ga-N(Al) (Å)	3.1697(8)	3.1241(6)	3.1130(7)	3.1086(2)
σ^2 (Å^2)	0.00044	0.00052	0.00073	0.00085
R-factor*	0.00641	0.00176	0.00952	0.00566

R-factor* is the parameter to evaluate the fitting quality.

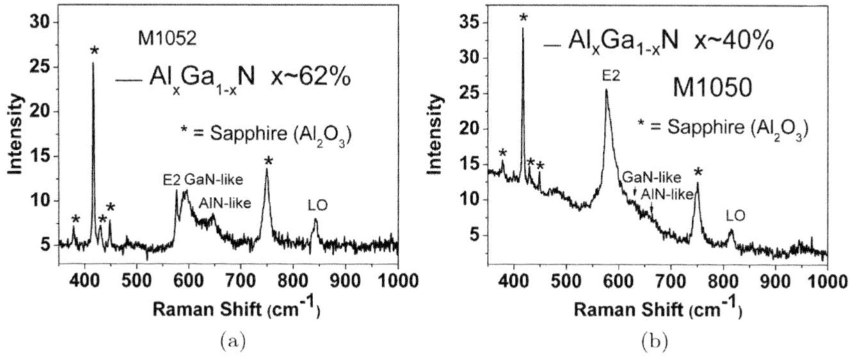

Fig. 26. 532 Raman spectra of two $Al_x Ga_{1-x}N$/Sapphires (a) M1052 ($x = 55\%$) (b) M1050 (38%). (Reprinted from Ref. [44] with permission from SPIE.)

3.1.3. *Raman scattering spectra of* $Al_x Ga_{1-x}N$

Figure 26 shows the Raman spectra of two AlxGa1-xN/sapphire samples. Four major bands from AlGaN alloys are observed: AlN-like and GaN-like A1 longitudinal optical (LO), and E2 transverse optical (TO) phonon mode, respectively, plus several peaks emanating from the substrate. When Al composition increases, the sample presents a higher free carrier concentration, i.e., a higher doping level.[44] Due to the deformation-potential mechanism and the electro-optic mechanism, affected by higher free carriers through their interactions, the lattice vibration will be stronger, which causes a stronger coupling of LO-phonon and plasma. So, it is observed that LO mode shifts to the higher energy side and the intensity of E2 Mode becomes weaker as the content of Al increases from 38% (Fig. 26(b)) to 55% (Fig. 26(a)).

3.1.4. *Temperature-dependent photoluminescence of* $Al_x Ga_{1-x}N$

Figures 27 (a–c) show the temperature-dependent photoluminescence (PL) spectra of the $Al_{0.2}Ga_{0.8}N$ samples with different In flow rates (M1710, In = 0 ccm; M1712, In = 50 ccm; M1714, In = 500 ccm). Label H and L indicat the high energy peak and low energy peak, respectively. To enable the low energy peak clearer, the drawing is enlarged (the inset). It is noted that the emission peak L of the AlGaN presents an anomalous temperature-dependent emission behavior (specifically an S-shaped shift, i.e. red-blue-red shift) with increasing temperature. This behavior is conventionally

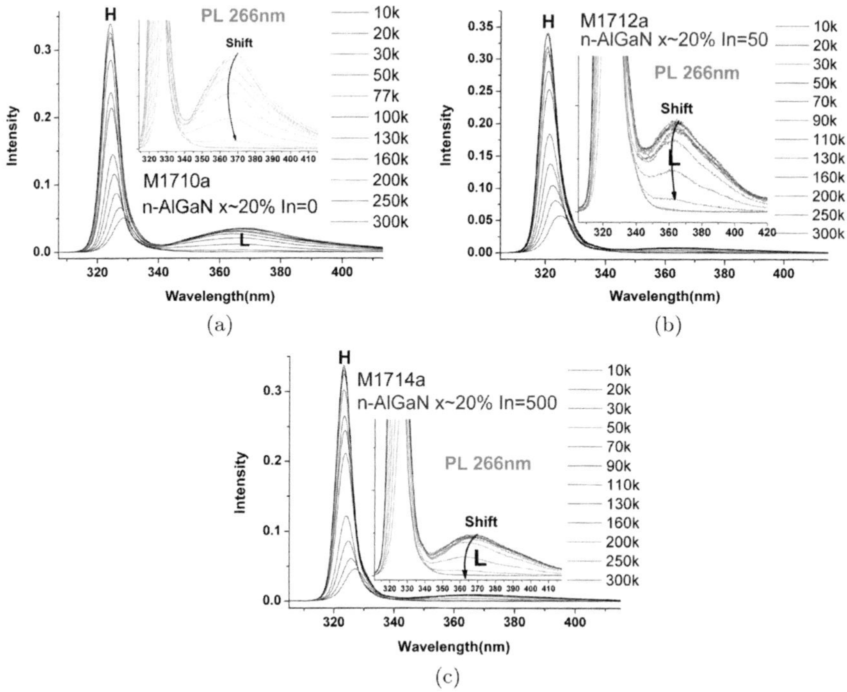

Fig. 27. PL spectra, excited by 266 nm, for the n-AlGaN/sapphire thin films with different In flow rates in the temperature range from 10 to 300 K. (a) M1710, (b) M1712 and (c) M1714. (Reprinted from Ref. [45] with permission from SPIE.)

interpreted as a sign of localization due to inhomogeneity of the potential and the carrier localization in its band-tail states such as the fluctuations in Al composition, layer thickness and the defects in the quantum wells.[45] Therefore, it is important to analyze how the behavior and these effects vary with emission energy.

Moreover, the anomalous temperature-dependent emission behaviors (this is a fingerprint of the excitation localization effect) are somewhat different for the $Al_{0.2}Ga_{0.8}N$ samples with different In flow rates. For the samples M1710, with temperature increasing, the emission energy increases from about 10 K to around 100 K, and then decreases up to room temperature (RT). In the cases of the sample M1712 and M1714, the emission energy increases from about 10 K to around 130 K, and then decreases up to RT. It should be noted that, when the temperature increases, the peak L only

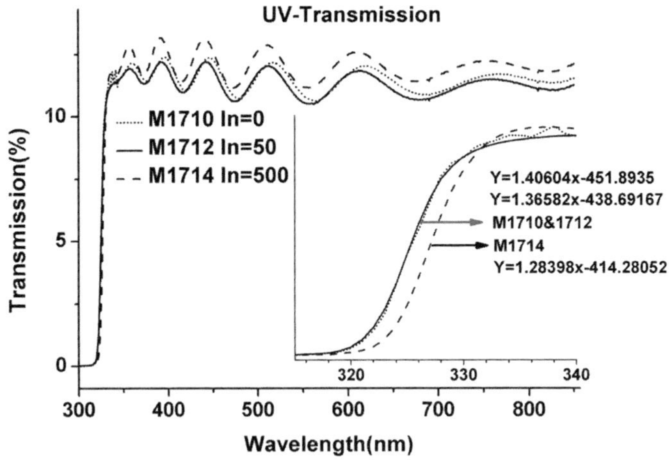

Fig. 28. Optical transmittance spectra of M1710, M1712 and M1714. (Reprinted from Ref. [45] with permission from SPIE.)

exists from 10 K to 200 K and the peak H shifts to the lower energy and its PL intensity decreased. Actually, the peak H is contributed to the content of Al in AlGaN/Sapphire thin film structure.

3.1.5. *Optical transmission*

Figure 28 shows the optical transmission spectra and the first order derivatives of the same $Al_xGa_{1-x}N$ films as Fig. 27 indicates. It is found that the optical transmittance of $Al_xGa_{1-x}N$ films is above 10% from UV wavelength to visible wavelength (300–800 nm). Moreover, these three samples nearly have a similar cut-off wavelength of transmission spectra, indicating that their energy band gaps are the same (near 320 nm, 3.875 eV), which corresponds to the PL result measured at RT. As a result of their Al compositions are a little different while In flow rate increased from 0 to 500 ccm, it can be observed that the spectrum of sample M1714 gradually shifted to the lower energy side (see the inset in Fig. 28).

3.2. Luminescence studies of $In_xGa_{1-x}N$

Figures 29(a) and (b) show the PL spectra of $In_xGa_{1-x}N$ films with different In contents (0.02, 0.04, 0.11, 0.15, 0.30 and 0.33) at 300 and 12 K, respectively. In Fig. 29, the peaks at 300 K are estimated to be at 3.35, 3.29,

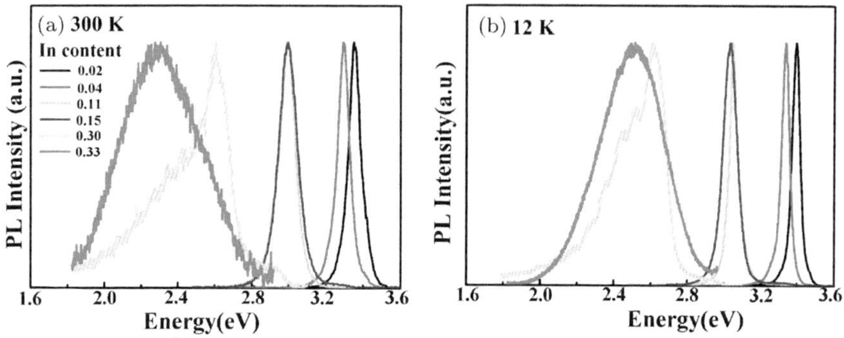

Fig. 29. PL spectra for the $In_xGa_{1-x}N$ films with different In contents measured at: (a) 300 K, and (b) 12 K. (Reprintedfrom Ref. [46] with permission from Elsevier.)

2.99, 2.99, 2.59, 2.31 eV, respectively–the peaks shift is caused by increasing In contents. Moreover, the PL shape tends to become asymmetric while the FWHM of the PL line shape increases. Again, there exists larger alloy compositional fluctuation in the In-rich InGaN films, which cause the phase separation and the excitonic localization effect.[46] This clearly explains as to why the phase separation in the InGaN films with higher In content plays a pronounced role in nitride-based material LEDs. In additions, the samples contain regions with higher In content (smaller band gap) embedded in regions with higher Ga content (higher band gap), which indicates that the broadening and asymmetry of the main PL shape originate from the accompanying of several lower PL peaks (as clearly indicated by the PL feature for In content of 0.3 and 0.33 in Fig. 29).[46]

In order to quantitatively investigate the relationship between the exciton localization effect and In content, the PL peak energy of InGaN samples with different In content is depicted as a function of temperature in Fig. 30. Obviously, the PL peak for the InGaN with In content of 0.02 exhibits a typical redshift with increasing temperature, as described by the Varshni expression[47]:

$$E(T) = E(0) - \frac{\alpha T^2}{T + \beta} \tag{4}$$

where $E(0)$ is the exciton energy at $T = 0\,\text{K}$, and α and β are fitting parameters. For example, setting $E(0) = 3.39\,\text{eV}$, $\alpha = 0.53\,\text{meV/K}$, and

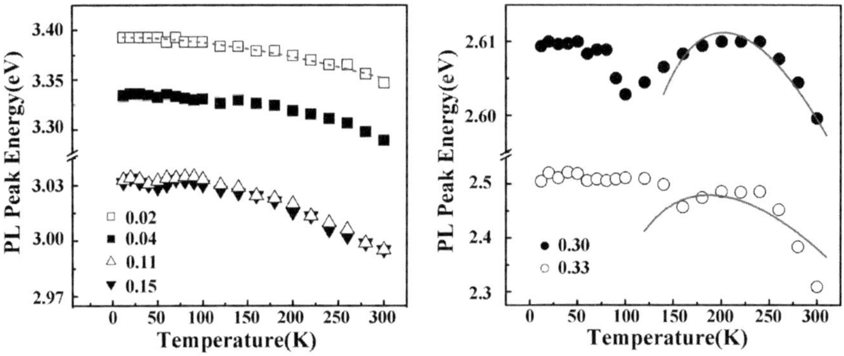

Fig. 30. Temperature dependence of PL peak position for the $In_x Ga_{1-x}N$ films with different In contents. (Reprinted from Ref. [46] with permission from Elsevier.)

$\beta = 850\,\text{K}$, one can get a good fitting result for the InGaN with In content of 0.02, as shown by the red dashed line in Fig. 30. However, for the InGaN with In content of 0.11 and 0.15, the PL peak redshifts slightly in the low temperature region, and then blueshifts in the intermediate temperature region, and again redshifts up to room temperature (the same PL behavior of AlGaN in Sec. 3.1.4), which indicates that the Varshni relation doesn't work as the In composition is larger than 0.11. The anomalous temperature-dependent behavior has been referred as S-shaped behavior.[47] Especially, when the In content increases, this behavior becomes more evident (as clearly indicated by the PL feature InGaN with In contents of 0.30 and 0.33 in Fig. 30).

Generally, S-shaped behavior (red-blue-red shift) in PL peak energy is interpreted by the exciton dynamic as a function of temperature. At first, a very low temperature froze the excitons out at the local minima. When the temperature increases, the excitons acquire sufficient thermal energy. They can overcome the small potential barriers in the local potential and relax to the absolute potential minimum, thus the PL peak energy redshifts. Then, the further increase in temperature can thermally transfer the excitons to higher energy states until reaching the conduction-band edge. The spectra exhibit blueshift. Again, increasing the temperature further, spectral redshift appears again due to the exciton localization is prevented by the thermal ionization effect.[47] Thus, the band gap as a function of temperature given by the Varshni's expression needs modification to include the exciton localization effect and can be written in the context of Eliseev's model as[48,49]:

$$E(T) = E(0) - \frac{\alpha T^2}{T+\beta} - \frac{\sigma^2}{K_B T} \tag{5}$$

where K_B is the Boltzmann's constant and σ is the excitonic localization energy. Therefore, after adopting appropriate σ values (about 29 and 65 meV for $In_xGa_{1-x}N$ with $x = 0.3$ and 0.33, respectively), the experimental PL peak S-shape behavior for InGaN with In-rich contents can be well fitted by Eq. (5), as plotted by the solid line in Fig. 30. It is emphasized that larger σ value means stronger exciton localization. The study shows large compositional fluctuation in In-rich InGaN — leading to phase separation, resulting in the observed S-shaped PL emission behavior.

4. The Group-III Nitrides Quaternary Alloys InAlGaN

4.1. *HRXRD spectra of $In_{1-x-y}Al_xGa_yN$*

Three $In_{1-x-y}Al_xGa_yN$ quaternary alloys with different In and Al compositions (listed in Table 8) were characterized by the HRXRD measurements to carry out the (004) $\omega - 2\theta$ scan, as shown as in Fig. 31. It is found that there exhibit two separated diffraction peaks associated with GaN and $In_{1-x-y}Al_xGa_yN$ for the M12 and M24 samples in Fig. 31(a), demonstrating that these quaternary $In_{1-x-y}Al_xGa_yN$ alloys were epitaxially grown on GaN. Moreover, one can observe that $In_{1-x-y}Al_xGa_yN$ peak become not quite evident as the Al to In ratio decreases. Especially, when the Al to In ratio decreases further to be around 4.9 (M28), only one GaN peak is observed in Fig. 31. In other words, the lattice of M28 matches to the underlying GaN layer, exhibiting the best structural and optical properties. Counter to that, the biggest lattice mismatch and strain exists in M12 — revealing the largest separated distance between GaN and InAlGaN peaks (dotted line in Fig. 31(a)) among the three samples.[50]

4.2. *Photoluminescence (PL) spectra of $In_{1-x-y}Al_xGa_yN$*

Figure 32 shows the varying-temperature PL spectra from the $In_{0.014}Al_{0.105}Ga_{0.881}N$ and the GaN substrate in the range of 10–300 K. At low temperature of 10 K, there exist two main emission features

Table 8. Sample number, composition, ratio of Al to In, the values of linewidth C, and correlation length L for the AlGaN-like A1 (LO) phonon modes in our InAlGaN/GaN heterostructure quaternary alloys. (Reprinted from Ref. [50] with permission from American Institute of Physics.)

Sample	In(%)	Al(%)	Al *to* In	$\Gamma(cm^{-1})$	L(Å)
M12	1.38	10.525	7.6	15.0	62
M14	2.73	13.977	5.1	14.0	66
M28	1.64	8.012	4.9	13.0	69

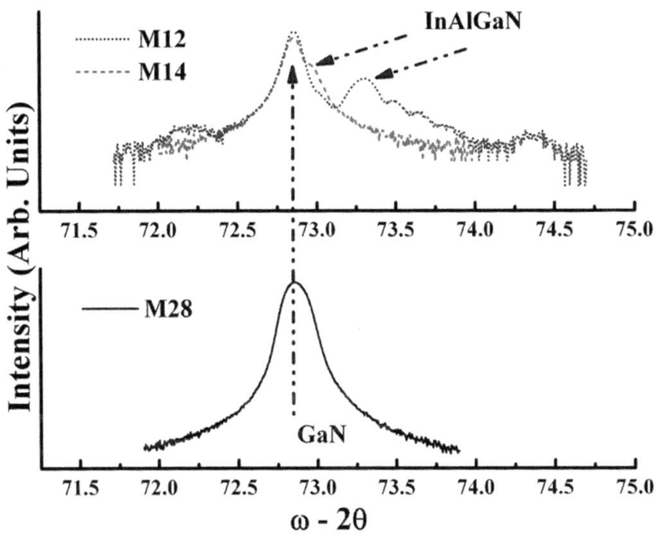

Fig. 31. Three $In_{1-x-y}Al_xGa_yN/GaN$ samples were characterized by HRXRD to carry out the (004) ω-2θ scan. (Reprinted from Ref. [50] with permission from American Institute of Physics.)

at around 3.60 eV and 3.46 eV, which correspond to the onset of the $In_{0.014}Al_{0.105}Ga_{0.881}N$ exciton band and the GaN substrate absorption, respectively. However, by increasing the temperature, the intensity of the left emission peak at the low-energy side decreases quickly and nearly becomes a shoulder at the temperature of 300 K when the right emission peak at the high energy side almost keeps its intensity unchanged.

Furthermore, the same as the case in Secs. 3.1.4 and 3.2, the positons of PL peak energy of $In_{0.014}Al_{0.105}Ga_{0.881}N$ thin film shift with the varying

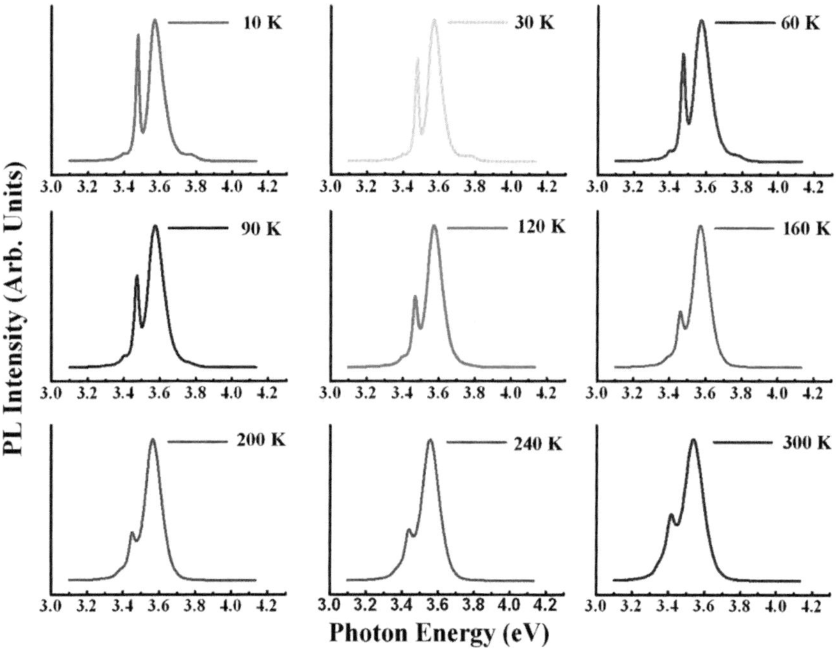

Fig. 32. PL spectra from the $In_{0.014}Al_{0.105}Ga_{0.881}N$ and the GaN substrate revealing the significant emissions at several different temperatures in the range of 10–300 K. (Reprinted from Ref. [51] with permission from Elsevier.)

temperature. Following the same approaches as used in Secs. 3.1.4 and 3.2, the shifts of the PL peak energy (EPL) of $In_{0.014}Al_{0.105}Ga_{0.881}N$ thin film are investigated as a function of the temperature varying from 10 K to 300 K in Fig. 33. Obviously, the anomalous luminescence peak energy called S-shaped behavior appears in Fig. 33, i.e., the EPL first redshifts, then blueshifts and further shows a monotonous redshift above 100 K. This behavior is quite different from that of the GaN band gap (shown as the inset). It is a typical behavior in disordered systems and originates from alloy potential fluctuations in the layers. In other words, when the distribution of InGaN-like clusters is inhomogeneous, larger InGaN-like cluster has the lower energy state, which will cause the stronger localized effect. The localized states will lead to the S-shaped behavior.[52]

It is useful to introduce the detailed process of the PL S-shaped emission shifts of InAlGaN even though Section 3.2 has the similar description for

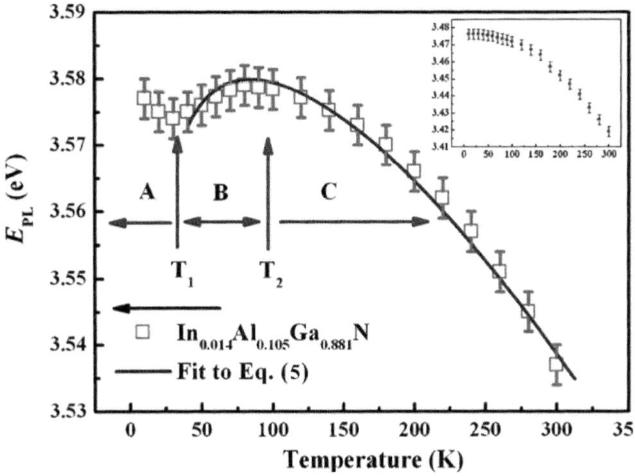

Fig. 33. The PL peak energy as a function of temperature for $In_{0.014}Al_{0.105}Ga_{0.881}N$ thin film (open squares) and the inset is the temperature dependence of the GaN band gap (closed circles). The solid curve is least squares fit to Eq. (5) according to Eliseev's model. (Reprinted from Ref. [51] with permission from Elsevier.)

InGaN. In this case, Fig. 33 are divided into three temperature regions labeled A, B and C to correspond to redshift, blueshift and redshift behavior of the EPL, respectively (T_1 and T_2 are the terminal temperatures of regions A and B). In region A from 10 to 30 K ($T < T_1$, T_1 is the temperature of maximum exciton localization), the charge carriers are initially trapped in the local minima of the potential. However, the rising temperature activates weak localized charge carriers to recombine nonradiatively or transit to the deeper localized states, resulting in a redshift of the PL peak. At higher temperatures of region B from 30 to 100 K, more and more thermal charge carriers occupy at higher-energy states, thus the PL peak begins to shift to higher energies and has a maximum at $T = T_2$. The temperature corresponding to the onset of the exciton delocalization.[51,53] In region C (above 100 K), by increasing the temperature further, the exciton localization energy becomes overshadow by the temperature-induced band shrinkage. Following the temperature dependence of the excitonic band gap, the PL peak will shift monotonously to lower energies (a redshift). Similar to Section 3.2, the temperature dependence of the PL peak position for $In_{0.014}Al_{0.105}Ga_{0.881}N$ sample can be described well by Eliseev's model

(Eq. (5), $E(T) = E(0) - \alpha T^2/(T+\beta) - \sigma^2/K_B T$),[53,54] as shown in Fig. 33. Here, the fitting parameters are adopted as follow: the exciton localization energy σ is around 10 ± 20 meV; the temperature coefficient $\alpha$is around 3.3×10^{-4} eV/K and the parameter related to the Debye temperature, β is about 730 ± 30 K.

Figure 34 shows the PL spectra of three quaternary InAlGaN samples with different Al and In compositions at 10 K and 300 K. Al (aluminum) is to be 13.97% for M14, 10.52% for M12, and 31.50% for M04, respectively, while In (indium) is to be 2.73% for M14, 1.38% for M12, and 3.50% for M04, respectively. In Fig. 34, the short-wavelength PL signals are the main PL features, representing the InAlGaN-based transition. Moreover, for the samples of M14 and M12, there are GaN-related features located at about 364 nm appearing on the right side of the InAlGaN-based PL peaks. Especially, the GaN-related PL peak of M12 becomes stronger from 10 K to 300 K while M14 exhibits opposite behavior. As often observed in the nitride- based semiconductor alloys, the PL emission bands have been

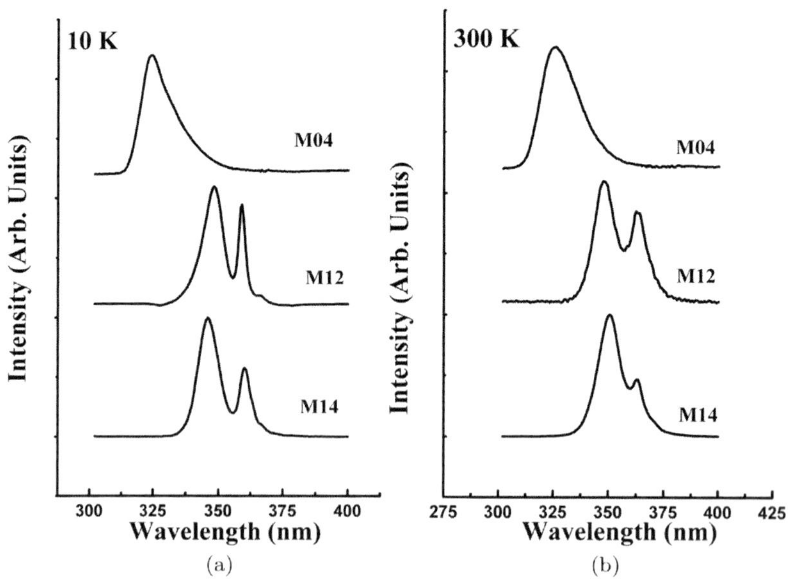

Fig. 34. PL spectra for the three quaternary $In_{1-x-y}Al_xGa_yN$ samples, namely M14, M12, and M04 as labeled in (a) for 10 K, and (b) for 300 K, respectively. (Reprinted from Ref. [55] with permission from Elsevier.)

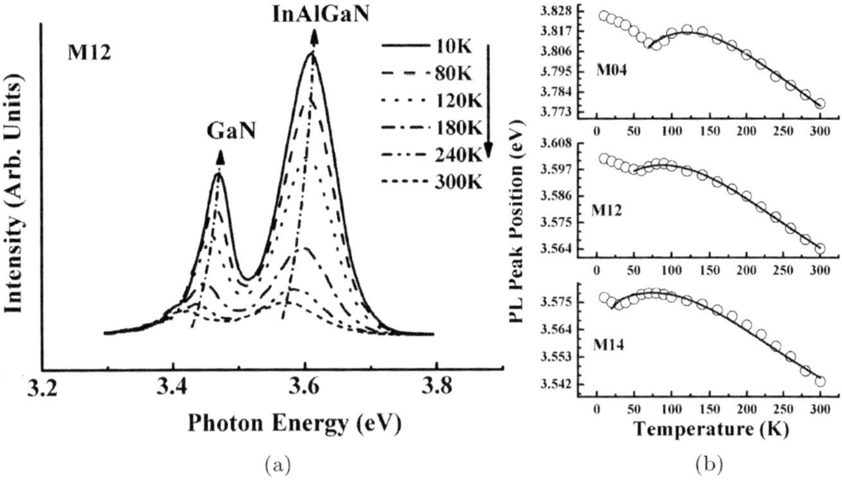

Fig. 35. The typical PL spectra of sample M12 at several different temperatures from 10 to 300 K is shown in (a), while the PL peak positions (open circles) as a function of temperature are shown in (b), respectively. (Reprinted from Ref. [55] with permission from Elsevier.)

attributed to a recombination of localized excitons.[55–59] Actually, it is also observed that the InAlGaN-based PL emission properties of these three InAlGaN samples are different.[55,60]

Figure 35(a) shows the temperature dependent PL spectra from 10 to 300 K of a quaternary alloyed $In_{1-x-y}Al_xGa_yN$ sample labeled M12, which has been described as a function of temperature (opened circles) in Fig. 35(b). Obviously, from Fig. 35(b), all the three $In_{1-x-y}Al_xGa_yN$ epilayers show a similar anomalous temperature-dependent emission behavior as Fig. 33 (S-shaped shift, i.e. red-blue-red shift), providing more evidences of exciton localization in potential minima caused by alloy fluctuations.[57,59] It is also observed that their PL peak position displaying S-shaped phenomena are different: ~30 K for M14, ~60 K for M12 and ~80 K for M04, contributing to their different localization degree.[57,59] Actually, while increasing Al/In ratio, the effect of localization of excitons becomes stronger and the S-curve behavior becomes more evident, as displayed in Fig. 35(b). Thus, quaternary $In_{1-x-y}Al_xGa_yN$ sample marked M04 shows the biggest S-curve shift because it has the biggest Al/In ratio.

To study the exciton localization effects in the $In_{1-x-y}Al_xGa_yN$ samples, based on the band-tail-filling model, a modified Bose–Einstein expression can be used to fit the PL peak position, which is somewhat different from Eq. (5) of the Eliseev's model but still takes the exciton localization effect into account[47,61–63]:

$$E_g(T) = E_B - a_B \times \left\{1 + \frac{2}{\left[\exp\left(\frac{\Theta_B}{T}\right) - 1\right]}\right\} - \frac{\sigma_E^2}{K_B T} \tag{6}$$

where $E_B = E_0 + a_B$(E_0 is the band gap at 0 K), a_B is the magnitude of the limiting slope of the corresponding $E_g(T)$ curve attributed to the strength of the electron/exciton- phonon interaction, K_B is the Boltzmann constant, $\Theta_B = \hbar\omega_{\text{eff}}/K_B$ corresponds to the average phonon temperature associated with the single ('phantom') oscillator, σ_{E} is defined to be close to the localization energy caused by the potential fluctuations.[55] Then, the redshift of PL peak with decreasing temperature is given by $\sigma_E^2/K_B T$.[64] Based on Gaussian distribution of localized states, Fig. 35(b) shows the corresponding results by fitting Eq. (6) in an appropriate range of temperature, which is valid above certain temperature and obtain the estimated values of E_B, a_B, Θ_B and σ_{E} as listed in Table 9 with the representative error bar.[47,61–63] From the results of Table 9, σ_{E} varies in the range from 4 ± 1 to 13 ± 1 meV as the a_B increases from 82 ± 2 to 156 ± 2 meV, and the Θ_B increases from 450 ± 25 to 530 ± 25 K. Obviously, the $In_{1-x-y}Al_xGa_yN$ samples with higher Al/In ratio has larger values of a_B, Θ_B and σ_{E}. It is simply because there exists stronger electron/exciton-phonon interaction enhanced by more alloy disorder (implying stronger exciton localization, i.e., larger σ_{E}) in $In_{1-x-y}Al_xGa_yN$ samples with higher Al/In ratio, which might lead to a higher value of a_B and Θ_B.[47]

Table 9. With the representative error bar, values of the parameters E_B, a_B, Θ_B, σ_E are obtained from a fit to Eq. (6). (Reprinted from Ref. [55] with permission from Elsevier.)

Sample Number	M14	M12	M04
E_B	3.58 ± 0.01	3.60 ± 0.01	3.84 ± 0.01
a_B	82 ± 2	94 ± 2	156 ± 2
Θ_{B}	450 ± 25	480 ± 25	530 ± 25
σ_B	4 ± 1	6 ± 1	13 ± 1

4.3. *Raman spectra of $In_{1-x-y}Al_xGa_yN$*

Figure 36 shows room temperature Raman spectra for three same $In_{1-x-y}Al_xGa_yN$ samples grown on GaN as shown in Fig. 31 over the range of 450–800 cm^{-1}. From Fig. 36, it is observed that there are three resolved phonon structures observed in each Raman spectrum of $In_{1-x-y}Al_xGa_yN$ quaternary alloys: the GaN E_2 (high) mode located at the region of 550–575 cm^{-1}, GaN A_1 (LO) mode is located 735 cm^{-1}, the AlGaN-like A_1 (LO) mode is located 750 cm^{-1} (described by the Gaussian fit (dotted line)

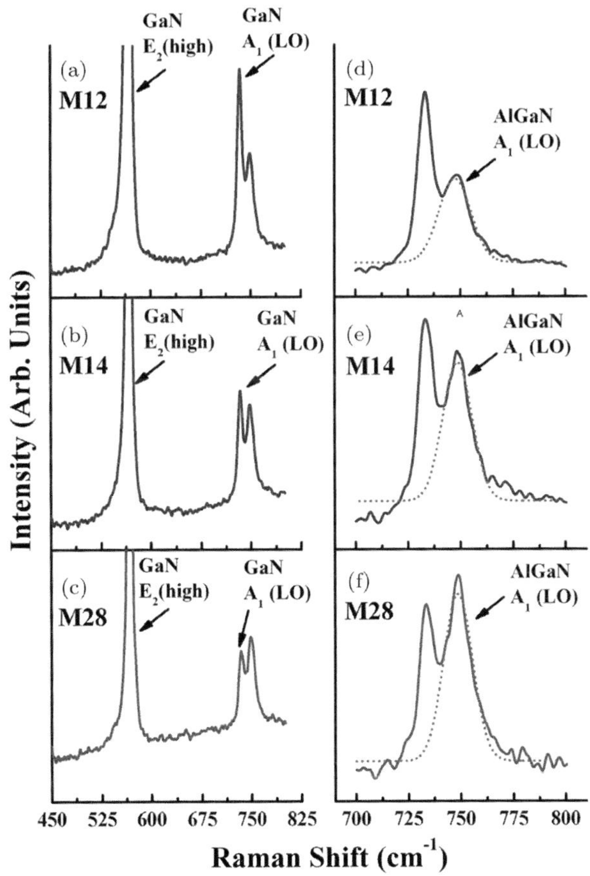

Fig. 36. Room temperature Raman spectra of the different InAlGaN samples over the range of 450–800 cm^{-1}. (Reprinted from Ref. [50] with permission from American Institute of Physics.)

of the Raman line shape and the spectra are displayed in (d)–(f) of Fig. 36). It should be noted that, both of the E_2 (high) mode of GaN and A_1 (LO) mode of GaN originate from the substrates, while the A_1 (LO) modes of AlGaN-like are similar to the Al-rich A_1 (LO) phonon lines from the earlier reports.[65–70] In additions, due to the In composition is much less than Al composition in these quaternary samples, the A_1 (LO) mode of InGaN-like, expected at around 700 cm^{-1} is not detectable.

Raman spectra truncated in the range of 675–800 cm^{-1} of the AlGaN-like A_1 (LO) modes from these three samples are shown in Fig. 37. As displayed in Fig. 37, increasing the ratio of Al to In leads to asymmetric

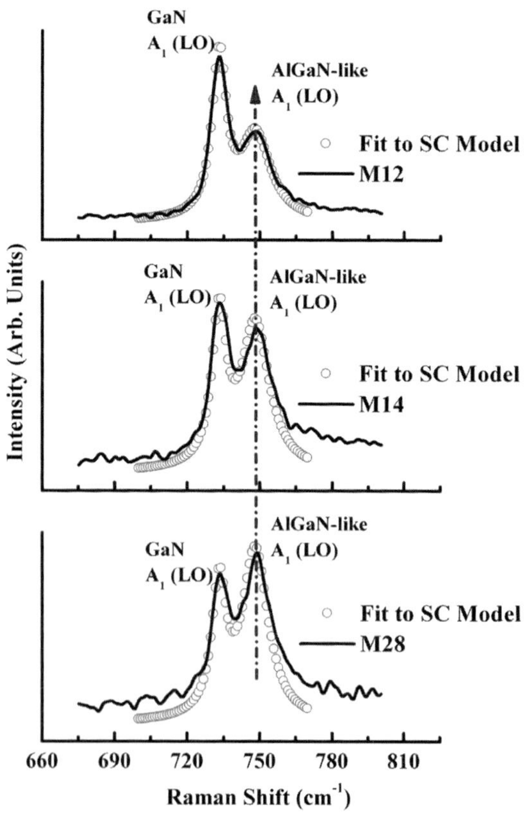

Fig. 37. Room temperature Raman spectra of the AlGaN-like A_1 (LO) modes from our quaternary samples M12, M14, and M28 are in the range of 675–800 cm^{-1}, respectively. (Reprinted from Ref. [50] with permission from American Institute of Physics.)

broadening of the spectra line shapes and redshift of the AlGaN-like A_1 (LO) mode. To the best of our knowledge, the process of atom substitution in alloy semiconductors results in topological and structural disorders. For allowed phonons, these disorders basically break the translational symmetry and cause $q \neq 0$ phonons affect the Raman line shape, corresponding to the so-called finite-size effects.[47] Due to the alloy compositional fluctuations, the correlation length of a propagating phonon is finite. Then, the spatial correlation model (SC) can be employed to study this line shape asymmetry,[68] as shown as the opened circles in Fig. 37, which has a good agreement with the experimental data.

5. Summary

In conclusion, we have reported the summary of our comprehensive investigations on the structural and optical properties of III-nitride semiconductors including the binary (GaN, AlN, InN), ternary $In_xGa_{1-x}N$, $Al_xGa_{1-x}N$, and quaternary InAlGaN alloy materials by exploiting a set of multiple spectroscopic techniques.

Synchrotron radiation X-ray absorption fine structure (SR-XAFS) approach has been a useful tool for extracting information about III-nitride materials of their crystal structure, electronic behavior including the lattice deformation and atomic bond length. As noted in Secs. 2.1 and 3.1.2, the lattice constants of different GaN films can be deduced from the interatomic distances in the second coordination shell around the Ga atom. In $Al_xGa_{1-x}N$ epilayers, the first shell Ga–N bond length exhibits only a weaker composition x dependence, while the second shell Ga-Ga bond length is strongly dependent on x and it is significantly longer than that of Ga-Al bond.

X-ray diffraction (XRD) is a non-destructive and rapid characterization tool for assessing the quality of semiconductor thin films and devices. In Secs. 2.2.1, 3.1.1, we have succinctly demonstrated the use of XRD for obtaining the growth orientation of InN films grown on LAO substrate which is affected by the growth conditions. We estimated the c- and a-lattice constants of various B-doped AlN films, Al compositions and the threading dislocation density in AlGaN epilayers, lattice mismatch of quaternary InAlGaN layers epitaxially grown on GaN, etc.

Raman scattering spectroscopy has been an equally effective and reliable method for probing the vibrational behavior and obtaining the informations on crystal phase, orientation and lattice dynamics of perfect/imperfect III-nitride materials. By using Raman scattering (cf. Secs. 2.2.4, 3.1.3, 4.3) spectroscopy and with a careful theoretical analysis we have demonstrated the effects of disorders or finite-size atoms responsible for the redshift and broadening of the Raman E_2 mode peak in GaN and InAlGaN films grown on different substrates, the phonon anisotropic properties of GaN wurtzite crystal and the LO-plasmon coupled modes of InN, AlGaN samples.

Photoluminescence is also an effective tool for obtaining the emission band gap of III-nitrides. The temperature dependent PL study revealed peaks of binary materials (cf. Sec. 3.2, 4.2) exhibiting redshifts while in the ternary and quaternary alloys with different compositions it exhibited different S-shaped behavior. The results are used to analyze exciton localization effects caused by the alloy compositional fluctuations.

By exploiting various characterization methods, we have noticed that the III-nitrides exhibit unique structural and optical properties — primarily affected by their growth conditions. We hope that the overview presented here on spectroscopic investigations is valuable for better understanding the optical, structural and luminescent traits in providing ways to improve the growth of these materials, which have demonstrated potential applications in the field of energy-optoelectronics technology needs.

Acknowledgments

This work was supported by the National Natural Science Foundation of China (NO. 61367004, NO. 11604058), and the Guangxi Natural Science Foundation (No. 2013GXNSFFA019001). One of the authors DNT wishes to thank Dr. Deanne Snavely, the Dean College of Natural Science and Mathematics for travel support and to the Graduate School at Indiana University of Pennsylvania for the Innovation grant.

References

1. Zhe Chuan Feng (ed.), *III-Nitride Semiconductor Materials*, Imperial College Press, London, 422pp (2006).
2. Zhe Chuan Feng(ed.), *III-Nitride Devices and Nanoengineering*, Imperia College Press, London, 462pp (2008).

3. Shuji Nakamura, Stephen Pearton, and Gerhard Fasol(ed.), *The blue laser diode: the complete story*, Springer Berlin, 343pp (2013).
4. Farid Medjdoub (ed.), *Gallium Nitride (GaN): Physics, Devices, and Technology*, CRC Press, Berlin, 372pp (2015).
5. Yong Huang, Andrew Melton, Balakrishnam Jampana, Muhammad Jamil, Jae-Hyun Ryou, Russell D Dupuis, and Ian T Ferguson, "Compositional instability in strained InGaN epitaxial layers induced by kinetic effects," *Journal of Applied Physics*, **110**, 064908 (2011).
6. M. Łopuszyński and Jacek A Majewski, "Composition dependence of elastic constants in wurtzite AlGaInN alloys," *Journal of Applied Physics*, **111**, 033502 (2012).
7. R. Oliva, J. Ibáñez, Ramón Cuscó, R. Kudrawiec, J. Serafinczuk, O. Martínez, Juan Jiménez, M. Henini, C. Boney, and A. Bensaoula, "Raman scattering by the E2h and A1 (LO) phonons of $In_xGa_{1-x}N$ epilayers ($0.25 < x < 0.75$) grown by molecular beam epitaxy," *Journal of Applied Physics*, **111**, 063502 (2012).
8. A. M. Fischer, Y. O. Wei, F. A. Ponce, M. Moseley, B. Gunning, and W. A. Doolittle, "Highly luminescent, high-indium-content InGaN film with uniform composition and full misfit-strain relaxation," *Applied Physics Letters*, **103**, 131101 (2013).
9. S. Alajlouni, Z. Y. Sun, J. Li, J. M. Zavada, J. Y. Lin, and H. X. Jiang, "Refractive index of erbium doped GaN thin films," *Applied Physics Letters*, **105**, 081104 (2014).
10. Yuya Hayakawa, Tomonori Fukuno, Katsuto Nakamura, Hideto Miyake, Kazumasa Hiramatsu, and Yoichi Yamada, "Binding energy of localized biexcitons in AlGaN-based quantum wells," *Applied Physics Express*, **7**, 122101 (2014).
11. M. Gladysiewicz, L. Janicki, M. Siekacz, G. Cywinski, C. Skierbiszewski, and R. Kudrawiec, "Theoretical and experimental studies of electric field distribution in N-polar GaN/AlGaN/GaN heterostructures," *Applied Physics Letters*, **107**, 262107 (2015).
12. Hyun Jeong, Hyeon Jun Jeong, Hye Min Oh, Chang-Hee Hong, Eun-Kyung Suh, Gilles Lerondel, and Mun Seok Jeong, "Carrier localization in In-rich InGaN/GaN multiple quantum wells for green light-emitting diodes," *Scientific reports*, **5** (2015).
13. Avinash Patsha, Sandip Dhara, and A. K. Tyagi, "Localized tip enhanced Raman spectroscopic study of impurity incorporated single GaN nanowire in the sub-diffraction limit," *Applied Physics Letters*, **107**, 123108 (2015).
14. Marco A. Sousa, Teresa C. Esteves, Nabiha Ben Sedrine, Joana Rodrigues, Márcio B. Lourenço, Andrés Redondo-Cubero, Eduardo Alves, Kevin P. O'Donnell, Michal Bockowski, and Christian Wetzel, "Luminescence studies on green emitting InGaN/GaN MQWs implanted with nitrogen," *Scientific reports*, **5** (2015).
15. Shibin Krishna, Neha Aggarwal, Monu Mishra, K. K. Maurya, Mandeep Kaur, Geetanjali Sehgal, Sukhveer Singh, Nita Dilawar, Bipin Kumar Gupta, and Govind Gupta, "Epitaxial growth of high In-content In 0.41 Ga 0.59 N/GaN heterostructure on (11–20) Al 2 O 3 substrate," *Journal of Alloys and Compounds*, **658**, 470–475 (2016).
16. Xiao Meng, Lai Wang, Zhibiao Hao, Yi Luo, Changzheng Sun, Yanjun Han, Bing Xiong, Jian Wang, and Hongtao Li, "Study on efficiency droop in InGaN/GaN light-emitting diodes based on differential carrier lifetime analysis," *Applied Physics Letters*, **108**, 013501 (2016).

17. Takuo Sasaki, Fumitaro Ishikawa, and Masamitu Takahasi, "Anomalous lattice deformation in GaN/SiC (0001) measured by high-speed in situ synchrotron X-ray diffraction," *Applied Physics Letters*, **108**, 012102 (2016).
18. Shuchang Wang, Xiong Zhang, Muchi Liu, Bowei Wang, Zhe Chuan Feng, and Yiping Cui, "Study of lattice deformation and atomic bond length for $Al_xGa_{1-x}N$ epi-layers with synchrotron radiation X-ray absorption spectroscopy," *Journal of Materials Science: Materials in Electronics*, **25**, 4800–4805 (2014).
19. P. Perlin, C. Jauberthie-Carillon, J. P. Itie, A. San Miguel, I. Grzegory, and A. Polian, "Raman scattering and X-ray-absorption spectroscopy in gallium nitride under high pressure." *Phys. Rev. B*, **45**, 83 (1992).
20. Y. L. Wu, Z. C. Feng, J. F. Lee, W. Tong, B. K. Wagner, I. Ferguson, and Weijie Lu, "X-ray absorption and Raman study of GaN films grown on different substrates by different techniques," *Thin Solid Films*, **518**, 7475–7479 (2010).
21. P. Parayanthal and Fred H. Pollak, "Raman scattering in alloy semiconductors:" spatial correlation" model," *Physical review letters*, **52**, 1822 (1984).
22. Hung Chiao Lin, Zhe Chuan Feng, Ming Song Chen, Ze Xiang Shen, Ian T. Ferguson, and Weijie Lu, "Raman scattering study on anisotropic property of wurtzite GaN," *Journal of Applied Physics*, **105**, 036102 (2009).
23. Cheng Chen, Xiang Ping Shu, Hua Yang Sun, Zhi Ren Qiu, Ting Wei Liang, Li Wei Tu, and Zhe Chuan Feng, "Temperature Dependence of Raman Scattering in M-Plane GaN with Varying III/V Ratios," *Advanced Materials Research*, **602–604**, 1453–1456 (2013).
24. Ching-Lien Hsiao, Jr-Tai Chen, Hsu-Cheng Hsu, Ying-Chieh Liao, Po-Han Tseng, Yen-Ting Chen, Zhe Chuan Feng, Li-Wei Tu, Mitch M. C. Chou, Li-Chyong Chen, and Kuei-Hsien Chen, "m-plane (1010) InN heteroepitaxied on (100)-γ -LiAlO2 substrate: Growth orientation control and characterization of structural and optical anisotropy," *Journal of Applied Physics*, **107**, 073502 (2010).
25. G. Koblmüller, A. Hirai, F. Wu, C. S. Gallinat, J. S. Speck, G. D. Metcalfe, H. Shen, and M. Wraback, "Molecular beam epitaxy and structural anisotropy of m-plane InN grown on free-standing GaN," *Applied Physics Letters*, **93** (2008).
26. Melvin B. McLaurin, Asako Hirai, Erin Young, Feng Wu, and James S. Speck, "Basal plane stacking-fault related anisotropy in X-ray rocking curve widths of m-plane GaN," *Japanese Journal of Applied Physics*, **47**, 5429 (2008).
27. T. Metzger, R. Höpler, E. Born, O. Ambacher, M. Stutzmann, R. Stömmer, M. Schuster, H. Göbel, S. Christiansen, and M. Albrecht, "Defect structure of epitaxial GaN films determined by transmission electron microscopy and triple-axis X-ray diffractometry," *Philosophical magazine A*, **77**, 1013–1025 (1998).
28. C. Stampfl and Chris G. Van de Walle, "Energetics and electronic structure of stacking faults in AlN, GaN, and InN," *Physical Review B*, **57**, R15052 (1998).
29. Dmitri N. Zakharov, Zuzanna Liliental-Weber, Brian Wagner, Zachary J. Reitmeier, Edward A. Preble, and Robert F. Davis, "Structural TEM study of nonpolar a-plane gallium nitride grown on (1120) 4 H-SiC by organometallic vapor phase epitaxy," *Physical Review B*, **71**, 235334 (2005).
30. Yee Ling Chung, Xingyu Peng, Ying Chieh Liao, Shude Yao, Li Chyong Chen, Kuei Hsien Chen, and Zhe Chuan Feng, "Raman scattering and Rutherford backscattering

studies on InN films grown by plasma-assisted molecular beam epitaxy," *Thin Solid Films*, **519**, 6778–6782 (2011).
31. Y. P. Varshni, "Temperature dependence of the elastic constants," *Physical Review B*, **2**, 3952 (1970).
32. R. Pässler, "Temperature dependence of exciton peak energies in multiple quantum wells," *Journal of applied physics*, **83**, 3356–3359 (1998).
33. Devki N. Talwar, Ying Chieh Liao, Li Chyong Chen, Kuei Hsien Chen, and Zhe Chuan Feng, "Optical properties of plasma-assisted molecular beam epitaxy grown InN/sapphire," *Optical Materials*, **37**, 1–4 (2014).
34. J. Serrano, A. Bosak, M. Krisch, F. J. Manjón, A. H. Romero, N. Garro, X. Wang, A. Yoshikawa, and M. Kuball, "InN thin film lattice dynamics by grazing incidence inelastic X-ray scattering," *Physical review letters*, **106**, 205501 (2011).
35. Devki N. Talwar, Zhe Chuan Feng, and Tzuen-Rong Yang, "Vibrational signatures of isotopic impurities and complexes in II-VI compound semiconductors," *Physical Review B*, **85**, 195203 (2012).
36. Devki N. Talwar, Zhe Chuan Feng, Jyh-Fu Lee, and P. Becla, "Structural and dynamical properties of Bridgman-grown $CdSe_xTe_{1-x}$ ($0 < x = 0.35$) ternary alloys," *Physical Review B*, **87**, 165208 (2013).
37. Deng Xie, Zhi Ren Qiu, Devki N. Talwar, Yi Liu, Jen–Hao Song, Jow–Lay Huang, Ting Mei, Chee Wee Liu, and Zhe Chuan Fang, "Investigation of optical parameters of boron doped aluminium nitride films grown on diamond using spectroscopic ellipsometry," *International Journal of Nanotechnology*, **12**, 97–110 (2014).
38. J. H. Song, Jow-Lay Huang, H. H. Lu, and J. C. Sung, "Investigation of wurtzite (B, Al) N films prepared on polycrystalline diamond," *Thin Solid Films*, **516**, 223–227 (2007).
39. M. Jayasakthi, R. Ramesh, P. Arivazhagan, R. Loganathan, K. Prabakaran, M. Balaji, and K. Baskar, "Structural and optical characterization of AlGaN/GaN layers," *Journal of Crystal Growth*, **401**, 527–531 (2014).
40. M. A. Moram and M. E. Vickers, "X-ray diffraction of III-nitrides," *Reports on Progress in Physics* **72**, 036502 (2009).
41. Bahadir Kucukgok, Na Lu, Ian T. Ferguson, Shu Chang Wang, Xiong Zhang, and Zhe Chuan Feng, "Structural and optical analyses of $Al_xGa_{1-x}N$ thin films grown by metal organic chemical vapor deposition," *Japanese Journal of Applied Physics*, **54**, 02BA05 (2015).
42. A. L. Ankudinov, B. Ravel, J. J. Rehr, and S. D. Conradson, "Real-space multiple-scattering calculation and interpretation of X-ray-absorption near-edge structure," *Physical Review B*, **58**, 7565 (1998).
43. Bruce Ravel and MATHENA Newville, "ATHENA, ARTEMIS, HEPHAESTUS: data analysis for X-ray absorption spectroscopy using IFEFFIT," *Journal of synchrotron radiation*, **12**, 537–541 (2005).
44. Ian Ferguson, Tse Yang Lin, Yee Ling Chung, Lin Li, Shude Yao, Y. C. Lee, Zhe Chuan Feng, Ian T. Ferguson, Weijie Lu, Matthew H. Kane, Nadarajah Narendran, and Tsunemasa Taguchi, "Optical, structural, and nuclear scientific studies of AlGaN with high Al composition," *SPIE*, **7784**, 778415-778415-778411 (2010).
45. Matthew H. Kane, T. Y. Lin, L. Li, C. Chen, Y. L. Chung, S. D. Yao, Y. C. Lee, Z. R. Qiu, I. T. Ferguson, D. S. Wuu, Z. C. Feng, Christian Wetzel, and Jian-Jang

Huang, "Material properties of MOCVD-grown AlGaN layers influenced by indium-incorporation," *SPIE*, **8123**, 812309-812309-812312 (2011).
46. T. Y. Wu, C. C. Chang, K. K. Tiong, Y. C. Lee, S. Y. Hu, L. Y. Lin, T. Y. Lin, and Z. C. Feng, "Luminescence studies in $In_xGa_{1-x}N$ epitaxial layers with different indium contents," *Optical Materials*, **35**, 1829–1833 (2013).
47. G. R. James, A. W. R Leitch, F. Omnes, and M. Leroux, "The effect of exciton localization on the optical and electrical properties of undoped and Si-doped $Al_xGa_{1-x}N$," *Semiconductor science and technology*, **21**, 744 (2006).
48. A. Bell, S. Srinivasan, C. Plumlee, H. Omiya, F. A. Ponce, J. Christen, S. Tanaka, A. Fujioka, and Y. Nakagawa, "Exciton freeze-out and thermally activated relaxation at local potential fluctuations in thick $Al_xGa_{1-x}N$ layers," *Journal of applied physics*, **95**, 4670–4674 (2004).
49. S. A. Lourenço, M. A. T Da Silva, I. F. L Dias, J. L. Duarte, E. Laureto, A. A. Quivy, and T. E. Lamas, "Correlation between luminescence properties of $Al_xGa_{1-x}As$/ GaAs single quantum wells and barrier composition fluctuation," *Journal of applied physics*, **101**, 113536 (2007).
50. S. Y. Hu, Y. C. Lee, Z. C. Feng, and Y. H. Weng, "Raman spectra investigation of InAlGaN quaternary alloys grown by metalorganic chemical vapor deposition," *Journal of Applied Physics*, **112**, 063111 (2012).
51. Sheng-Yao Hu, Yueh-Chien Lee, Zhe-Chuan Feng, and Shi-Hong Yang, "Anomalous luminescence behavior in the InAlGaN thin film," *Journal of Alloys and Compounds*, **509**, 2300–2303 (2011).
52. C. H. Chen, L. Y. Huang, Y. F. Chen, H. X. Jiang, and J. Y. Lin, "Mechanism of enhanced luminescence in $In_xAl_yGa_{1-x-y}N$ quaternary alloys," *Applied physics letters*, **80**, 1397 (2002).
53. Yatendra Pal Varshni, "Temperature dependence of the energy gap in semiconductors," *Physica*, **34**, 149–154 (1967).
54. P. G. Eliseev, "The red σ^2/kT spectral shift in partially disordered semiconductors," *Journal of applied physics*, **93**, 5404–5415 (2003).
55. S. Y. Hu, Y. C. Lee, Y. H. Weng, I. T. Ferguson, and Z. C. Feng, "Characterization of temperature-dependent photoluminescence properties of InAlGaN quaternary alloys," *Journal of Alloys and Compounds*, **587**, 153–157 (2014).
56. M. E. Aumer, S. F. LeBoeuf, F. G. McIntosh, and S. M. Bedair, "High optical quality AlInGaN by metalorganic chemical vapor deposition," *Applied physics letters*, **75**, 3315–3317 (1999).
57. E. Monroy, N. Gogneau, F. Enjalbert, F. Fossard, D. Jalabert, E. Bellet-Amalric, Le Si Dang, and B. Daudin, "Molecular-beam epitaxial growth and characterization of quaternary III–nitride compounds," *Journal of applied physics*, **94**, 3121–3127 (2003).
58. R. J. Potter, and N. Balkan, "Optical properties of GaNAs and GaInAsN quantum wells," *Journal of Physics: Condensed Matter*, **16**, S3387 (2004).
59. Sergio Fernández-Garrido, J. Pereiro, F. González-Posada, E. Muñoz, E. Calleja, Andrés Redondo-Cubero, and R. Gago, "Photoluminescence enhancement in quaternary III-nitrides alloys grown by molecular beam epitaxy with increasing Al content," *Journal of Applied Physics*, **103**, 046104 (2008).
60. T. N. Oder, J. Li, J. Y. Lin, H. X. Jiang, "Photoresponsivity of ultraviolet detectors based on $In_xAl_yGa_{1-x-y}N$ quaternary alloys," *Appl. Phys. Lett.* **77**, 791 (2000).

61. H. P. D. Schenk, M. Leroux, and P. De Mierry, "Luminescence and absorption in InGaN epitaxial layers and the van Roosbroeck–Shockley relation," *Journal of Applied Physics* **88**, 1525–1534 (2000).
62. H. X. Jiang and J. Y. Lin, "AlGaN and InAlGaN alloys-epitaxial growth, optical and electrical properties, and applications," *Optoelectronics Review*, 271–286 (2002).
63. S. O. Usov, A. F. Tsatsul'nikov, V. V. Lundin, A. V. Sakharov, E. E. Zavarin, M. A. Sinitsyn, and N. N. Ledentsov, "Energy characteristics of excitons in structures based on InGaN alloys," *Semiconductors*, **42**, 720–725 (2008).
64. J. Christen and D. Bimberg, "Line shapes of intersubband and excitonic recombination in quantum wells: Influence of final-state interaction, statistical broadening, and momentum conservation," *Physical Review B*, **42**, 7213 (1990).
65. V. Yu Davydov, I. N. Goncharuk, A. N. Smirnov, A. E. Nikolaev, W. V. Lundin, A. S. Usikov, A. A. Klochikhin, J. Aderhold, J. Graul, and O. Semchinova, "Composition dependence of optical phonon energies and Raman line broadening in hexagonal $Al_xGa_{1-x}N$ alloys," *Physical Review B*, **65**, 125203 (2002).
66. C. H. Chen, Y. F. Chen, Z. H. Lan, L. C. Chen, K. H. Chen, H. X. Jiang, and J. Y. Lin, "Mechanism of enhanced luminescence in $In_xAl_yGa_{1-x-y}N$ quaternary epilayers," *Applied physics letters*, **84**, 1480–1482 (2004).
67. A. Cros, A. Cantarero, N. T. Pelekanos, A. Georgakilas, J. Pomeroy, and M. Kuball, "Resonant Raman characterization of InAlGaN/GaN heterostructures," *physica status solidi (b)*, **243**, 1674–1678 (2006).
68. Jiejun Wu, Jiemin Li, Guangwei Cong, Hongyuan Wei, Panfeng Zhang, Weiguo Hu, Xianglin Liu, Qinsheng Zhu, Zhanguo Wang, and Quanjie Jia, "Temperature dependence of the formation of nano-scale indium clusters in InAlGaN alloys on Si (111) substrates," *Nanotechnology*, **17**, 1251 (2006).
69. Tongjun Yu, Yaobo Pan, Zhijian Yang, Ke Xu, and Guoyi Zhang, "Influence of V/III flux ratio on electrical properties of AlInGaN/Gan heterostructures grown by MOCVD," *Journal of crystal growth*, **298**, 211–214 (2007).
70. Nagarajan Subramaniyam, Oh Tae Su, Kumar Muthusamy Senthil, Hong Chang-Hee, and Suh Eun-Kyung, "Structural and Optical Properties of In-Rich InAlGaN/InGaN Heterostructures for White Light Emission," *Japanese Journal of Applied Physics*, **47**, 4413 (2008).

Part III

III-Nitride Devices and Nano-Structures

CHAPTER 8

III-NITRIDE NANO-STRUCTURES AND IMPROVING THE LUMINESCENCE EFFICIENCY FOR QUANTUM WELL LEDS

Peng Chen

School of Electronic Science and Engineering, Nanjing University
22 Hankou Road, Nanjing, Jiangsu Province, P. R. China
pchen@nju.edu.cn

Novel low-dimensional structure and lower polarization effect will bring GaN-based optoelectronic devices more excellent physical properties and great potential applications. This chapter systematically explores the morphology change and luminescence properties of InGaN quantum well as the dimension reduced, also includes the study from polar plane to semipolar plane.

1-dimensional MQW nanorod arrays were fabricated by inductively coupled plasma (ICP) etching by using a self-assembled Ni nanoparticle mask. At room temperature, an enhancement of 3.1 times in total integrated photoluminescence intensity is achieved in comparison to that of the as-grown MQW structure. The internal quantum efficiency of the nanorod arrays reaches 61.2%.

0-dimensional InGaN nanostructures have been demonstrated to grow on SiO_2-pretreated GaN surface. At room temperature, an enhancement of 4.2 times in total integrated photoluminescence intensity is achieved in comparison to that of the same InGaN layer grown on the untreated GaN surface. Highly uniform and periodic InGaN Nanostructures have been demonstrated by Selective Area Epitaxy (SAE).

The key influence factors of 2-dimensional semipolar GaN facets grown on stripe and cross masks by SAE is discussed. It is found that the thermal stability of {1-101} semi-polar facet is better than {11-22} facet.

The internal quantum efficiency of {1-101} and {11-22} InGaN MQWs are 65.6% and 55.7% respectively, which are much higher than the (0001) MQWs.

1. Self-assembling High-efficiency 1-Dimensional Nanostructure: InGaN MQW Nanorod Arrays

1.1. *Introduction*

InGaN multiple quantum well (MQW) structures based on the wide-energy-gap group-III nitrides have become particularly attractive for the fabrication of light-emitting diodes (LEDs) and laser diodes (LDs).[1–5] However, the large lattice mismatch between GaN and InN, high-density dislocations, In-rich clusters and other crystalline defects generated in the InGaN/GaN MQW structure, can lead to a reduction of the internal quantum efficiency.[6–9] Threading dislocations (TDs) also reduce the emission efficiency, increase the leakage current, and shorten the life time of devices.

Thus, various methods have been proved beneficial for enhancing internal quantum efficiency for In-GaN based MQWs by reducing the density of dislocation and other defects, such as homoepitaxy with thick GaN substrates,[10] epitaxial layer overgrowth,[11,12] and using a multiple intermediate layer.[13] However, using these techniques, it is difficult to reduce the dislocation density to be less than $10^6\ \text{cm}^{-2}$. GaN substrates with low dislocation density of $\sim 10^6\ \text{cm}^{-2}$ are beginning to available, while they are still very expensive. Therefore, developing an effective affordable technique for fabricating low-TD-density GaN-based materials and device structures is an important issue for improving the qualities of nitride-based optoelectronic devices, particularly for LEDs.

On the other hand, by reducing the size of the emitting area to nanoscale, dislocation-free structures can be obtained and then the luminescence degradation caused by the dislocation can be avoided. In addition, nanostructures are also helpful for increasing the light extraction efficiency. Thus, by using a nanoscale emitting structure, both internal quantum efficiency and light extraction efficiency can be improved, resulting in a significant improvement of the luminescence.

In this chapter, the fabrication of the InGaN MQW nanorod arrays will be demonstrated and optical characterizations will be discussed.

1.2. *Experimental process*

The MQW nanorod arrays were fabricated by inductively coupled plasma (ICP) top-down etching using a self-assembled Ni nanoparticle mask on a plane InGaN/GaN MQW structure. Detailed photoluminescence (PL) measurements at all temperatures have been performed on the nanorod arrays and the corresponding as-grown MQW structure. From the PL measurements, the effect of a significant reduction of defects in the nanorod arrays on their optical properties is investigated. All the samples used in this study were grown on *c*-plane sapphire substrates by a metalorganic chemical vapor deposition (MOCVD) system with trimethylgallium, trimethylindium, and ammonia as the precursors of Ga, In, and N, respectively. Nitrogen was used as the carrier gas for the sources. The MQW structure was grown on a 2-μm GaN template, containing five periods of $In_{0.15}Ga_{0.85}N$/GaN quantum wells (2-nm well and 8-nm barrier). Finally, the MQWs were covered by a 10-nm thick GaN cap layer.

After the MOCVD epitaxial growth, the planar InGaN/GaN MQW samples were fabricated into the nanorod arrays by using ICP etching via a self-assembled nickel nanomask. However, the dry etching could cause lattice defects near etching surface, which degenerates the electrical and optical properties in the etch damage region. In order to minimize the damages caused by the dry etching, the etching condition must be optimized by using small RF power during the etching process.[14] For further removal of the surface damage caused by the dry etching, the MQW nanorod arrays were immersed in HCl solvent at 70°C for 15 min.

The optical properties of the nanorod arrays were investigated by temperature-dependent PL measurements. The spectra were recorded as a function of temperature ranging from 15 to 300 K by using a 325-nm cw laser. The fabrication process for the InGaN MQW nanorod arrays is schematically illustrated in Fig. 1.

1.3. *Results and discussions*

Figure 2 shows the geometrical aspects of the sample observed by using a field emission scanning electron microscope (FE-SEM). From the SEM image, it can be seen that the lateral sizes of the MQW nanorods are about 200–300 nm, which result in an area of the nanorods that accounts for 25.1%

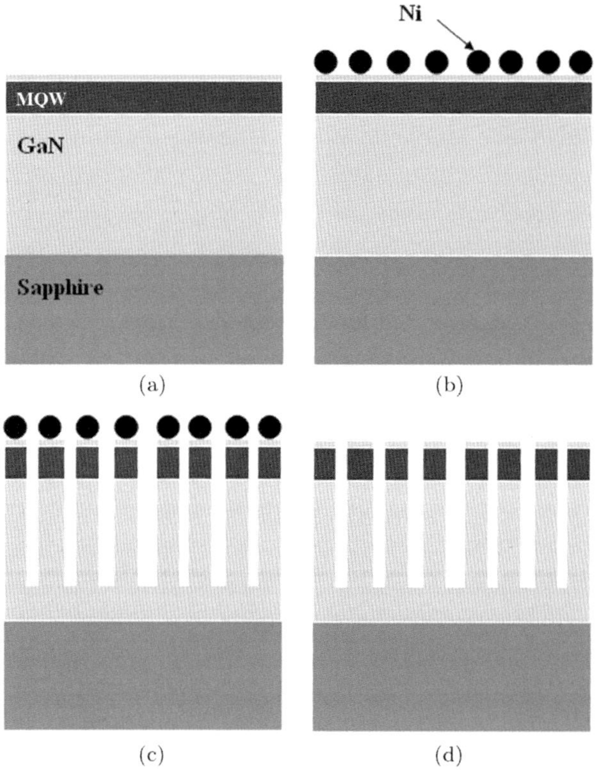

Fig. 1. A schematic illustration of the fabrication process for the InGaN MQW nanorod arrays. (a) planar MQW structure grown by MOCVD, (b) self-assembled Ni nanomask, (c) the nanorod arrays formed by ICP etching, (d) the MQW nanorod arrays after final HCl etching.

of the total area. The etching depth is about 400 nm and the inclination angle of the etched sidewalls is higher than 85° as shown in Fig. 2.

A 325-nm helium-cadmium laser was used as the excitation source for the InGaN MQW nanorod arrays and the planar InGaN MQW structure. Figure 3 shows the photoluminescence spectra of the two structures at room temperature with the laser excitation power of 1.5 mW.

It can be clearly seen that the integrated PL intensity of the nanorod arrays is 3.1 times that of the as-grown planar MQWs. The peak width of the PL spectrum from the nanorod arrays is slightly narrower than that from the planar MQWs. The full width at half maximum (FWHM) of the

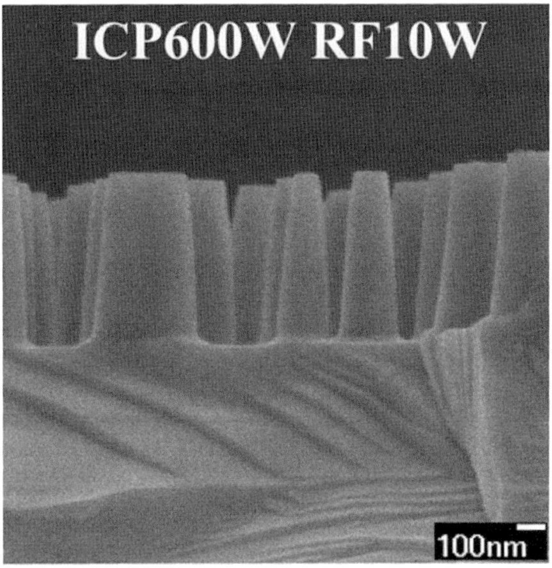

Fig. 2. SEM image of InGaN MQW nanorod arrays after ICP etching.

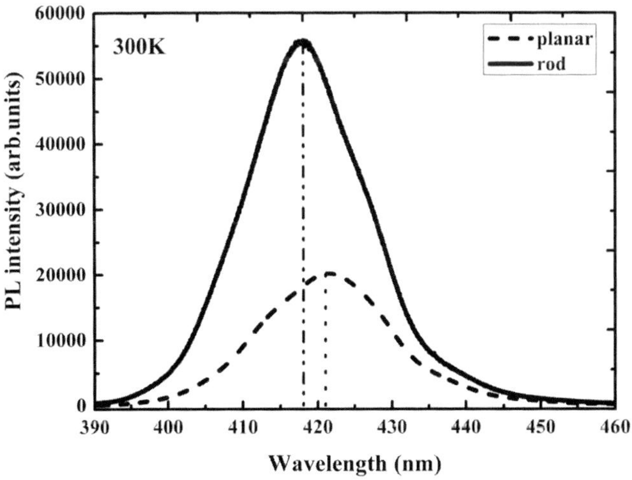

Fig. 3. PL spectra of the MQW nanorod arrays and as grown planar MQWs at room temperature.

spectra are 19.17 nm for the nanorod arrays and 20.26 nm for the planar MQWs, respectively. Furthermore, the PL peak position of the nanorod arrays shows a 3.4-nm blue-shift compared with that of the as-grown planar MQW sample.

Thus, the nanorod array structure has a significant impact on the quantum well light emitting characteristics. To understand the phenomena, the changes in the internal quantum efficiency (η_{int}) and the light extraction efficiency (η_{ext}) should be investigated separately due to the different sample structures. For an MQW sample, its PL intensity is determined by the light emitting area, the absorption efficiency of the pumping laser (η_{abs}), η_{int} and η_{ext}. Accordingly, the PL intensity can be expressed as[15]

$$PL\ intensity \propto (emission\ area) \times \eta_{abs} \times \eta_{int} \times \eta_{ext}.$$

Because the excitation laser beam is just to the right of the samples' surface, we assume that η_{abs} for the as-grown MQW and the nanorod arrays can be equal for the effective united area, which has been proved by Kuo *et al.*[16] Thus, as the area of the nanorods accounts for 25.1% of the total surface area, and the integrated PL intensity of the nanorods is 3.1 times than that of the planar MQWs, $\eta_{int} \times \eta_{ext}$ of the nanorod arrays reaches 12.3 times than that of the as-grown planar MQW sample.

It is obvious that this huge enhancement should be contributed by both η_{int} and η_{ext}. In order to determine η_{int} for both the samples, the temperature-dependent PL measurements were performed at temperatures from 15 to 300 K. The normalized PL intensities as a function of temperature are shown in Fig. 4. It is well known that the internal quantum efficiency of a semiconductor sample for luminescence is defined as the ratio of the integrated PL intensity at RT (300 K) over that at the lowest temperature to be reached (15 K in our case). Therefore, from Fig. 4, it can be seen that η_{int} of the as-grown planar MQW sample is 33.8% and $\eta_{int}$of the MQW nanorod arrays is 61.2%, which clearly indicate the higher emission efficiency for the nanorod arrays.

Meanwhile, it is worth noting here that η_{ext} of the MQW nanorod arrays is much better than that of as-grown planar MQWs. For conventional as-grown planar MQWs, a large fraction of the generated light is trapped because of total internal reflection and, hence, cannot escape from the as-grown planar MQWs surface. According to previous reports,[16–18] η_{ext}

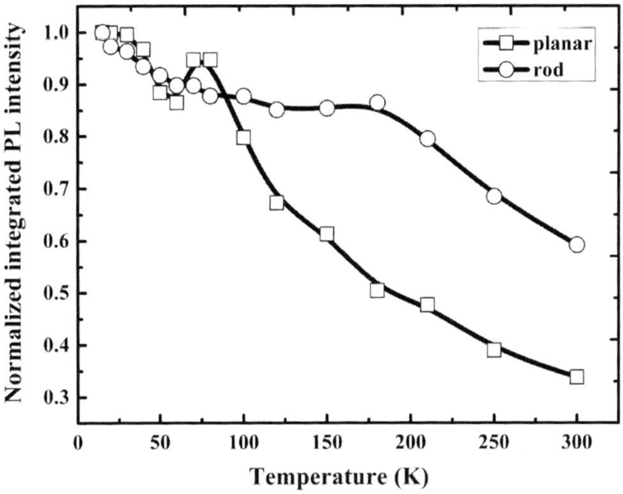

Fig. 4. Temperature dependent normalized integrated PL intensity for the MQW nanorod arrays and as-grown planar MQWs from 15 K to 300 K.

through a top surface rough process which breaks the total internal reflection can greatly increase the output emission intensity. From Fig. 1, it is clear that the diameters and the spaces of the nanorods are smaller than the light wavelength of the MQW emission, thus strong diffusion scattering is taking place in the nanorod array region. Due to the strong light scattering from the MQW nanorod arrays, η_{ext} of our MQW nanorod arrays can be greatly increased. After a simple calculation, it is known that for GaN-based planar optical waveguides, η_{ext} is theoretically estimated to be 4.5% for single-side light extraction. In principle, if there is no total reflection, the maximum value of η_{ext} is 50% for single-side light extraction. In our work, when η_{int} of 33.8% and 61.2% are adopted, we can derive that η_{ext} of our MQW nanorod arrays is 30.5% for single-side light extraction, i.e., 61.0% for total light extraction, which is near 7 times η_{ext} of the planar MQWs sample, comparable to previous reports.

The increase of η_{int} can be originated from several aspects. In our work, the strain relaxation may occur in the MQW nanorod arrays when the planar MQWs have been nanotextured. Most of the strain relaxation releases the thermal mismatch strain which is induced by the high temperature growth of GaN, not the lattice mismatch strain. The formation of the nanorods can make the planar MQWs into small disks. However, the diameters (about 100nm or a little more) of the nanorods are still much larger than the thickness of the InGaN layers (less than 10 nm). Thus, the strain relaxation on the multiple quantum wells is small. Although the strain relaxation can increase η_{int}, in our experiment it seems not to be the dominant reason for the increase of η_{int}. Considering the major difference between both samples is the different emitting area, a significant reduction in dislocations and other defects enhances the radiative recombination in the MQWs. It has been known that the density of dislocations in the GaN layer normally is 10^8–10^9 cm^{-2}. Thus the average distance between the dislocations is about 1nm–300 nm, which is much larger than the diameter and spacing of the nanorod arrays. Thus, the nanorod is nearly dislocation free, the non-radiative recombination on the dislocations is greatly avoided. Although η_{int} reaches 61.2%, it is still far from perfect. This shows that the nanorods cannot avoid all dislocations and some surface defaces may still remain in our structure, which act as the nonradiative centers to reduce η_{int}.

To further understand the increase of the η_{int}, the finite-difference time-domain (FDTD) simulations is adopted to simulate the electromagnetic energy and field distributions in the MQW nanorod arrays to theoretically explore the emission and extraction mechanisms. Figure 5(a) shows the electromagnetic energy distribution at the emission wavelength of in the

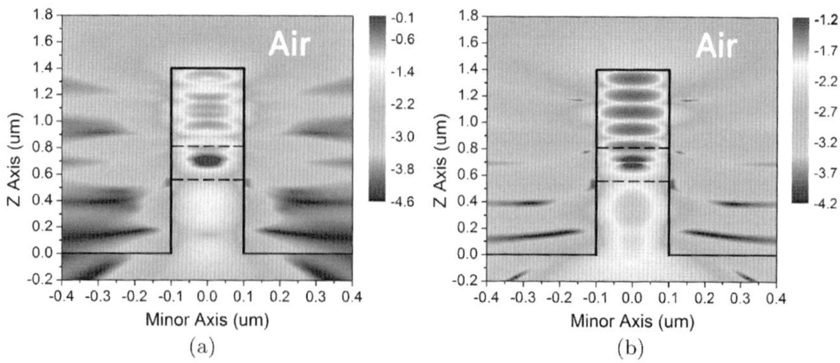

Fig. 5. (a) The electromagnetic field energy and (b) H-component distribution in the a-axis-z-axis plane for a single InGaN MQW nanorod. The nanorod is shown the solid black lines, and the dashed line marks the MQW location.

MQW nanorod arrays. The MQW location is labeled as the black lines. This simulated result demonstrates that some part of the electromagnetic energy density is concentrated at the location of MQWs in the nanorod due to the reflection from the top interface, which results in the enhancement of the radiative recombination emission at these places. Figure 5(b) exhibits a typical magnetic field component H distribution at the MQW emission wavelength. It can be noted that the nanorod is able to act as an optical waveguide along the z axis, supporting at least one guided mode. Thus the radiative recombination emission can be coupled to these guided modes and can be well directed to extract from the nanorod,[19] leading to a higher light extraction.

1.4. *Conclusions*

In conclusion, self-assembling high-efficiency InGaN MQW nanorod arrays have shown a great improvement of the internal quantum efficiency and light extraction efficiency to be 1.75 times and near 7 times than that of as-grown planar MQW structure. The efficiency improvement is contributed by the low damage dry etching process, significant reduction of dislocation and optical waveguide effect in the nanorods. These indicated that the MQW nanorod array structure has great potential to prepare high efficiency optoelectronic devices for future advanced device design.

References

1. S. Nakamura and G. Fasol, *Blue Laser Diode*, (Berlin:Springer) p 4 (2000).
2. J. X. Wang, L. Wang, Z. B. Hao, and Y. Luo, *Chin. Phys. Lett.*, **28**, 118105 (2011).
3. W. C. Peng and Y. S. Wu, *Appl. Phys. Lett.*, **88**, 181117 (2006).
4. Z. L. Xie, R. Zhang, B. Liu, X. Q. Xiu, H. Su, Y. Li, X. M. Hua, H. Zhao, P. Chen, P. Han, Y. Shi, and Y. D. Zheng, *Chin. Phys. Lett.*, **28**, 087102 (2011).
5. Y. K. Kuo, T. H. Wang, and J. Y. Chang, *Appl. Phys. Lett.*, **100**, 031112 (2012).
6. Y. S. Lin, K. J. Ma, C. Hsu, Y. Y. *et.al. Appl. Phys. Lett.*, **80**, 2571 (2002).
7. S. W. Feng, Y. C. Cheng, Y. Y. Chung, C. C. Yang, Y. S. Lin, C. Hsu, K. J. Ma, and J. I. Chyi, *J. Appl. Phys.*, **92**, 4441 (2002).
8. M. K. Chen, Y. C. Cheng, J. Y. Chen, C. M. Wu, C. C. Yang, K. J. Ma, J. R. Yang, and A. Rosenauer, *J. Cryst. Growth*, **279**, 55 (2005).
9. Y. C. Cheng, E. C. Lin, C. M. Wu, C. C. Yang, J. R. Yang, A. Rosenauer, K. J. Ma, S. C. Shi, L. C. Chen, C. C. Pan, and J. I. Chyi, *Appl. Phys. Lett.*, **84**, 2506 (2004).
10. X. A. Cao, J. M. Teetsov, M. P. D'Evelyn, D. W. Merfeld, and C. H. Yan, *Appl. Phys. Lett.*, **85**, 7 (2004).

11. A. Bell, R. Liu, F. A. Ponce, H. Amano, I. Akasaki, and D. Cherns, *Appl. Phys. Lett.*, **82**, 349 (2003).
12. S. Nagahama, N. Iwasa, M. Senoh, T. Matsusgita, Y. Sugimoto, H. Kiyoku, T. Kozaki, M. Sano, H. Matsumura, H. Umemoto, K. Chocho, and T. Mukai, *Jpn. J. Appl. Phys.*, **39**, L647 (2000).
13. A. Kikuchi, T. Yamada, S. Nakamura, K. Kusakabe, D. Sugihara, and K. Kishino, *Jpn. J. Appl. Phys.*, **39**, L330 (2000).
14. Z. G. Yu, P. Chen, G. F. Yang, B. Liu, Z. L. Xie, X. Q. Xiu, *et.al. Chin. Phys. Lett.*, **29**, 078501 (2012).
15. H. Y. Ryu, J. K. Hwang, D. S. Song, I. Y. Han, and Y. H. Lee, *Appl. Phys. Lett.*, **78**, 1174 (2001).
16. M. L. Kuo, Y. J. Lee, and T. C. Shen, *Opt. Lett.*, **34**, 2078 (2009).
17. C. Huh, K. S. Lee, E. J. Kang, and S. J. Park, *J. Appl. Phys.*, **93**, 9383 (2003).
18. T. Fujii, Y. Gao, R. Sharma, E. L. Hu, S. P. DenBaars, and S. Nakamura, *Appl. Phys. Lett.*, **84**, 855 (2004).
19. A.-L. Henneghien, B. Gayral, Y. Désières and J-M. Gérard, *J. Opt. Soc. Am. B.*, **26**, 2396 (2009).

2. High Efficiency, Highly Uniform and Periodic InGaN 0-Dimensional Nanostructures

2.1. *Introduction*

Low dimensional semiconductor structures give rise to attractive physical phenomena that have been applied in optoelectronic and electronic devices with improved functionalities, such as quantum dot (QD) laser diodes (LDs)[1] and single-electron transistors.[2] Research into nitride semiconductor (GaN, AlGaN and InGaN) nanodots is attracting a lot of attention. At present, most of semiconductor nanodots are grown by the Stranski-Krastanow (SK) growth mode through self-organized processes.[3] However, lattice mismatch induced strain is not sufficient to result in SK growth mode in the nitride system. In order to reach a 3-dimensional growth mode in nitride system, Si as antsurfactant is used to modify GaN or AlGaN surface states to realize nitride nanodots. The fabrication of GaN-related nanodots has been reported by metalorganic chemical vapor deposition (MOCVD)[4] and molecular beam epitaxy.[5] Generally, high-density defects ($\sim 10^9\,cm^2$) in the nitride materials severely affect device performance, such as luminescence efficiency and lifetime. The diameter of the nitride nanodots can be much smaller than the interspaces of those extended defects. By using such nanodots as the active layer, recombination of electrons and holes could be confined in the nanodots, not degraded by the defects. Thus, reducing the dimension of the active layer can improve the performance of the GaN-based devices.[6]

In the self-organized processes described above, random distribution of the QD sizes and locations usually occurs,[7] which limits the benefits from low-dimensional structures. Usually, the dynamic of exciton decay is found to be monoexponential and the emission should very narrow in a single dot. But, the disorder of the QD system governs the recombination dynamics to give a nonexponential photoluminescence (PL) decay of the entire QD ensemble.[8]

In this chapter, the growth of dense InGaN nanodots on pretreated GaN surfaces is presented, and the fabrication and optical properties of long-range ordered InGaN nanodot arrays grown on GaN/sapphire substrates by SAE are discussed.

2.2. *High-density InGaN nanodots grown on pretreated GaN surfaces*

To fabricate the InGaN nanodots, first a GaN film was grown on a c-plane sapphire up to a thickness of $2\mu m$ by MOCVD. Next 100-nm of SiO_2 was deposited onto the GaN film by plasma-enhanced chemical vapor deposition. Before the epitaxial growth of InGaN, the sample was dipped in dilute hydrofluoric acid and dilute hydrochloric acid to remove the SiO_2 film and native oxide of GaN. After all the cleaning processes, a few InGaN was grown on the pretreated GaN in the MOCVD reactor for 60, 120 and 180 seconds respectively. The nominal growth rate of the InGaN is 0.08 nm per second. The growth temperatures are selected as 740°C and 755°C. By control of the growth duration and temperature, InGaN nanodots with various sizes and density were achieved. For reference, another InGaN with the same growth condition was grown on an untreated GaN surface. All samples were investigated by means of atomic force microscopy (AFM) and photoluminescence (PL) at room temperature.

Figure 6(a) and (b) are AFM images of the InGaN nanodots grown on the GaN surface at 740°C with different growth durations. The InGaN nanodots with the highest density and smallest size are obtained by the shortest growth duration (60 seconds), as shown in Fig. 6(a). The lateral sizes of the dots are mostly in the range of 37 ± 11 nm. The average height of the dots is 3.6 nm

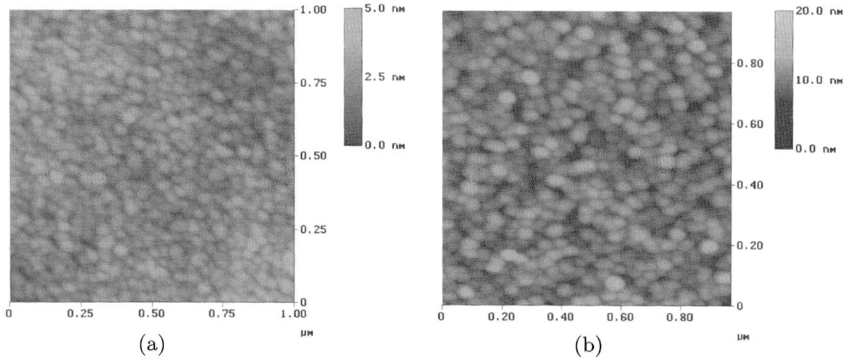

Fig. 6. AFM images of the InGaN nanodots grown on the SiO_2 treated GaN surfaces at 740°C for (a) 60 seconds; (b) 180 seconds.

and the density is around $9 \times 10^{10}\,cm^{-2}$. When the growth duration of the InGaN is extended to 120 seconds, the InGaN nanodots become bigger. The lateral sizes of the dots are in the range of 48 ± 14 nm. The average height of the dots is 11 nm and the density is around $5 \times 10^{10}\,cm^{-2}$. The InGaN nanodots grown for the longest duration (180 seconds) are shown in Fig. 6(b). The lateral sizes of the dots are even bigger, in the range of 53 ± 15 nm. The height of the dots becomes less uniform than that of those by grown shorter time. The dots density reduces to about $3 \times 10^{10}\,cm^{-2}$. For comparison, the same InGaN was grown on an untreated GaN surface for 180 seconds, shown in Fig. 7. The morphology of the InGaN is smooth as normal InGaN epilayer. There are some small pits in the film, which are common structure in InGaN film grown on sapphire and induce by threading dislocations.[9,10] These results show that the treatment of SiO_2 changes the properties of the GaN surface. After the treatment, the subsequent growth of InGaN tends to 3-dimensional mode, instead of 2-dimensional growth mode as usual.

It has been known that Si atoms can change the surface states of GaN or AlGaN.[5,6,11] The exposure of a SiH_4/NH_3 gas mixture to the AlGaN or GaN surfaces results in the formation of a SiN nano-mask. The surface states were modified, and then the growth of GaN nanodots occurred. In our experiments, during the deposition and coverage of SiO_2 on the GaN surface, a few mount of Si can occupy the surface sites of the GaN and remain on the surface after removal of the SiO_2. Considering the reference

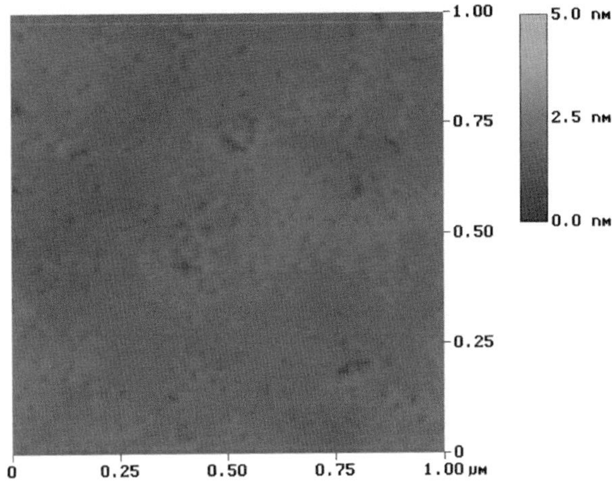

Fig. 7. AFM image of the InGaN thin film grown on the untreated GaN surface at 740°C for 180 seconds.

InGaN thin film grown on untreated GaN surface shown in Fig. 7, it is clear that the same growth conditions give different morphologies because of the different GaN surfaces. The InGaN nanodots formation shows that the SiO_2 treatment can also modify the GaN surface states, which results in the growth of InGaN nanodots.

The growth temperature is another crucial parameter for the InGaN nanodots formation. When the growth temperature is increased, the average dot size is reduced. However, if the growth temperature is too high, the initial 3-dimensional growth behavior will quickly turn to 2-dimensional mode. One sample grown at 755°C for 120 seconds is shown in Fig. 8. From this image, it can be seen that most nanodots already coalesced to form large flat terraces. If growing for longer time, a flat surface will be formed (not shown here).

The optical properties of the nanodots were investigated by means of PL at room temperature, as shown in Fig. 9. The sample was excited by a He-Cd laser ($\lambda_{ex} = 325\,\text{nm}$). Comparing with the PL spectrum from the thin InGaN film, stronger PL emissions can be observed from all InGaN nanodots without cap layer. The fringe patterns in the spectra are due to the light interference in the 2-μm GaN film. From Fig. 9, the emission from the InGaN thin film grown on the untreated GaN surface shows the

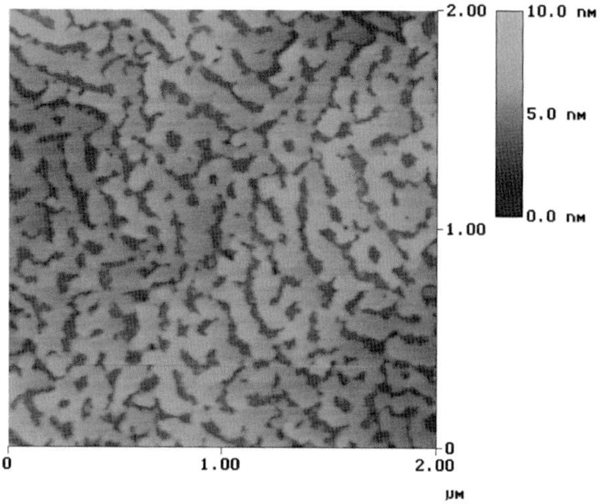

Fig. 8. AFM image of the InGaN thin film grown on the untreated GaN surface at 755°C for 180 seconds.

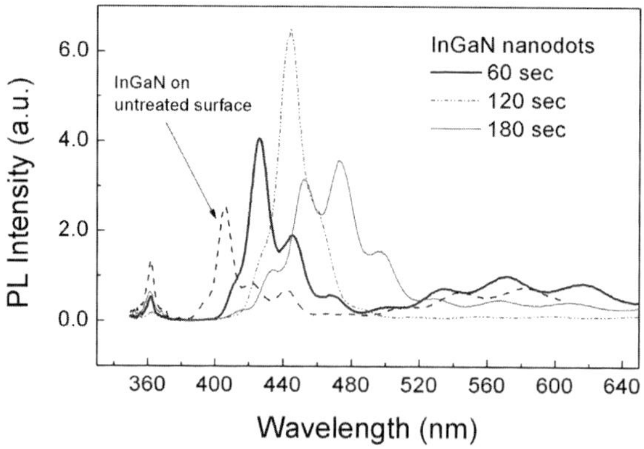

Fig. 9. Photoluminescence spectra of InGaN nanodots and thin film at room temperature.

shortest peak wavelength, indicating the smallest indium composition. The InGaN nanodots with longer growth duration shows the PL emission at longer wavelength, which indicates more indium incorporated in the bigger nanodots. Higher indium incorporation efficiency in the InGaN nanodots could be induced by strain relaxation during the nanodot growth. The strain relaxation is also thought to contribute to the strong PL emission. These results show that the SiO_2 treatment process provides an alternative method to produce high quality InGaN nanodots.

In summary, we have demonstrated a method for the fabrication of InGaN nanodots with controllable size and density. The InGaN nanodots grown on SiO_2 treated GaN surface have lateral sizes from 26 nm to 68 nm and density up to $9 \times 10^{10}\,cm^{-2}$. The strong PL from the uncapped nanodots at room temperature indicates the good optical quality of the InGaN nanodots. Furthermore, the nanodots can be incorporated into technologically important applications for quantum and optoelectronic devices.

2.3. *Growth and optical properties of highly uniform and periodic InGaN nanostructures*

In order to achieve nanodots with homogeneous size and controlled positioning, various methods have been carried out to promote nucleation at expected sites. These methods can be divided into two categories in principle. The first is by surface states control,[12–14] such as by self-organized anisotropic strain engineering,[12] by using vertical elastic interaction in QD superlattices,[13] by ion sputtering-induced surface instability,[14] and so forth. Locally ordered nanodot arrays were produced by these methods. The other sort of approaches relies on nanoscale selective area epitaxy (NSAE) on a pre-patterned substrate or growth template. The nano-patterning of substrates/templates can be realized by a variety of lithographic or non-lithographic techniques, such as interference lithography,[15,16] anodic oxidation of aluminum to form nano-porous alumina,[17–19] nanosphere lithography,[21] focused ion beam sputtering,[22] and electron-beam lithography (EBL).[23–26] At present, state-of-the-art EBL allows the creation of patterns with versatile physical geometry/localization, and homogeneous/tunable size. Rigorously-aligned InAs/GaAs nanodot arrays with a standard deviation of less than 5% for both dot diameter and height was achieved by molecular beam epitaxy on an EBL-patterned GaAs substrate.[25] Well-resolved excited states of the homogeneous QDs were observed by power-dependent PL measurements.[26] Due to the compatibility with conventional semiconductor device fabrication processes, EBL has been used for the fabrication of patterned QD edge-emitting lasers operating at room temperature.[27]

Due to the high temperature involved in the growth of III-nitride materials, it is difficult to realize dot formation through S-K growth mode. Thus, Si is used as antisurfactant to modify GaN or AlGaN surface states to realize nitride nanodot formation, which has been reported by the means of metalorganic chemical vapor deposition (MOCVD)[28] and molecular beam epitaxy.[29] The fabrication and evaluation of a UV light-emitting diode (LED) incorporating GaN quantum dots as the active layer is demonstrated.[30] The nitride nanodot formation by using SiO_2 to supply Si as the antisurfactant has been reported.[31] However, those are random nanodots. The first report of the quasi-order InGaN nanodots and nanorings grown on the patterned GaN surfaces by the NSAE is by nanopatterns from the porous aluminum oxide.[32] However, the long-range ordered nanostructure arrays are hard to obtain due to the nature of the self-assembly porous aluminum oxide.

Here, the fabrication and optical properties of long-range ordered InGaN nanodot arrays grown on GaN/sapphire substrates by NSAE will be described. The dense periodic nanohole patterns with different hole diameters and a pitch of 200 nm were defined by EBL and transferred into a SiO_2 layer which serves as the growth template. InGaN nanodots and nanorings were grown through the nano-porous SiO_2 template by MOCVD. The detailed processing conditions have been reported in Ref. 33.

Typical SEM images of the grown InGaN nanodot and nanoring arrays were shown in Fig. 10. The digitized images were processed to create three-dimensional profiles for a single dot/ring, as shown in the insets. With a precise replication of the patterns in the SiO_2 template, the InGaN nanodots/nanorings have good size uniformity and well-defined positioning. Our best result of the lateral nanodot size inhomogeneity was found to be as small as 3%,[33] which is much smaller than those created by using anodized alumina templates.[19,32] The nanodot and nanoring diameters vary from 50 to 100 nm, as defined by different pattern sizes on the SiO_2 template. As illustrated by the three-dimensional views, different physical geometries were developed for different pattern sizes. The nanodots with an average diameter of 60 nm possess a cone shape (Fig. 10a), whereas 100-nm nanorings have a volcano-like profile (Fig. 10b). As discussed in Ref. 32, the selective area growth process is primarily controlled by the migration of In and Ga adatoms on the sample surface. In the NSAE, the

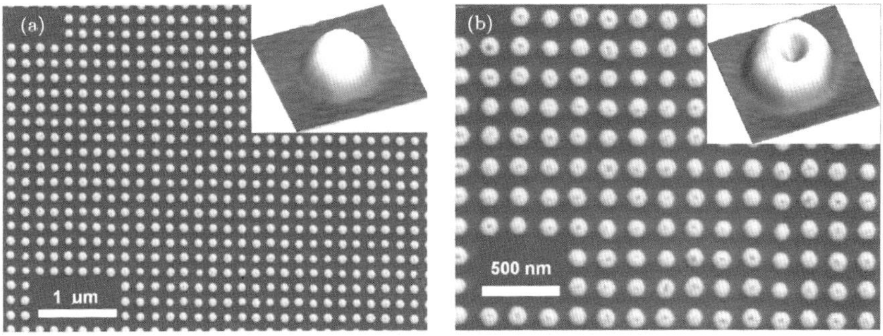

Fig. 10. SEM images of 200-nm-spaced (a) InGaN nanodot and (b) nanoring arrays. The average diameter is 60 nm for nanodots and 100 nm for nanorings respectively. The insets show three-dimensional views of a single nanodot/nanoring: the 60-nm nanodots have a cone shape, and the 100-nm nanorings have a volcano-like profile.

nucleation and growth of InGaN nano-islands starts preferably from the edges of the nanoholes, resulting in the formation of nanorings at the first stage. The nanoholes with smaller diameter will be quickly filled as the reactants supply is continued, leading to the formation of nano-cones. The time required to form cone-shape nanodots depends on the nanohole size, i.e. it takes longer time to fill up a larger hole. As a result, the variation of the morphology from nanoring to nanodot can be observed on the different patterns for the same growth duration.

The high uniformity of the nanodot size results in distinct improvements about the nanodots emission properties. At room temperature, the PL spectra from the highly uniform/periodic 60-nm InGaN nanodots and the random InGaN nanodots grown by using Si as antisurfactant (showed in Fig. 1c and Fig. 4 of Ref. 31) were compared in Fig. 11 with normalized intensity. For both samples, the raw curves reflect some fringes induced by light interference in the GaN layers, as shown by dash lines in the figure. By a proper data processing, the fringes were eliminated for the evaluation of the emission linewidth, as indicated by solid lines in Fig. 11 respectively. The PL spectrum from the ordered InGaN nanodots has a full width at half maximum (FWHM) of 240 meV, much smaller than that from the random InGaN nanodots (323 meV), **in spite of a longer peak wavelength.** In $In_xGa_{1-x}N$ material system, it has been known that higher In composition in the InGaN results in longer emission wavelength and larger compositional

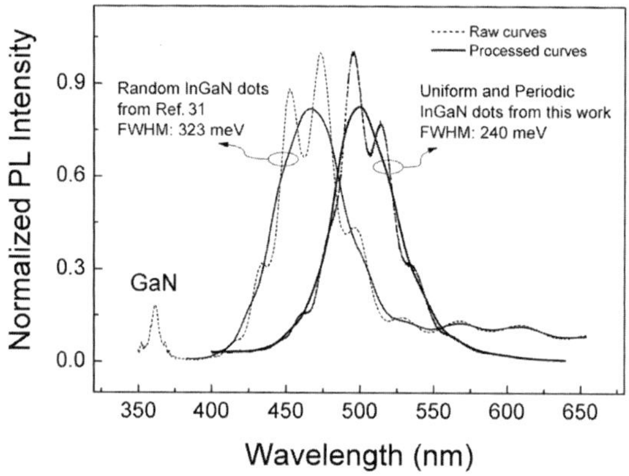

Fig. 11. Normalized room-temperature PL spectra from ordered InGaN nanodots grown by NSAE and random InGaN nanodots grown by using Si antisurfactant (Ref. 31). The raw curves with interference fringes (shown in dashed lines) were processed for the comparison of PL emission width (shown in solid lines).

fluctuation. The compositional fluctuation can lead to broader emission spectrum. Thus, the reduction of emission peak width is clearly attributed to the great improvement in the nanodot size homogeneity. In fact, the present sample grown by NSAE has a lateral size deviation of 6%, while the random nanodots present a much larger size deviation of 30%.[31] This indicates that the inhomogeneous broadening of the emissions from the nanodot ensembles can be greatly reduced by one third when the dispersion of dot sizes is reduced from 30% to 6%. However, the FWHM of the ordered InGaN nanodots is not as small as expected considering such high size uniformity, probably due to the non-uniform distribution of the In composition induced by the selective growth.

Figure 12 shows typical PL spectra from 60-nm InGaN nanodot arrays under different pumping levels of $0.01I_0$, $0.1I_0$, $0.8I_0$ and I_0 at 77 K, where the highest excitation power density I_0 is around 1.2×10^3 W/cm^2. A striking result is that there is no shift of the PL peak position in spite of the excitation power density increased by two orders of magnitude. This is unlike those in other InGaN-based heterostructures. As a matter of fact, a blue shift of the PL peak with the increase of the excitation power can often be observed for random InGaN nanodots, as well as for strained InGaN quantum well structures. For random InGaN nanodots, with the increase of

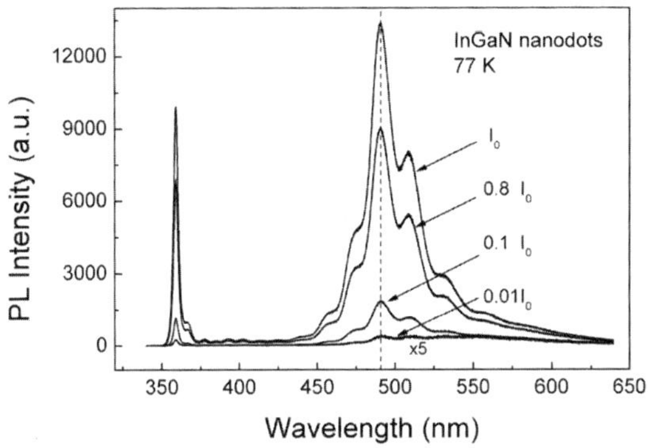

Fig. 12. PL spectra from 60-nm InGaN nanodot arrays under different excitation levels of $0.01I_0, 0.1I_0, 0.8I_0$ and I_0 at the temperature of 77 K, where I_0 denotes the highest excitation power.

the excitation power density, a blue shift can be induced by more photo-generated carriers migrating from larger dots to smaller dots.[34] When there is little difference in the nanodot sizes as in the case of the present InGaN nanodot arrays, the carrier redistribution does not occur. Therefore, the PL peak position of the size-uniform InGaN nanodots will not shift with the increase of the excitation power. The blue shift also is familiar in strained InGaN/GaN quantum well, in which the quantum-confined Stark effect due to piezoelectric filed causes the shift of the emission peak.[35]

Figure 13a shows the PL emissions from the 60-nm InGaN nanodot arrays at various temperatures ranging from 77 K to 270 K. For a better inspection of the emission properties, the curves have been processed to remove those interference fringes. The inset is the spectra from the GaN part in this sample. With the increase of the temperature from 77 K to 270 K, the emission linewidth of the nanodots increased about 11.5%, i.e. from 217 meV to 242 meV. At the same time, the emission linewidth of the GaN increased 42%, i.e. from 33meV to 47meV, which is within expected – a lot of thermo-activated energy levels contributing to the emission at higher temperature in the GaN. Based on the results from the InGaN nanodot arrays, it can be concluded that the photo-generated carriers were highly localized in the nanodots and those recombination centers did not change very much. On the other hand, the carry diffusion among different nanodots is greatly suppressed by the highly size homogeneity and the 200-nm space between

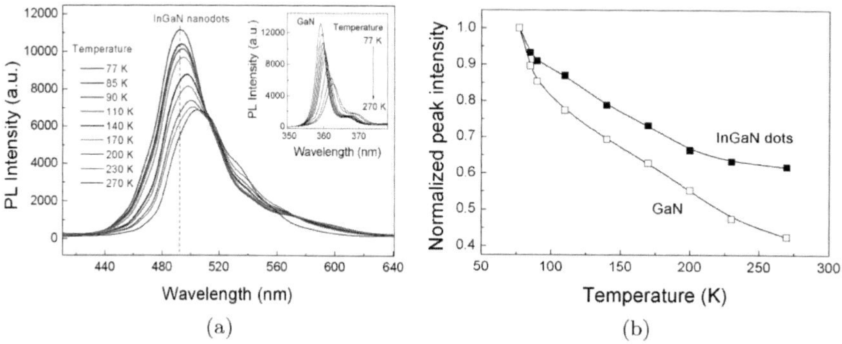

Fig. 13. (a) Temperature-dependent PL spectra from ordered InGaN nanodots and the GaN layer on the same sample. The spectra have already been processed to eliminate the fringes caused by light interference. (b) PL peak intensities of the InGaN nanodot arrays and the GaN layer plotted as a function of the temperature extracted from (a). The peak intensities have been normalized to their maximum values at 77 K for the InGaN nanodots and the GaN layer respectively.

the dots. The highly localized carriers give higher emission efficiency at higher temperature. The emission peak intensities from both the nanodots and the GaN are plotted in Fig. 4b with normalization to the values at 77 K individually. Although both intensities decreased with the increase of temperate, the nanodots shows much smaller in intensity loss than the GaN. The emission peak intensity at 270 K from the nanodots is about 64% of the value at 77 K, whereas it is 42% for the GaN. This retained ratio of 64% is also much higher than the value reported in self-assembly InGaN nanodots which is not more than 30 % within the same temperature range.[36] These results suggest a remarkable PL emission efficiency improvement from the highly uniform and periodic InGaN nanodots at higher temperatures.

In summary, highly uniform and periodic InGaN nanostructure arrays by NSAE using EBL-patterned templates have been fabricated. The InGaN nanodot/nanoring arrays produced by this approach have a very narrow distribution of diameters (3% in the best sample) and extremely good array symmetry. Improved optical properties resulting from the size uniformity of the InGaN nanodots were attained. The study on the room-temperature PL shows that the InGaN nanodot arrays with good uniformity can give much narrower PL emission compared to the random InGaN nanodots. Owing to the strong localization of the photon-generated carriers, a remarkable

PL emission efficiency at room temperature was achieved up to about 60% (with respect to the intensity at 77 K) in the InGaN nanodot arrays, which is much higher than those in the GaN and the self-assembled InGaN nanodots. These results reveal that InGaN nanostructures with highly uniform size/composition and good symmetry can be fabricated by the NSEA, and the nanostructures present good optical properties. Furthermore, this technology is promising for the realization of the optoelectronic devices with high efficiency and high sensitivity.

References

1. S. Fafard, K. Hinzer, S. Raymond, M. Dion, J. McCaffrey, Y. Feng, and S. Charbonneau, *SCIENCE*, **274**, 1350 (1996)
2. M. H. Devoret and R. J. Schoelkopf, *NATURE*, **406**, 1039 (2000).
3. D. J. Eaglesham and M. Cerullo, *Phys. Rev. Lett.*, **64**, 1943 (1990).
4. S. Tanaka, S. Iwai, and Y. Aoyagi, *Appl. Phys. Lett.*, **69**, 4096 (1996).
5. X. Q. Shen, S. Tanaka, S. Iwai, and Y. Aoyagi, *Appl. Phys. Lett.*, **72**, 344 (1998)
6. S. Tanaka, J. S. Lee, P. Ramvall, and H. Okagawa, *Jpn. J. Appl. Phys.*, **42**, L885 (2003).
7. G. Springholz, V. Holy, M. Pinczolits, and G. Bauer, *Science*, **282**, 734 (1998)
8. T. Bartel, M. Dworzak, M. Strassburg, A. Hoffmann, A. Strittmatter, and D. Bimberg, *Appl. Phys. Lett.*, **85**, 1946 (2004).
9. I. H. Kim, H. S. Park, Y. J. Park, and T. Kim, *Appl. Phys. Lett.*, **73**, 1634 (1998).
10. N. Sharma, P. Thomas, D. Tricker, and C. Humphreys, *Appl. Phys. Lett.*, **77**, 1274 (2000).
11. Hideki Hirayama, Satoru Tanaka, Peter Ramvall, and Yoshinobu Aoyagi *Appl. Phys. Lett.*, **72**, 1736 (1998).
12. T. Mano, R. Notzel, G. J. Hamhuis, T. J. Eijkemans, and J. H. Wolter, *Appl. Phys. Lett.*, **81**, 1705 (2002).
13. H. Brune, M. Giovannini, K. Bromann, and K. Kern, *Nature*, **394**, 451 (1998).
14. S. Facsko, T. Dekorsy, C. Koerdt, C. Trappe, H. Kurz, A. Vogt, and H. L. Hartnagel, *Science*, **285**, 1551 (1999).
15. S. C. Lee, K .J. Malloy, and S. R. J. Brueck, *J. Appl. Phys.*, **90**, 4163 (2001).
16. S. C. Lee, K. J. Malloy, L. R. Dawson, and S. R. J. Brueck, *J. Appl. Phys.*, **92**, 6567 (2002).
17. X. Mei, D. Kim, H. E. Ruda, and Q. X. Guo, *Appl. Phys. Lett.*, **81**, 361 (2002).
18. X. Mei, M. Blumin, M. Sun, D. Kim, Z. H. Wu, H. E. Ruda, and Q. X. Guo, *Appl. Phys. Lett.*, **82**, 967 (2003).
19. J. Liang, H. Luo, R. Beresford, and J. Xu, *Appl. Phys. Lett.*, **85**, 5974 (2004).
20. W. L. Xu, M. J. Zheng, G. Q. Ding, and W. Z. Shen, *Chem. Phys. Lett.*, **411**, 37 (2005).
21. Y. B. Zheng, S. J. Chua, C. H. A. Huan, and Z. L. Miao, *J. Cryst. Growth.*, **268**, 369 (2004).
22. S. W. Kim, M. Ueda, M. Funato, Shigeo Fujita, and Shizuo Fujita, *J. Appl. Phys.*, **97**, 104316 (2005).

23. T. Ishikawa, S. Kohmoto, and K. Asakawa, *Appl. Phys. Lett.*, **73**, 1712 (1998).
24. S. Kiravittaya, H. Heidemeyer, and O.G. Schmidt, *Physica E.*, **23**, 253 (2004).
25. S. Kiravittaya, A. Rastelli, and O. G. Schmidt, *Appl. Phys. Lett.*, **88**, 043112 (2006).
26. S. Kiravittaya and O. G. Schmidt, *Appl. Phys. Lett.*, **86**, 206001 (2005).
27. V. C. Elarde, R. Rangarajan, J. J. Borchardt, and J. J. Coleman, *IEEE Photon. Techno. Lett.*, **17**, 935 (2005).
28. S. Tanaka, S. Iwai, and Y. Aoyagi, Appl. *Phys. Lett.*, **69**, 4096 (1996).
29. X. Q. Shen, S. Tanaka, S. Iwai, and Y. Aoyagi, *Appl. Phys. Lett.*, **72**, 344 (1998).
30. S. Tanka, J. S. Lee, P. Ramvall, and H. Okagawa, *Jpn. J. Appl. Phys.*, **42**, L885 (2003).
31. P. Chen, S. J. Chua, and J. N. Tan, *Appl. Phys. Lett.*, **89**, 023114 (2006).
32. P. Chen, S. J. Chua, Y. D. Wang, M. D. Sander, and C. G. Fonstad, *Appl. Phys. Lett.*, **87**, 143111 (2005).
33. A. Chen, S. J. Chua, P. Chen, X. Y. Chen, and L. K. Jian, *Nanotechnology.*, **17**, 3903 (2006).
34. U. H. Lee, D. Lee, H. G. Lee, S. K. Noh, J. Y. Leem, and H. J. Lee, *Appl. Phys. Lett.*, **74**, 1597 (1999).
35. T. Takeuchi, S. Sota, M. Katsuragawa, M. Komori, H. Takeuchi, H. Amano, and I. Akasaki. *Jan. J. Appl. Phys.*, **36**, L382 (1997).
36. C. Adelmann, J. Simon, G. Feuillet, N. T. Pelekanos, B. Daudin, and G. Fishman, *Appl. Phys. Lett.*, **86**, 206001 (2005).

3. Morphology Evolution and Emission Properties of Semipolar InGaN Multiple Quantum Wells Grown on GaN Microfacets by Selective Area Epitaxy

3.1. *Introduction*

Optoelectronic materials and devices based on InGaN multiple quantum wells(MQWs) have been developed across the blue and green wavelength range using metal–organic chemical vapor deposition(MOCVD).[1–3] However, for longer wavelengths the efficiency of devices based on GaN grown in the commonly used c-direction is continuously decreasing with increasing indium content in the active region.[4] This fact is thought to be mainly caused by strong built-in electrical fields as a result of the biaxially compressively stressed InGaN quantum wells. These piezoelectric fields (PFs)[5] lead to a local separation of electrons and holes within the QWs because of quantum confined stark effect (QCSE), and consequently lead to less efficient devices due to a reduced overlap of the electron and hole wave functions. Therefore, the researchers began looking for ways to avoid or reduce the internal electric field. The most promising proposal is to develop GaN materials and devices on non-polar and semi-polar plane surface.

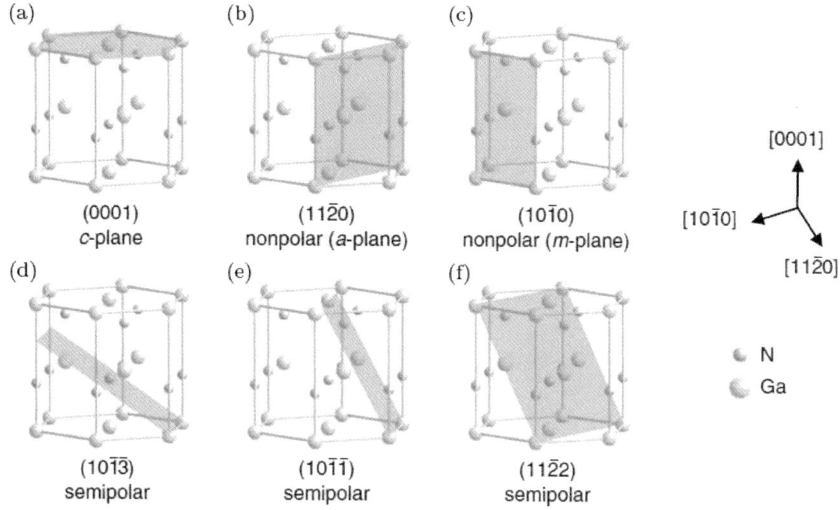

Fig. 14. The schematic diagrams of c-plane surface, non-polar surfaces, as well as typical semi-polar surface faces.

Figure 14 shows the schematic diagrams of polar c-plane, nonpolar a-plane and m-plane as well as some typical semi-polar crystal planes. It can be seen that a-plane {11-20} and m-plane {10-10} are perpendicular to the c-plane. The figure also lists some typical semi-polar planes, such as {11-22} and {1-101} plane, which show 58 degrees and 62 degrees to c-plane, respectively.

It has been known that different crystalline planes present different polarization electric field, resulting in different wave function overlap for electrons and holes.As seen from Fig. 15, the polar c-plane has a maximum polarization electric field. The polarization electric field alone nonpolar a-plane and m-plane is zero. Some semi-polar crystal planes, such as {11-22}, {1-101} planes, the polarization field is also significantly reduced, almost only one third of c-plane.

Compared to the optoelectronic devices grown on c-plane, the wave function overlap for electrons and holes can be increased for those devices on nonpolar and semi-polar surfaces because of the smaller polarization electric field, allowing higher internal quantum efficiency, as shown in Fig. 15. Although the growth of non-polar and semi-polar GaN material is not mature, but many researchers have been working on this novel structure,

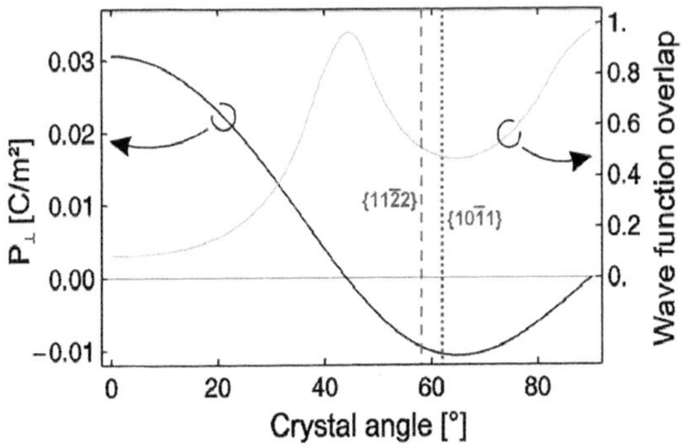

Fig. 15. Polarization electric field strength and electron hole wave functions overlap for different planes.

hoping to weigh the cost and device efficiency to realize commercial devices.

Non-polar GaN-based LED research begans in 2000, a study of m-plane GaN/AlGaN heterojunction structure has been reported on LiAlO2 substrate by using MBE.[6] Then in 2002, a-plane GaN film has been obtained on the r-plane sapphire substrate by MOCVD growth, and then non-polar InGaN/GaN optoelectronic devices were achieved.[7] Non-polar and semi-polar LEDs have become hot topics.

Numerous groups have tried to fabricate InGaN/GaN MQWs on nonpolar planes to avoid QCSE,[8,9] but the layers contain numerous nonradiative recombination centers since it is difficult to achieve a high crystallinity on nonpolar planes. Another approach to suppress the PFs is to use selective area epitaxy(SAE)[10] to obtain semipolar facets of high crystalline quality. Selective area epitaxy has shown great success in the growth of InGaN alloys. In selective MOCVD, growth occurs in openings or "windows" produced in dielectric mask such as SiO_2, SiN_x and bare sapphire when uniform GaN is grown and etched.[11–13] Lateral growth extending over the mask material reduces the vertical propagation of threading dislocations to very low level, and the generation of sidewall crystal facets can realize wide latitude in composition by controlling the mask design and growth conditions.[14] Up to now many researchers have reported growth of InGaN MQWs on the semipolar facets of GaN using the stripes pattern parallel to the [1-100] or [11-20] direction.[15,16]

3.2. *Semi-polar InGaN multiple quantum wells on GaN microfacets using parallel stripe patterns*

All samples were grown by MOCVD using trimethylgallium (TMGa), triethylgallium (TEGa), trimethylindium (TMIn), and ammonia (NH_3) as precursors. First, about $2 - \mu$m thick high quality GaN on c-plane sapphire substrates were grown. A 200-nm SiO_2 layer was then deposited as a mask material using plasma enhanced chemical vapor deposition (PECVD). The parallel stripe SiO_2 mask pattern with a width of 12 μm and a spacing of 4 μm was formed along the direction of [1-100] or [11-20] by conventional photo lithography and wet etching. The second epi-taxial GaN with side facets was grown in the open window at different temperatures of 890°C, 920°C, 955°C and 990°C (denoted as samples #1, #2, #3, #4) to investigate the influence of growth temperature on the morphology. Next, five-period InGaN/GaN quantum wells were grown. Finally, the MQWs were capped with a 20 nm thin GaN layer. For comparison, the same MQW structure was grown on c-plane GaN templates under the same conditions, which was denoted as the planar reference MQW structure without any mask pattern.

The geometrical aspects of the samples were investigated by field emission scanning electron microscope (FE-SEM). Information about the local composition of quantum wells was extracted from cathodeluminescence (CL) and photoluminescence (PL).

Usually, GaN is wurtzite structure by MOCVD growth. According to the distribution of crystal planes in the wurtzite structure, as well as differences growth rate of each crystal plane, obtained microfacets topography will be a huge difference by the selective area epitaxial growth when the mask pattern alignment is along different crystallographic orientation.

Figure 16 show the cross-sectional SEM micrographs of InGaN/GaN MQWs grown on GaN side facets with the SiO_2 stripe in [11-20] and [1-100] directions, respectively. As shown in Fig. 16(a), triangular GaN cross section mesas are obtained for stripes parallel to the [11-20] direction, while inclined sidewalls with flat (0001) surface have achieved for mask stripes oriented along the [1-100] direction as shown in Fig. 16(b). The triangular mesas are bounded by {1-101} side facets with the inclined angle of $\sim$62°, and the trapezoidal mesas {11-22} side facets with the inclined angle of $\sim$56°. It can be obviously seen from the insets of Fig. 16(a) and (b)

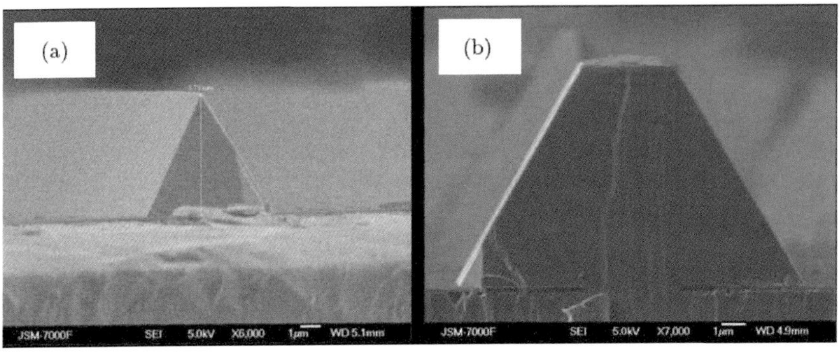

Fig. 16. Cross-sectional SEM images of stripes in [11-20] (a), and [1-100] (b) directions.

that GaN grown along the [11-20] stripes forming large area uniform triangular mesas with smooth sidewalls, as the morphology is independent of reactor pressure and growth temperature, and the lateral and vertical growth rates are constant, leading to the smooth surfaces consist of specific {1-101} facets with nitrogen polarity.[17] Whereas morphology for selective grown GaN in the [1-100] direction is poor, the mesas cannot coalesce completely and consist of step edges on the sidewalls. That's because the shapes of GaN selectively grown in the [1-100] are sensitive to the reactor pressure and growth temperature, and the (0001) and {11-22} facets are the most stable facets in this direction owing to their low surface energy.[18]

Figure 17 shows the SEM images of samples grown at 890°C, 920°C, 955°C and 990°C with stripes in [11-20] direction. The size of the mask is $4 - 4\mu$m (window-mask). It can be seen from the Fig. 2.5, as the temperature increases, the epitaxial growth always has a smooth {1-101} facet, indicating {1-101} facet is independent to the growth temperature under these growth conditions.

Figure 18 shows the SEM images of samples grown at 890°C, 920°C, 955°C and 990°C with stripes in [1-100] direction. It can be seen from the figure, only the {11-22} facet is formed by the epitaxial growth at the lower temperature. When the temperature rises to 955°C, flat (0001) facet appeared in the junction of two {11-22} facets. Further increasing the growth temperature, it was found that the {11-22} facets become narrower, and (0001) facets become wider, which indicates the increase of the lateral

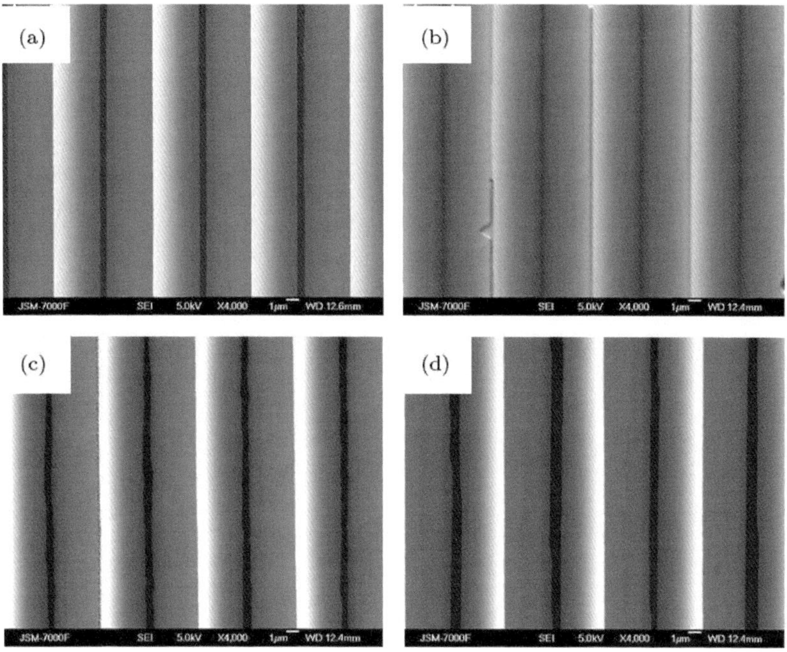

Fig. 17. SEM images of stripes in [11-20] directions grown at 890°C (a), 920°C (b), 955°C (c) and 990°C (d).

growth rate with the increase of the temperatures.

$$\varepsilon_{zz} = -\frac{2C_{13}}{C_{33}}\varepsilon_{xx}$$

$$P_{PE} = P_z \cos\theta + P_x \sin\theta$$

According to the above results, it can be concluded that the relationship between the facets formation and the growth temperatures with the stripes in [11-20], and [1-100] directions, as shown in Fig. 19.

InGaN/GaN MQWs grown on the two GaN stripe directions differ in the CL peak positions as plot in Fig. 20. The peak on the position of 365nm for both directions is arose from near-band-edge emission in GaN template. The MQWs on the triangular mesas of mask pattern along the [11-20] direction reveals monochromatic emission at 420~440 nm.

For the sample with the stripe at [1-100] direction, there are two facets, i.e. {11-22} facet and (0001) facet. CL from the two facets show huge

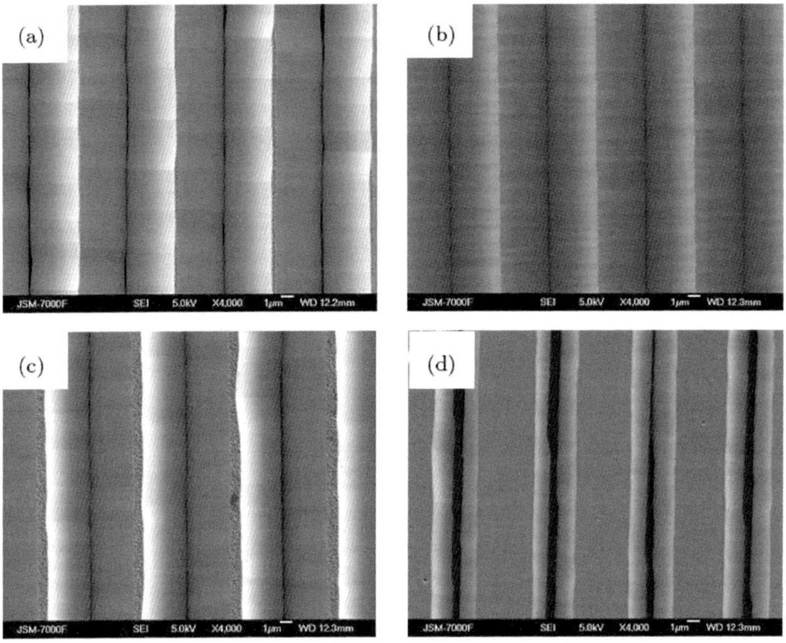

Fig. 18. SEM images of stripes in [1-100] directions grown at 890°C (a), 920°C (b), 955°C (c) and 990°C (d).

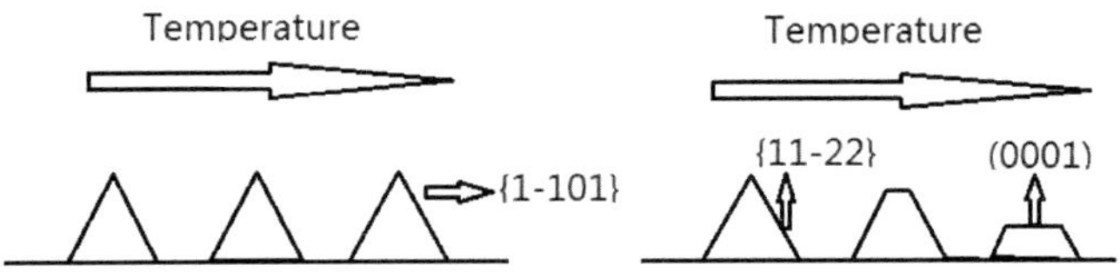

Fig. 19. The relationship between the facets formation and the growth temperatures with the stripes in [11-20] (left), and [1-100] (right) directions.

difference. The emission peak is 412 nm for {11-22} facet, while 518 nm for (0001) facet.

In order to compare the η_{int} of different facets, the temperature-dependent PL measurements were performed at temperatures from 15 to 300 K. The normalized PL intensities as a function of temperature are shown in Fig. 21. It is well known that the internal quantum efficiency of a semiconductor sample for luminescence is defined as the ratio of the integrated PL intensity at RT (300 K) over that at the lowest temperature

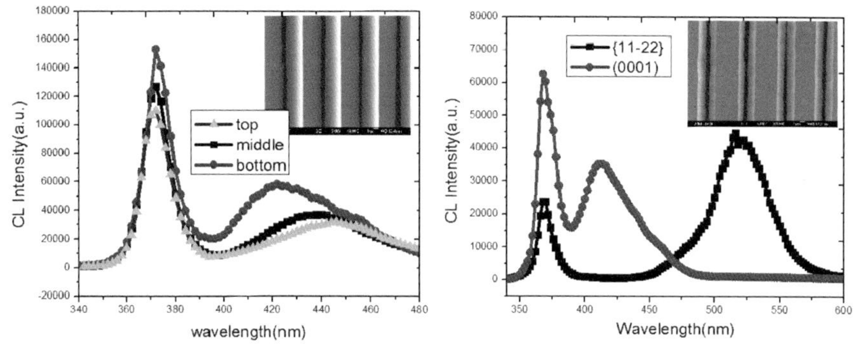

Fig. 20. The CL spectra of the stripes in [11-20] (left), and [1-100] (right) directions.

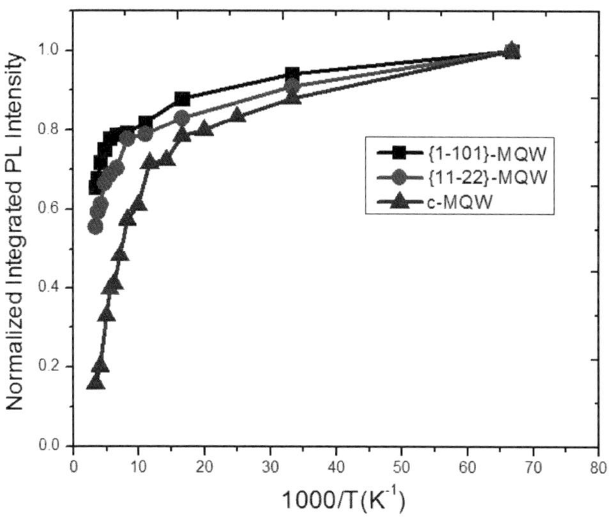

Fig. 21. The normalized integrated PL intensity of the MQWs on the semi-polar planes and c-planes variation with temperatures

to be reached (15 K in our case). Therefore, from Fig. 21, it can be seen that η_{int} of the c-plne planar MQW sample is 15.9%, 55.7% for the MQW on {11-22} facet and 65.6% for the MQW on {1-101} facet, which clearly indicates the higher emission efficiency for the semi-polar MQW. These results demonstrate that the elimination of the electric field polarization and QCSE effect can significantly increase InGaN/GaN multiple quantum wells luminescent properties.

3.3. *Semi-polar InGaN multiple quantum wells on GaN microfacets using crossover stripe patterns*

All samples were grown with conditions similar to above. The crossover stripe SiO_2 mask pattern with a width of 12μm and a spacing of 4μm was formed along the direction of [1-100] and [11-20] by conventional photo lithography and wet etching, as shown in Fig. 22. The second epi-taxial GaN with side facets was grown in the open window at different temperature of 890°C, 920°C, 955°C and 990°C.

Figure 23 shows the top-view SEM images of the investigated InGaN/GaN MQWs microfacet structures grown at temperatures of 890°C (a), 920°C (b), 955°C (c), and 990°C (d). As a result of the mask design, the structures are accurately ordered and exhibit a perfect homogeneity. As is shown in Fig. 23 the morphology of the microfacet structures changes with the GaN growth temperatures. The shapes formed by the outer edges of selectively overgrown GaN on crossover stripe patterns gradually change from nearly rectangular to hexagonal when the GaN growth temperature elevates from 890°C to 990°C. The boundaries of these shapes are composed of a series of surfaces with c-plane surface on top. The c-plane surface becomes larger with increasing growth temperature due to the surface migration, and the lateral vapor-phase diffusion is enhanced. As it is known from previous experiments stripes oriented along the direction of [11-20] form smooth triangular {1-101} facets, while stripes in the

Fig. 22. Optical microscope image of mask design in the shape of crossover stripe patterns.

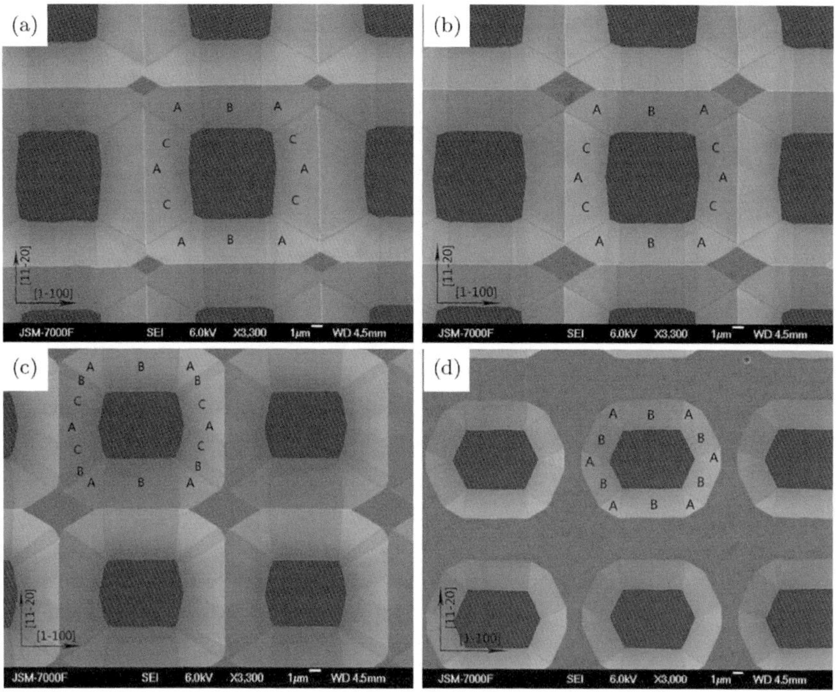

Fig. 23. Top-view SEM images of SAE InGaN/GaN MQWs on GaN microfacets grown at different temperatures of 890°C (a), 920°C (b), 955°C (c), and 990°C (d).

[1-100] direction produce {11-22} side facets with planar (0001) surface.[19] However, by using crossover stripe patterns, the semipolar facets formed along the [11-20] direction are not only {1-101} facets but also other types of semipolar facets. The growth rates at different positions along the [11-20] direction between crossover stripe masks gradually change because of the variation of surface migration and lateral vapor-phase diffusion of group III atoms at different locations. Moreover, this difference of growth rates along the [11-20] direction increases with growth temperatures, which can account for the changed shapes of inner rings in Fig. 23.

Upon close inspection, the varying shapes are defined by three types of sidewall facets which are denoted as A, B and C. Facets given common designations belong to the same family. Figure 24 shows the standard equivalent stereographic (0001) projection diagram of hexagonal structures, which is useful for identifying the crystal facets since it corresponds to the

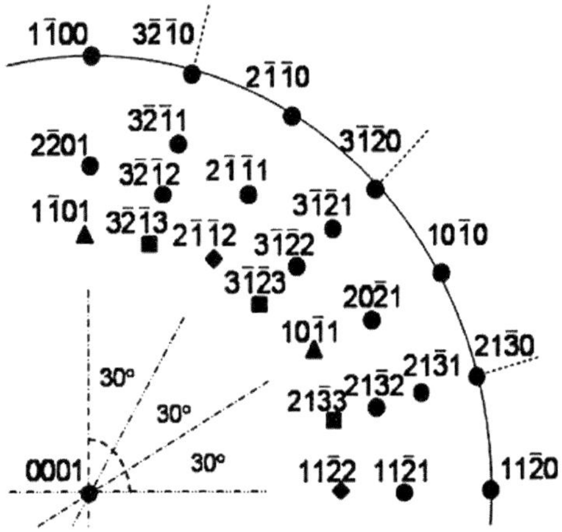

Fig. 24. Standard equivalent stereographic (0001) projection of hexagonal structures.

angular relationship between directions and the lattice planes.[20] All the semipolar facets are confirmed by measuring the inclined angles between the semipolar facets and the c-plane in the cross-sectional SEM images (not shown here). Based on the standard projection, the A facets are identified as equivalent semipolar {1-101} facets since the A facets are found to be inclined at an angle of 62° to the (0001) plane. All the B and C facets have an angle of 58° to the (0001) plane. The B facets which can be identified as {11-22} facets have been observed for InGaN/GaN stripes oriented along the [1-100] direction.[19] It can be explicitly observed from Fig. 23 that the C facets are formed between the {1-101} and {11-22} facets, so they are best identified as semipolar {21-33} facets based on Fig. 3. Different growth rates along the [1-100] and [11-20] directions and the variation of surface stability with elevated temperatures would be attributed to the formation of these semipolar facets by selective area epitaxy.[21] After identifying the types of sidewall facets surrounding the uncovered masks, it can be found that a obvious change of types of semipolar facets formed along the [11-20] direction when the GaN growth temperature elevates to 990°C. Specifically, besides {1-101} facets, the semipolarfacets formed along the [11-20] direction change from {21-33} facets to {11-22} facets. This is because that the {11-22} plane formation is more stable than that of the {21-33} plane when the GaN growth temperature is high.

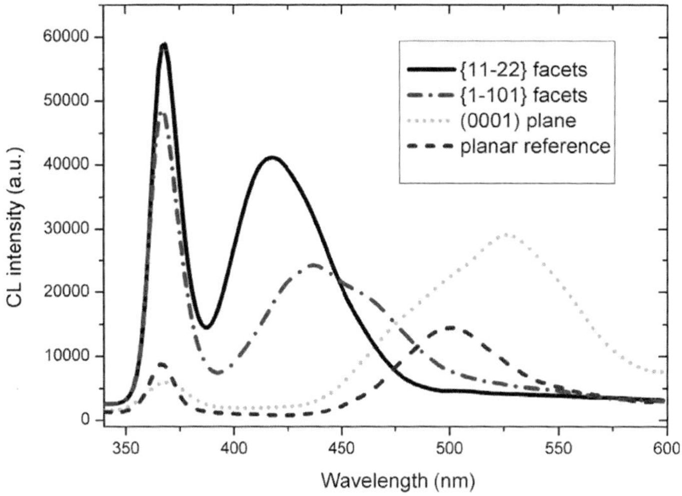

Fig. 25. Room temperature CL spectra of InGaN/GaN MQWs grown on microfacet structures and on planar reference structure without any mask patterning.

CL measurement is performed to investigate the emission properties of the InGaN/GaN MQWs grown on the aforementioned 990°C-grown GaN semipolar facets. Typical CL spectra of InGaN/GaN MQWs grown on different microfacets are displayed in Fig. 25, showing that the CL peaks at about 525, 440, and 420 nm are coming from MQWs on c-plane (0001) top facet, semipolar {1_101} and {11_22} side facets, respectively. Each spectrum has a peak near 368 nm corresponding to the GaN underlying the microstructural MQWs. The emission peak of MQWs grown on {11_22} facets presents a blueshift by 20 nm compared to the MQWs grown on {1_101} facets, which can be explained by variations in the growth rate on different crystal planes and more efficient incorporation of In on the {1_101} facet planes as compared to the {11_22} facet planes since the different growth rates influence the In incorporation efficiency.[22] For the SAE of InGaN, growth rates on different crystal facets are ordered as {1_101} >{11_22},[23] which is consistent with the sequence of the emission wavelength of MQWs on these semipolar facets. As is shown in Fig. 25, the intensity of underlying GaN (IGaN) is ordered as IGaN({1_101}) < IGaN ({11_22}), which can verify the variation of growth rate on these plane. The faster growth rate of InGaN and GaN on {1_101} facets compared to

{11_22} facets gives rise to thicker well and barrier layers. Since the MQWs material can attenuate the incident electrons prior to reaching the subsequent GaN, these thicker layers will cause greater attenuation, resulting in reduced intensity. In Addition, a narrower FWHM of CL spectrum of {11_22} facet compared to {1_101} facet can be discovered in Fig. 4. The FWHM of CL spectrum is influenced by the In spatial distribution,[24] indicating weaker indium fluctuation on {11_22} facets.

Furthermore, as shown in Fig. 25, a large redshift in emission wavelength for (0001) plane MQWs compared to its inclined {1_101} and {11_22} plane MQWs is attributed to the thicker InGaN well thickness and indium enrichment in the selectively grown InGaN/GaN MQWs. The composition increased from about 20% on the {11_22} and {1_101} plane to about 40% on the (0001) plane considering an increased well thickness from 2 nm to 3 nm, and barrier thickness from 10 nm to 15 nm, as the position changed from semipolar to polar plane.

The indium enrichment originates from additional source supply due to the surface migration effect and lateral vapor-phase diffusion during selective area epitaxy. The migration length of the group III atoms on the mask and their diffusion lengths in the vapor phase are ordered according to In > Ga > Al.[25] Therefore, indium enrichment during selective area epitaxy occurs due to the enhanced supply of In species from the inclined facets to the top surface. Hence, this type of SAE microstructure has potential applications on light-emitting devices with multi-wavelengths on single chip due to the large difference of wavelength between top and sidewall facets.

In comparison with the planar grown referential MQWs on the c-plane template, the (0001) plane MQWs of microfacet structures in this work show a longer emission wavelength. As it is well-known from selective growth, the local growth rate of the selectively grown structures increases due to gas phase diffusion of the precursor gases from the masked areas to the growth windows. Therefore, the longer emission wavelength can be attributed to a larger thickness of InGaN well and a higher incorporation efficiency of In.

To check for the uniformity of emission properties for MQWs on each microfacet, monochromatic CL intensity images at select wavelengths were acquired over the InGaN/GaN MQWs microfacet structure. It can be seen in Fig. 26 that {11-22} facets emit at 420 nm; light is notably absent from

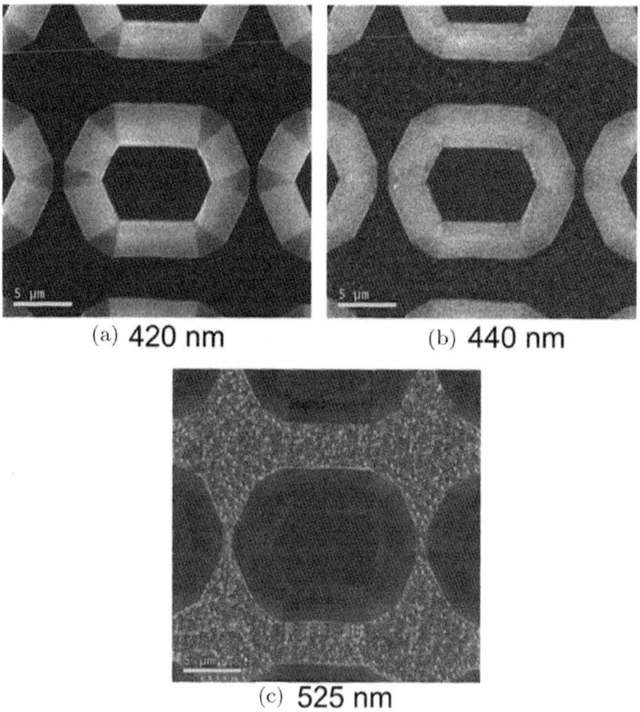

Fig. 26. Monochromatic CL intensity images at select wavelengths of InGaN/GaN MQWs microfacet structure.

the {1-101} facets and the top (0001) planar surface at this wavelength. This agrees with the emission peak in the CL spectrum of {11-22} facets and low intensity in the spectra at this wavelength from {1-101} facets in Fig. 4. Along with the consistent result in emission wavelength, the image exhibits that uniform emission can be seen from the entire inclined {11-22} facets at around 420 nm. The uniform emission supports the fact that the MQWs growth is homogeneous on the {11-22} GaN side facets. However, strong and nearly uniform emission intensity can be seen from all the sidewall facets of the microfacet structure at 440 nm. At 525 nm, we note that all the sidewall facets are dark, while emission intensity is only seen from the top area due to the planar c-plane MQWs with more In content than that of the sidewall MQWs. The results are in agreement with the CL spectra in Fig. 25.

In summary, selective area epitaxy have been applied to grow the GaN microfacet structures. The selectively overgrown GaN on a template masked with crossover stripe patterns form periodic hexagonal shapes with the boundary composed of six large rectangular semipolar {11-22} and six small triangular semipolar {1-101} side facets, while c-plane (0001) is formed on top. CL spectra observed InGaN/GaN MQWs on the microfacets show a large redshift of emission wavelength for (0001) plane MQWs of selectively grown compared to its inclined {1-101} and {11-22} plane MQWs, while the emission peak of MQWs grown on {11-22} facets presents a blueshift by 20 nm compared to the MQWs grown on {1-101} facets. The variation of emission wavelength on different planes is attributed to variations in growth rate and indium incorporation during InGaN growth. Furthermore, MQWs on microfacets exhibits uniform emission.

Acknowledgments

This work was supported by the national key technology research and development program (2016YFB0400100, 2016YFB0400602), Hi-tech Research Project (2015AA033305), the nature science foundation of Jiangsu province (BE2015111), Collaborative innovation center of solid state lighting and energy-saving electronics and Key Laboratory of Advanced Photonic and Electronic Materials.

References

1. S. Nakamura, T. Mukai, and M. Senoh, *Appl. Phys. Lett.*, **64**, 1687 (1994).
2. A. K. Viswanath, J. I. Lee, S. T. Kim, and D. Kim, *J. Cryst. Growth*, **260**, 322 (2004).
3. K. S. Ranaiah, Y. K. Su, S. J. Chang, C. H. Chen, and F. S. Juang, *Appl. Phys. Lett.*, **85**, 401 (2004).
4. G. Chen, M. Craven, A. Kim, A. Munkholm, S. Watanabe, M. Camras, W. Gotz, and F. Steranka, *Phys. Status Solidi A*, **205**, 1086 (2008).
5. F. Bernardini, V. Fiorentini, and D. Vanderbilt, *Phys. Rev. B.*, **56**, R10024 (1997).
6. P. Waltereeit, O. Brandt, M. Ramsteiner, A. Trampert, H. T. Grahn, J. Menniger, M. Reiche, R. Uecker, P. Reiche, and K. H. Ploog, *Phys. Status Solidi A*, **180**, 133 (2000).
7. M. Craven, S. H. Lim, F. Wu, J. S. Speck, and S. P. Denbaars, *Appl. Phys. Lett.*, **81**, 1201 (2002).
8. P. Waltereit, O. Brandt, A. Trampert, H. T. Grahn, J. Menniger, M. Ramsteiner, M. Reiche, and K.H. Ploog, *Nat. Lond.*, **406**, 865 (2000).

9. T. Onuma, A. Chakraborty, B. A. Haskell, S. Keller, S. P. DenBaars, J. S. Speck, S. Nakamura, U. K. Mishra, T. Sota, and S. F. Chichibu, *Appl. Phys. Lett.*, **86**, 151918 (2005).
10. T. Wunderer, J. Wang, F. Lipski, S. Schwaiger, A. Chuvilin, U. Kaiser, S. Metzner, F. Bertram, J. Christen, S. S. Shirokov, A. E. Yunovich, and F. Scholz, *Phys. Status Solidi C*, **7**, 2140 (2010).
11. X. Zhang, P. D. Dapkus, and D. H. Rich, *Appl. Phys. Lett.*, **77**, 1496 (2000).
12. T. S. Zheleva, O. H. Nam, W. M. Ashmawi, J. D. Griffin, and R. F. Davis, *J. Cryst. Growth*, **222**, 706 (2001).
13. P. de Mierry, N. Kriouche, M. Nemoz, S. Chenot, and G. Nataf, *Appl. Phys. Lett.*, **90**, 231918 (2010).
14. T. Tsuchiya, J. Shimizu, M. Shirai, and M. Aoki, *J. Cryst. Growth*, **276**, 439 (2005).
15. Y. Kawakami, K. Nishizuka, D. Yamada, A. Kaneta, M. Funato, Y. Narukawa, and T. Mukai, *Appl. Phys. Lett.*, **90**, 261912 (2007).
16. S. Khatsevich, D. H. Rich, X. Zhang, W. Zhou, and P. D. Dapkus, *J. Appl. Phys.*, **95**, 1832 (2004).
17. H. Miyake, A. Motogaito, and K. Hiramatsu, *Jpn. J. Appl. Phys.*, **38**, L1000 (1999).
18. K. Hiramatsu, K. Nishiyama, M. Onishi, H. Mizutani, M. Narukawa, A. Motogaito, and H. Miyake, *J. Cryst. Growth*, **221**, 316 (2000).
19. G. F. Yang, P. Chen, Z. G. Yu, H. Zhao, B. Liu, X. M. Hua, Z. L. Xie, X. Q. Xiu, Z. L. Wu, F. Xu, Z. Xu, X. M. Hua, P. Han, Y. Shi, R. Zhang, and Y. D. Zheng, Selective epitaxy of InGaN/GaN multiple quantum wells on GaN side facets, *Physica E*, **45**, 61 (2012).
20. B. D. Cullity and S. R. Stock, Elements of X-ray Diffraction, Prentice-Hall, Englewood Cliffs, NJ, 2001.
21. K. Hiramatsu, K. Nishiyama, M. Onishi, H. Mizutani, M. Narukawa, A. Motogaito, and H. Miyake, Fabrication, and characterization of low defect density GaN usingfacet-controlled epitaxial lateral overgrowth (FACELO), *J. Cryst. Growth*, **221**, 316 (2000).
22. S. Keller, B. P. Keller, D. Kapolnek, A. C. Abare, H. Masui, L. A. Coldren, U. K. Mishra, and S. P. DenBaars, *Appl. Phys. Lett.*, **68**, 3147 (1996).
23. J. Park, P. A. Grudowski, C. J. Eiting, and R. D. Dupuis, *Appl. Phys. Lett.*, **73**, 333 (1998).
24. K. Nishizuka, M. Funato, Y. Kawakami, Y. Narukawa, and T. Mukai, *Appl. Phys. Lett.*, **87**, 231901 (2005).
25. T. Tsuchiya, J. Shimizu, M. Shirai, and M. Aoki, *J. Cryst. Growth*, **276**, 439 (2005).

CHAPTER 9

FABRICATION AND CHARACTERIZATION OF GREEN RESONANT-CAVITY LIGHT-EMITTING DIODES PREPARED BY WAFER TRANSFER TECHNOLOGIES

Shih-Yung Huang* and Ray-Hua Horng

Department of Industrial Engineering and Management,
Da-Yeh University, 168 University Rd.,
Changhua 51591, Taiwan R.O.C.

Department of Electronics Engineering,
National Chiao Tung University,
1001 University Rd., Hsinchu 300, Taiwan R.O.C.
**syh@mail.dyu.edu.tw*

Green resonant cavity light-emitting diodes (RCLEDs) were demonstrated using a combination of wafer transfer and laser lift-off techniques. RCLED fabricated through all dielectric DBRs, hybrid mirrors, H^+ implantation, and vertical electrodes structures. The chapter discussed the effects of the Fabry-Pérot cavity involving the spectral purity, superior emission directionality, inherent temperature stability, and enhanced light extraction efficiency. Moreover, the RCLEDs for plastic optical fiber communication applications are also reported.

1. Introduction

Resonant cavity light-emitting diodes (RCLEDs) have been developed for optical communication along plastic optical fibers (POFs), optical interconnect, optical signal processing, and local-area network applications.[1] Compared to LEDs, RCLEDs offer many advantages, including spectral purity, superior directionality of the emitted light, inherent temperature stability, and improving the light extraction efficiency.[2–5] The design of the RCLED structure must be consider that (1) the reflectance of the Fabry-Pérot cavity mirrors which is relative to the light mode density. The high reflectance of the mirrors can result in the more the number of spontaneously emitted photons increasing[6]; (2) the cavity length can modify the spontaneous emission lifetime, the spectral purity, the light directionality, and the light extraction efficiency. These characteristics have been demonstrated in the other groups[7–9]; (3) the current confinement layers were demanded in order to decreasing lateral emitted light for fiber-coupled. (4) the thermal effect on RCLED devices is also an important issue. It is well known that the thermal issue can decrease the performance of devices. Currently, the raw materials of POFs are mostly using the polymethyl methacrylate, that of the signal transmission windows give a minimum attenuation loss at wavelength of 510∼530 and 570 nm.[10] These wavelengths can be achieved easily through using III-V nitride materials, which light emission properties are less temperature dependent than those of the commercial AlGaInP light source apply to POFs..[11] An RCLED structure consists of an active region was placed in two distributed Bragg reflector (DBR) mirrors that form a Fabry-Pérot cavity. However, obtaining highly reflective DBRs for the GaN-based RCLED is problematic partly due to AlGaN-based DBR and typical substrate choice. The high Al content GaN/AlGaN DBR layer is applied, but the lattice mismatch between AlN and GaN results in a large strain within the DBR structure and causes high defect densities which cause poor electrical conductivity[12–14] and substrates such as sapphire have a low thermal conductivity. These properties result in high Joule heating for the operating GaN-based device, thus reducing its performance. For this reason, dielectric DBRs offer a meaningful direction for exploring the route toward RCLEDs. Some authors have reported advantages of using dielectric DBRs for nitride light emitters.[15,16]

In this chapter, we propose the fabrication and characterization of green RCLEDs prepared by wafer transfer technologies. This technique can avoid the lattice mismatch from epitaxial-DBRs result in low reflectivity and narrow optical stopband width. Furthermore, it also solves the poor electrical conductivity issues through vertical-conducting structures. In Sec. 2, we describe the experimental procedures of our research on the green RCLEDs, including the epi-LED wafer growth, RCLEDs fabrication processes, and measurement methods. In section 3, the characterization of green RCLEDs fabrication by all dielectric DBRs and hybrid mirrors have been investigated. Furthermore, the optical and electrical properties of green RCLEDs with hydrogen ion-implantation were also investigated. We propose an approach in the improvement of thermal dissipation and increase the light-extraction efficiency by a combination of laser lift-off (LLO) and electroplating techniques. Details of the performance of green RCLEDs with vertical electrodes will be described. Finally, we summarize this dissertation and discuss the future development.

2. Experimental

2.1. *Fabrication of green RCLEDs with all dielectric DBRs and hybrid mirrors*

The GaN-based LEDs epilayers structure prepared using a metalorganic chemical vapor deposition (MOCVD). The epilayers were produced using standard epitaxial procedures. The RCLED processes proceeded as follows: a size $356 \times 356\,\mu m^2$ p-mesa was etched by inductively coupled plasma (ICP). An indium tin oxide (ITO) transparent conductive layer was deposited on the p-GaN layer to form the current spreading layer. A SiO_2 insulation layer was deposited as a current confinement layer to define the aperture region and the mesa region. The Cr/Au metals were deposited to form p- and n-contact pads. A total of five pairs of TiO_2/SiO_2 DBRs were evaporated as the top mirrors through using the electron beam evaporator. The thicknesses of the TiO_2 and SiO_2 layer were estimated through using an in-situ optical reflectance trace and quartz monitor. Then, a LLO process begins. The sample was bonded to a temporary glass carrier. An excimer ($\lambda = 248\,nm$) pulse laser was irradiated from the back surface of the sapphire substrate. After the entire wafer was illuminated by the laser beam, the two-inch sapphire substrate was separated from the sample. A diluted HCl solution was used to dissolve the residual Ga droplets on the exposed n-GaN surface. Subsequently, 7.5 pairs of TiO_2/SiO_2 DBRs were evaporated on an n-GaN layeras as a bottom mirror, to form a Fabry-Pérot cavity. The sample was then fused to the silicon substrate through using a wafer transfer technique. Finally, the sample was removed from the temporary glass carrier and a complete RCLED structure was obtained. This type of RCLED is called RCLED-I. Another structures of the RCLED devices have the same process

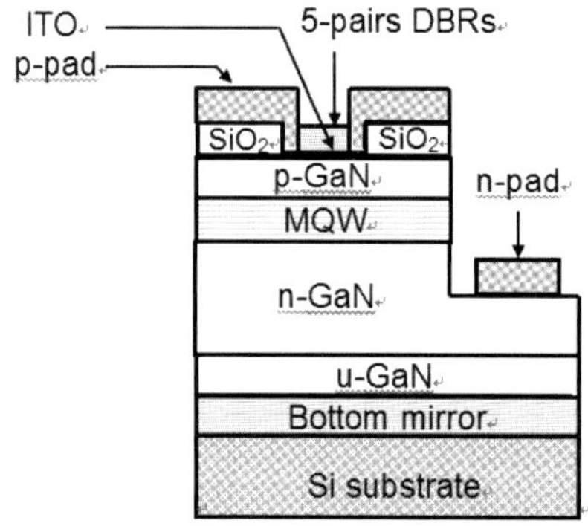

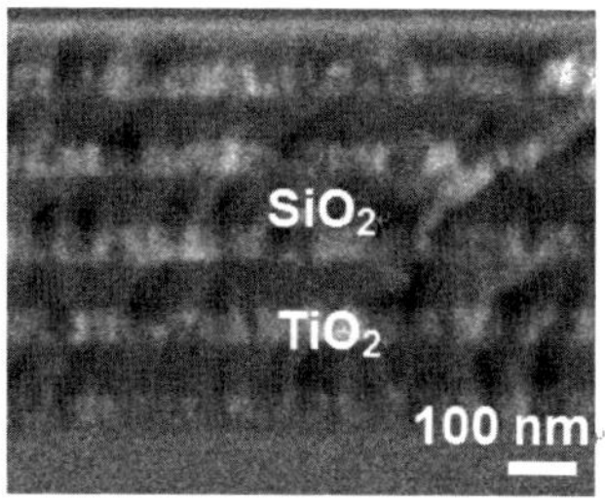

Fig. 1. Schematic diagram of a green RCLED structure. A field emission scanning electron microscopy micrograph of 5-pairs dielectric DBR is also illustrated.

procedures, except for the bottom mirrors. An Ag layer (thickness: 3000 Å) was utilized, instead of 7.5-pairs dielectric TiO_2-SiO_2 DBR, as the bottom mirrors. Here, RCLED-II denotes the type of RCLEDs with hybrid mirrors. A cross-sectional diagram of the Green RCLED structures prepared by wafer transfer techniques is shown in Fig. 1. Also given in this figure is the field emission scanning electron microscopy (FE-SEM) micrograph of the 5-pairs TiO_2/SiO_2 dielectric DBR in this structure.

2.2. *Fabrication of green RCLEDs using hydrogen Ion-Implantation techniques*

2.2.1. *Fabrication of different H-Ion implantation concentrations on green RCLEDs*

In order to ascertain the optimum implantation concentration for the current confinement layer of RCLEDs, three sets of samples are prepared by MOCVD technology, involving the *p*-GaN, epi-LED wafer and RCLED devices, all of which consist of as-grown one and samples with H^+ implanted with 10^{13}, 10^{14} and 10^{15} ions/cm^2, respectively. Before the RCLED devices fabrication, we first prepared the as-grown *p*-GaN of $\sim$0.1 μm on un-doped GaN ($\sim$1 μm)/sapphire, as shown in Fig. 2(a), and H-ion-implanted *p*-GaN with the 10^{13}, 10^{14} and 10^{15} ions/cm^2, respectively, to determine the tendency of the electrical and optical properties. Second, the material properties of the as-grown epitaxial LED wafer including the multi-quantum wells (MQW) structure and H^+-implanted (with 10^{13}, 10^{14}

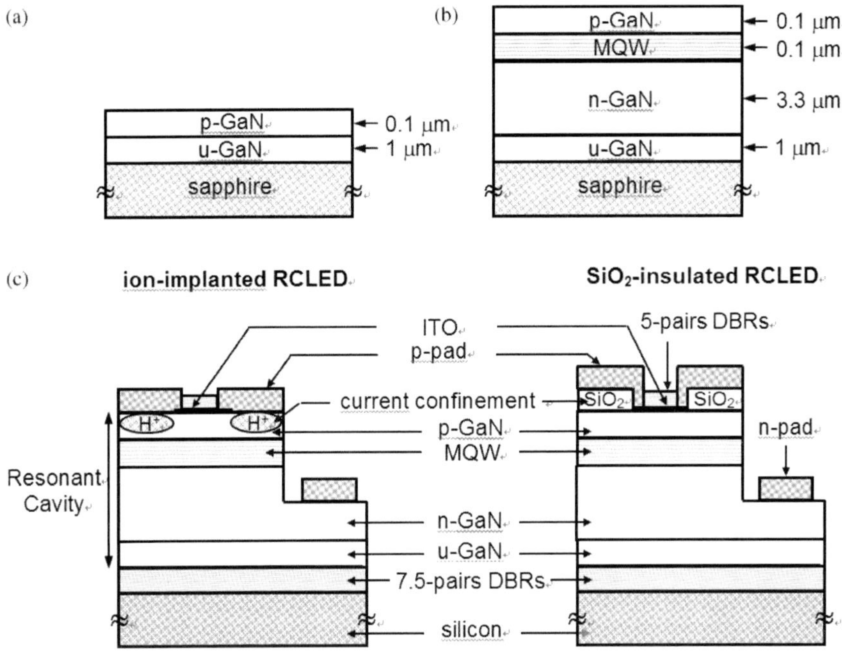

Fig. 2. Schematic diagrams of (a) p-GaN, (b) epi-LED wafer structures, and (c) ion-implanted and SiO_2-insulated RCLED devices. Reprinted with permission from Ref. 31. Copyright (2010) Semiconductor Science and Technology.

and 10^{15} ions/cm^2) epi-LED wafers are also investigated. The schematic drawing of epi-LED wafer structure is shown as in Fig. 2(b). Third, the as-grown, 10^{13}, 10^{14} and 10^{15} ions/cm^2 implanted RCLEDs were fabricated in order to demonstrate the influence of different hydrogen implantation concentration on characteristics of green RCLEDs. All RCLED devices have similar epilayer and process procedure (as mentioned in Sec. 2.1), except for the current confinement layers. The ion-implanted process involved the following procedures. An amorphous thin film, SiN_X, as the protective layer to prevent the channeling effect, was deposited via plasma enhanced chemical vapor deposition (PECVD) at a temperature of 350°C.[17] The ion-implanted region is defined using photolithography. The hydrogen ion implantation concentrations used in this work are 10^{13}, 10^{14} and 10^{15} ions/cm^2, respectively, and the implantation energy is 70 KeV. The implantation profiles were designed by the simulations via an ion transport in matter (TRIM) software to allow the implantation ions not penetrating beyond the top p-GaN layer. Subsquently, we fabricated the SiO_2-insulated (as-grown), 10^{13}, 10^{14} and 10^{15} ions/cm^2 implanted RCLED using wafer transfer and LLO technologies. The schematic diagrams of the ion-implanted and SiO_2-insulated RCLED are shown as in Fig. 2(c).

2.2.2. *Fabrication of green RCLEDs with H-Ion implantation and SiO_2 insulation layers*

In order to demonstrate the effect of current confinement on the RCLED, two types of devices were fabricated. The RCLEDs with and without H^+ implantation had similar epi-layer and process structures (as mentioned in Sec. 2.1), except for the current confinement layers. The ion-implanted process involved the following procedures. A Si_3N_4 film was deposited on the p-GaN layer and the bombardment region was masked either with or without photoresist using common photolithography. The photoresist was used to protect the epilayer from H^+ damage. Subsequently, the epi-wafer was bombarded with H^+ (energy: 70 KeV, dose: 1×10^{14} ions/cm^2) through the Si_3N_4 film. After the implantation, the chip processes involved the p-mesa of a size $356 \times 356\,\mu m^2$, current spreading layer, and p- and n-contact pads were also fabricated. Then the sample was transferred to the Si substrate with top and bottom DBR mirrors by the LLO and wafer transfer process to complete RCLED with H^+ implantation on silicon substrate specimen. This type of RCLED is called RCLED-III. For comparison, a RCLED without H^+ implantation was also fabricated. The only difference was to adopt the SiO_2 film as insulation layers form the RCLEDs without H^+ implantation. The RCLED-IV denotes the type of RCLEDs with SiO_2 insulation layers. The cross-sectional diagrams of both types of RCLEDs with H^+ implantation and SiO_2 insulation layers samples are shown in Fig. 3. The insert is a SEM micrograph of the GaN surface after LLO process.

2.3. *Fabrication of green RCLEDs with vertical electrodes*

In order to analyze the temperature dependence of the RCLEDs, RCLEDs with Cu and sapphire substrates were fabricated. The planar electrodes were also applied to an RCLED/Cu device in order to maintain the consistence of process condition and emission light area with RCLED/sapphire. The

Fig. 3. Schematic diagrams of (a) the RCLEDs without H^+ implantation (i.e. RCLED-IV devices), and (b) the RCLEDs with H^+ implantation (i.e. RCLED-III devices). The insert is a SEM micrograph of the GaN surface after LLO process. Reprinted with permission from Ref. 32. Copyright (2007) Journal of The Electrochemical Society.

GaN-based LEDs epitaxial structure were grown on 2-inch (0001)-oriented sapphire substrates. The epilayer and chip process structures as mentioned in Sec. 2.1, except for the top and bottom mirrors process. The *p*-mesa, current spreading layer, a SiO_2 insulation layer, and *p*- and *n*-contact pads were fabricated. $TiO_2/SiO_2/TiO_2$ multilayers were evaporated as the top mirrors using the electron beam evaporator. It presents ~51% reflectivity at 492–530 nm. Afterwards, the LED epiwafer was bonded to a temporary glass carrier using an adhesive layer, and a UV pulse laser was irradiated from the backside of the sapphire substrate to separate the sapphire from the LED sample. A diluted HCl solution was used to remove the residual Ga droplets on the exposed *n*-GaN surface. Subsequently, a Ti (30 Å)/Al (3000 Å) bilayer was evaporated on an exposed *n*-GaN surface as a bottom mirror. The bottom mirrors serve not only as a reflector, but also as a conduction layer. The bottom mirror presents ~87% reflectivity at 496–508 nm. The Pt film with 50 nm thickness was then deposited on the Ti/Al bilayer to serve as the seed layer for plating a Cu substrate. Then, a 130-μm thick Cu metal was electroplated as a permanent substrate for the RCLED/Cu device. Finally, a complete RCLED/Cu device was obtained after it was removed from the silica carrier. For comparison, a GaN-based epilayer with the Fabry-Pérot cavity structure was also fabricated for the RCLED/sapphire device. The RCLED/Cu and the RCLED/sapphire had the same epilayer and process structures except for the substrates. The RCLED epilayers with a bottom reflector were also bonded to the sapphire

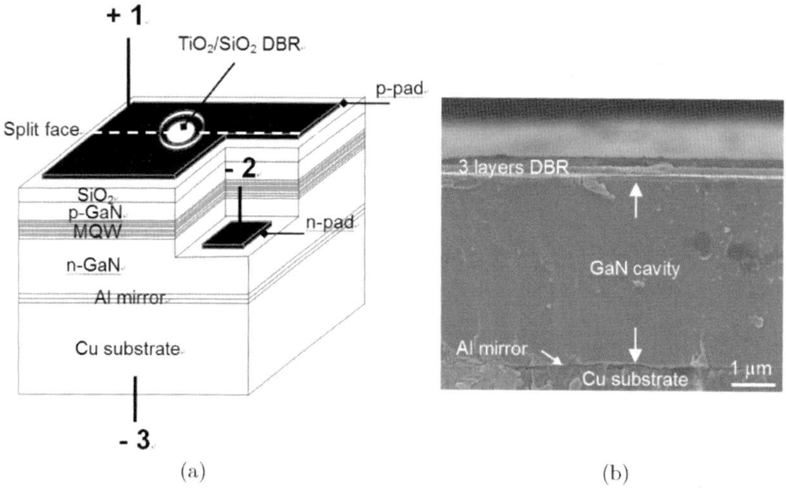

Fig. 4. (a) Schematic diagram of the RCLED/Cu, and (b) a SEM cross-section micrograph picture of the RCLED/Cu.

substrate using the thermal pressure wafer bonding. A schematic drawing of the RCLED/Cu device is shown in Fig. 2.4. All RCLEDs with Cu and sapphire substrates were probed on the electrodes of +1/-3 and +1/-2 labels, respectively. A cross-sectional micrograph of the RCLED/Cu device by SEM is also shown in Fig. 4(b). These chips were mounted on the gold-coated TO-cans without epoxy.

3. Results and Discussion

3.1. *Characteristics of green RCLEDs fabrication by all dielectric DBRs and hybrid mirrors*

3.1.1. *Characterization of green RCLEDs with all dielectric DBRs*

Figure 5 shows the reflectivity spectra of the top 5-pairs and bottom 7.5-pairs of TiO_2/SiO_2 dielectric DBRs. It was found that the high reflectivity of top and bottom DBRs were very close to the target wavelength at 525 nm. They also presented a broad stopband width of ~100 nm as the reflectivity of above 80%. Furthermore, the reflectivity of the top five pairs of TiO_2/SiO_2 DBRs (85%) was designed lower than that of the bottom dielectric DBRs (99.9%) at 525 nm, which forms a mirror of

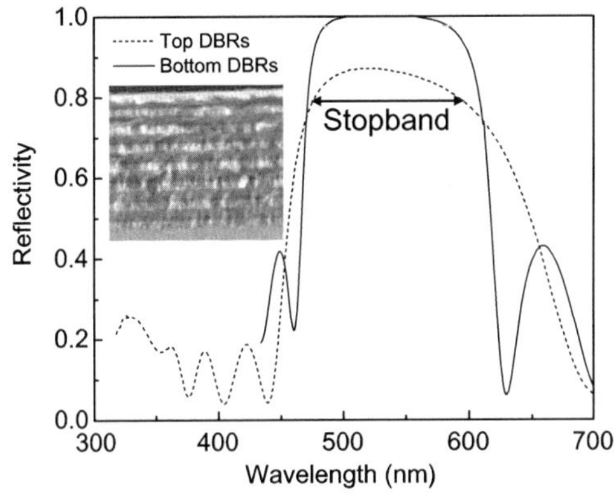

Fig. 5. Reflectivity spectra of the top 5-pairs and bottom 7.5-pairs of TiO_2/SiO_2 dielectric DBRs. The inset presents a FE-SEM image of 7.5-pairs of dielectric DBRs. Reprinted with permission from Ref. 32. Copyright (2007) Journal of The Electrochemical Society.

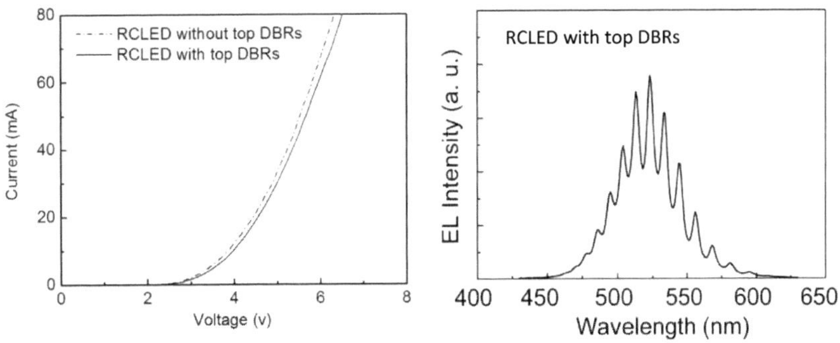

(a) (b)

Fig. 6. (a) Forward $I - V$ characteristics of the all dielectric DBRs RCLED with and without top DBRs. (b) Room-temperature EL spectra of the all dielectric DBRs RCLED with top DBRs.

semi-transmittance and semi-reflectivity. The FE-SEM image of 7.5-pairs of dielectric DBRs are also shown in the inset of Fig. 5. The thickness of the TiO_2 and SiO_2 are 40 nm and 95 nm, respectively. The result indicated that the thickness of DBRs is conforming to the designed thickness.

The room temperature (RT) forward $I - V$ characteristics of the all dielectric DBRs RCLED with and without top DBRs are shown in Fig. 6(a). Under an injection current of 20 mA, they present 4.55 and 4.45 V forward voltages for the all dielectric DBRs RCLED with and without top DBRs, respectively. The result indicated that the RCLED with top DBRs has similar $I - V$ characteristics as compared with that of the RCLED without top DBRs. The optical properties of the RCLED with all dielectric DBRs were studied using EL measurements, as shown in Fig. 6(b). The emission light was examined from the top of the nonencapsulated device at RT. The emission peak wavelength was located at 525 nm under an injection current density of 0.6 KA/cm^2. It was found that the emission spectrum of the RCLED with all dielectric DBRs was periodically modulated with several narrow resonant wavelength peaks due to the Fabry-Pérot oscillation with a cavity thickness of about 5 μm. The number of periodically modulated peak can be further reduced by decreasing the thickness of the resonant-cavity. In addition, it can be seen that cavity modes are well defined and exhibit a linewidth of an approximately 5 nm. The quality (Q) factor for this resonant cavity structure can be obtained, through the following Eq. (1)

$$Q = \left(\frac{2n_{eff}L}{\lambda_0}\right) \times F \tag{1}$$

where the F is the mode spacing/cavity mode. The Q factor was estimated to be approximately 105.

Figure 7 shows the injection current density versus the full width at half maximum (FWHM) of the emission peak from both types of all dielectric

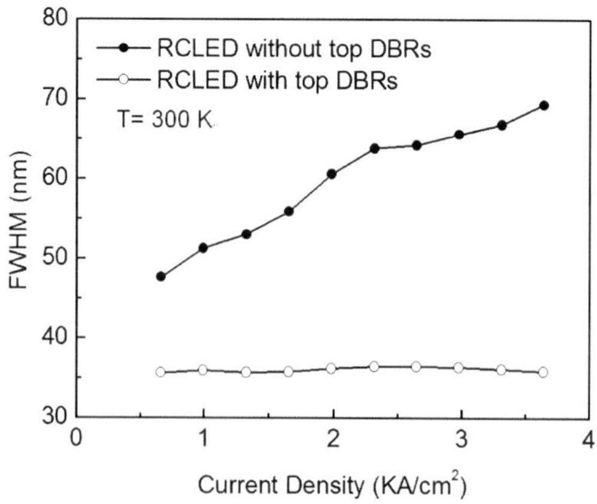

Fig. 7. FWHMs of EL spectra versus current density for the all dielectric DBRs RCLED with and without top DBRs.

DBRs RCLEDs at RT. It was found that the FWHM of the RCLED with top DBRs almost kept at 35 nm as the injection current density increased. The improvement in FWHM stability could be due to the superior resonant-cavity effect. On the contrary, a much broader EL emission was observed in the RCLED structure without top dielectric DBRs when the injection current density increased. Such a result could be attributed to the inferior resonant-cavity effect. To study the resonant-cavity effect on the light emission characteristics of the RCLED, the angle-resolved EL spectra were shown in Fig. 8. In these plots, the 0° of EL spectrum corresponds to the emission normal to the surface. We clearly observed a shorter resonant wavelength at higher emission angle, which is a typical microcavity effect and agrees well with the previous nitride-based RCLED report.[18] We demonstrate this through the following relation (2)

$$\frac{\lambda_\theta}{\lambda_0} = \sqrt{1 - \left(\frac{\sin^2(\theta)}{n_{cav}^2}\right)} \tag{2}$$

where θ is the angle with regard to the emitted normal to the surface, λ_θ and λ_0 are the EL peak wavelength at θ and normal to the surface, and n_{cav} is the refractive index of the resonant-cavity. If n_{cav} is a constant, the value of λ_θ decreased as the value of θ increase.

Figure 9 shows the peak wavelength of EL emission taken from both types of all dielectric DBRs RCLEDs as a function of injection current

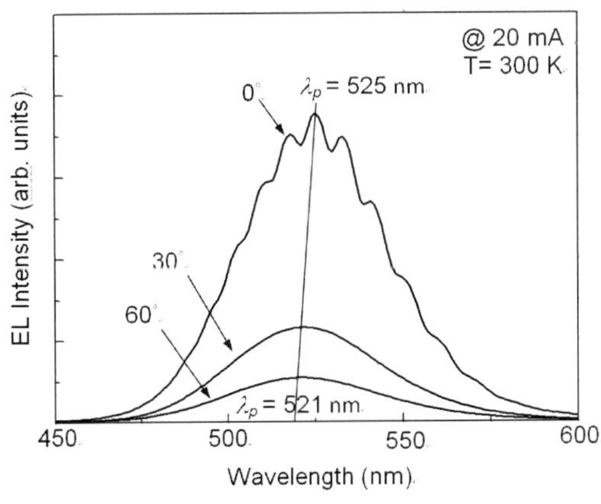

Fig. 8. Angle-resolved EL spectra of an InGaN all dielectric DBRs RCLED with top DBRs.

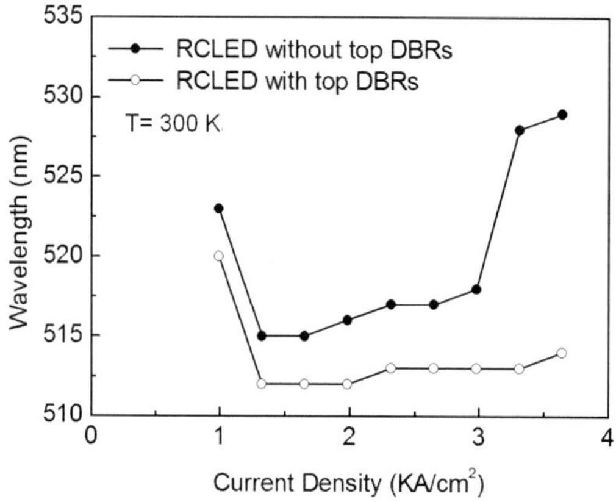

Fig. 9. EL wavelength shift versus current density for the all dielectric DBRs RCLED with and without top DBRs.

density. A strong EL blue shift was observed when the injection current density increased. It could be due to the localization effect caused by the high indium mole composition fluctuation in the green InGaN well and/or filling the band-tail states. Besides, when the injection current density increased, the all dielectric DBRs RCLED without top DBRs revealed a large red shift than the one with top DBRs. On the contrary, a more stable EL emission peak was observed in the all dielectric DBRs RCLED structures due to the superior resonant-cavity effect.

3.1.2. *Characterization of green RCLEDs with hybrid mirrors*

Figure 10 shows the electroluminescence (EL) spectrum of the hybrid mirrors RCLEDs with top DBRs. It was found that the emission spectrum of the RCLED with top DBRs structure was periodically modulated with several narrow resonant wavelength peaks due to the Fabry-Pérot oscillation. Moreover, the mode spacing about 9 nm between 525 and 534 nm resulted in the vertical cavity formed during two mirrors structures. The mode spacing $\Delta\lambda$ can be expressed as Eq. (3)

$$\Delta\lambda = \frac{\lambda_0^2}{2n_{eff}L} \tag{3}$$

where L is the cavity length, n_{eff} is the effective refractive index of the cavity, and λ_0 is the wavelength of the peak emission at normal to the

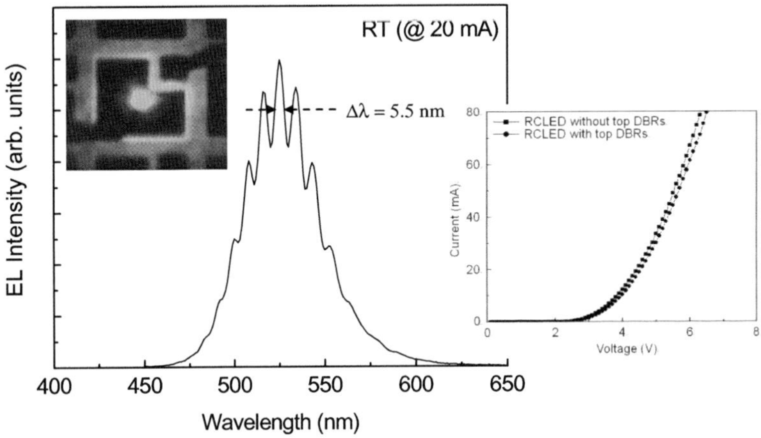

Fig. 10. The EL spectra and a plane view optical microscope image of the hybrid mirrors RCLED with top DBRs. The forward $I - V$ characteristics of both types of RCLEDs are also shown in the inset.

surface. Furthermore, the mode spectrum of this resonant-cavity structure is well defined and exhibits a cavity mode of approximately 5.5 nm for the dominant mode at $\lambda = 525$ nm, which gives a quality (Q) factor of about 95. The inset of Fig. 10 shows the forward $I - V$ characteristics of the hybrid mirrors RCLEDs with and without top DBRs. The both types of RCLEDs have similar $I - V$ characteristics. This result indicates that the processes of LLO and wafer bonding do not degrade electric properties. The plane view optical microscope image of the hybrid mirrors RCLEDs with top DBRs is also shown in the inset. Obviously, the RCLED structures can provide the circular light pattern. However, some light emission still occurs from the device edge. This problem can be overcome in the device fabrication and is under study of the Sec. 3.2.4.

Figure 11(a) shows the RT output power of both types of RCLEDs as a function of the injection current density. The output power was measured by a Si photodetector from the normal direction. It was found that the output power of the both RCLEDs increased proportionally to the injection current density. Under an injection current density of 0.6 KA/cm^2, the EL intensity of the hybrid mirrors RCLED with top DBRs structure shows 1.5 times the magnitude of the one without top DBRs structure. The improvement of optical output was attributed to the superior resonant-cavity enhancing the

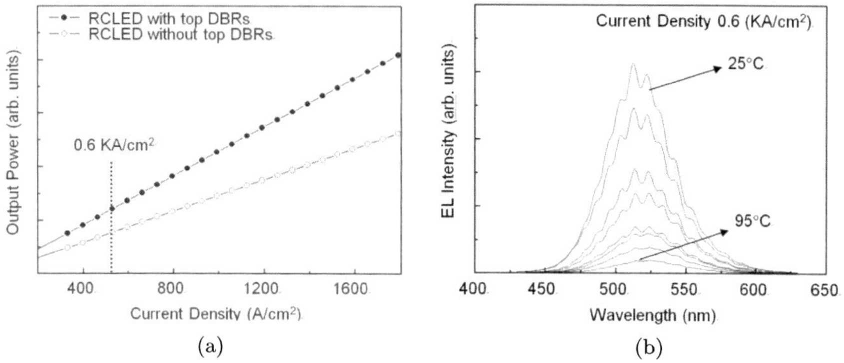

Fig. 11. (a) Light output power as a function of injection current density for hybrid mirrors RCLEDs with and without top DBRs. (b) Temperature-dependent EL spectra of hybrid mirrors RCLEDs with top DBRs. Reprinted with permission from Ref. 33. Copyright (2006) IEEE Photonics Technology Letters.

spontaneous emission at its resonant wavelength. A prime concern in the thermal stability of the hybrid mirrors RCLED with top DBRs was revealed by the temperature-dependent EL spectra, as shown in Fig. 11(b), where the chip was not encapsulated. It was found that the EL intensity decreased and the emission spectra broadened with the increase in operating temperature. When the temperature increased from 25°C to 95°C under an injection current density 0.6 KA/cm^2, the EL intensity decreased to about 6% of the RT one. Moreover, the resonant peak wavelength showed a red shift at a rate of 0.157 nm/°C. Such a phenomenon could be attributed to band gap shrinkage at high temperature.

3.1.3. *Comparison of green RCLEDs fabrication by all dielectric DBRs and hybrid mirrors*

In this section, we report on the comparison of green RCLEDs fabrication by all dielectric DBRs and hybrid Mirrors. The both types of RCLED structures were described in Sec. 2.1. The RCLED with all dielectric DBRs is called RCLED-I. Another structures of RCLED-II denotes the type of RCLEDs with hybrid mirrors. The simulation results of both types of RCLED devices will be described. The simulation results of spontaneous emission rate spectra of both types of RCLEDs at 20 mA were shown in Fig. 12 It is clearly seen that a main emission peak wavelength (525 nm)

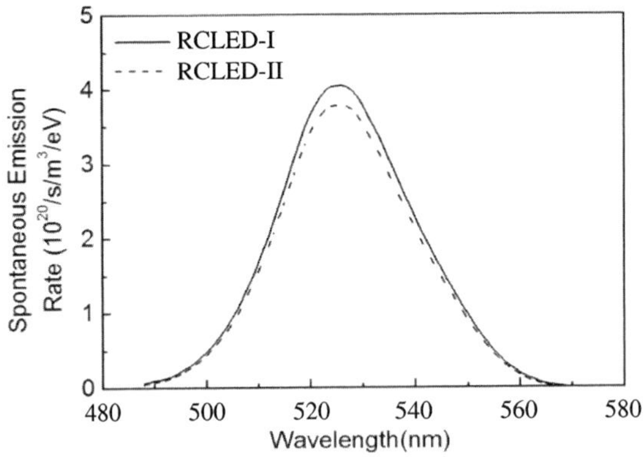

Fig. 12. Simulation results of spontaneous emission rate spectra of both types of RCLEDs at 20 mA.

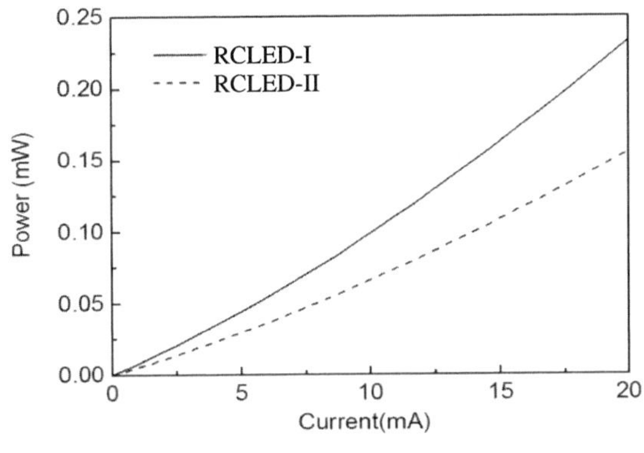

Fig. 13. Simulation results of light output power of both types of RCLEDs as a function of injection current.

of the numerical spontaneous emission rate spectra and the experimental EL spectra are of good agreement. The numerical spontaneous emission rate of the RCLED-I is higher than that of the RCLED-II. The phenomenon could be attributed to the absorption of dielectric DBRs was lower than that of the metal mirrors result in the optical performance enhancement of RCLED-I. Fig. 13 shows the simulation results of light output power of both types of RCLEDs as a function of injection current. The output power of both types of RCLEDs increases with the rise in injection current. The light output power of the RCLED-I is higher than that of the RCLED-II under the same injection current. It could be attributed to the dielectric DBRs have higher reflectivity than the metal mirrors result in the resonant cavity enhancing the spontaneous emission at its resonant wavelength.

3.2. *Characteristics of green RCLEDs with hydrogen ion-implantation*

3.2.1. *Effects of different H-Ion implantation concentrations on P-GaN films*

Monitoring the change in the sheet resistivity and carrier mobility is a common strategy for the electrical assessment of ion-induced damage to a semiconductor material. We adopt the four-point probe and Hall measurement method to analyze the relations between the sheet resistivity and carrier mobility characteristics, respectively. Fig. 14(a) shows the sheet resistivity and carrier mobility characteristics of the *p*-GaN/u-GaN/sapphire under different implantation concentrations. The sheet resistivity values of as-grown, 10^{13}, 10^{14} and 10^{15} ions/cm^2 *p*-GaN samples are 32.4, 340, 2.54×10^7, $8.8 \times 10^7\ \Omega \cdot$ cm, respectively. It can be seen clearly that the sheet resistivity increases while an increase of the implantation concentration, which may be caused by defects and various hydrogen complexes caused due to the hydrogen implantation.[19] The carrier mobility

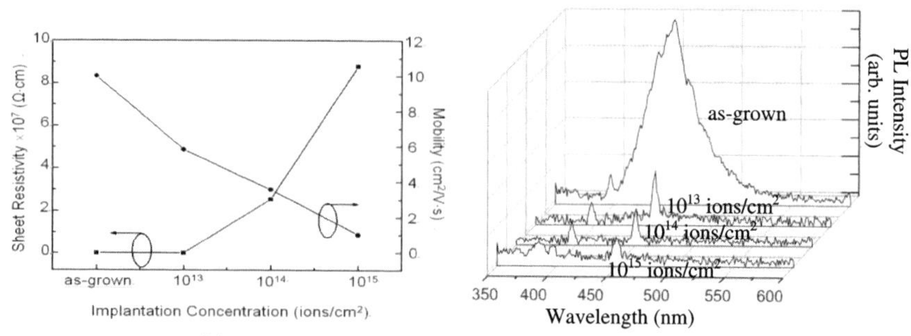

Fig. 14. (a) Sheet resistivity and carrier mobility of the p-GaN under different implantation concentrations. (b) Photoluminescence spectra of as-grown, 10^{13}, 10^{14} and 10^{15} ions/cm^2 p-GaN samples. Reprinted with permission from Ref. 31. Copyright (2010) Semiconductor Science and Technology.

of as-grown, 10^{13}, 10^{14} and 10^{15} ions/cm^2 p-GaN samples are 10, 5.82, 3.57 and 1.02 cm^2/V·s, respectively. The carrier mobility reduces with an increase of the implantation concentration because the implantation-induced defects blocked the carriers in p-GaN layers. The relations between the sheet resistivity and carrier mobility indicate that the sheet resistivity increases while the carrier mobility decreases. This could be attributed to the creation of defect states within the p-GaN acting as traps for the holes. From the above phenomena, it can be found that the implantation concentration of 10^{14} ions/cm^2 is a tailored concentration to acts as the current confinement layer for the RCLED devices. Fig. 14(b) shows the PL spectra of as-grown, 10^{13}, 10^{14} and 10^{15} ions/cm^2 implanted p-GaN samples. The blue luminescence (BL) band peak is at about 450 nm due to transitions from a deep donor to the shallow acceptor.[20] The BL band in these samples did not shift with increasing implantation concentration but its intensity decrease is dependent on implantation concentration from as-grown up to 10^{15} ions/cm^2. This could be attributed to the increasing amount of ion implantation inducing more defects in p-GaN layers. Also, it is worthy to note that there also exists extra ultraviolet luminescence (UVL) band peak centered at 385 and 396 nm due to the donor-acceptor-pair transitions.[20]

In order to understand the structural properties of the damaged layer in this implantation concentration range, HRXRD was performed to monitor the change of the GaN (0002) peak. Figure 15 shows the HRXRD spectra of the GaN (0002) peak for as-grown, 10^{13}, 10^{14} and 10^{15} ions/cm^2 implanted p-GaN samples. Each spectrum was fitted with Gaussian to obtain its FWHM value, which is shown in the inset of Fig. 3.11. The FWHM values of as-grown, 10^{13}, 10^{14} and 10^{15} ions/cm^2 implanted p-GaN samples are 18.0, 18.0, 21.6, 28.8 seconds, respectively. These results indicate that the FWHM of 10^{15} ions/cm^2 p-GaN samples is broadest due to the defect accumulated by high implantation concentration. On the contrary, the FWHM of 10^{13} ions/cm^2 implanted p-GaN sample is same as the as-grown one, indicating that the low dose H-implantation (10^{13} ions/cm^2) did not cause too much defects to result in the crystal lattice damaged. The phenomenon agrees with the results of the sheet resistivity and a tailored concentration with 10^{14} ions/cm^2 dose to acts as the current confinement layer.

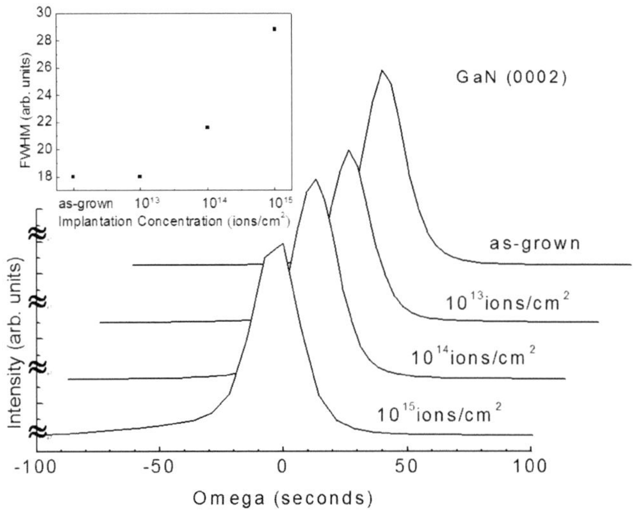

Fig. 15. XRD spectra of as-grown, 10^{13}, 10^{14} and 10^{15} ions/cm^2 p-GaN samples. The FWHM of as-grown, 10^{13}, 10^{14} and 10^{15} ions/cm^2 p-GaN samples are also shown in the inset.

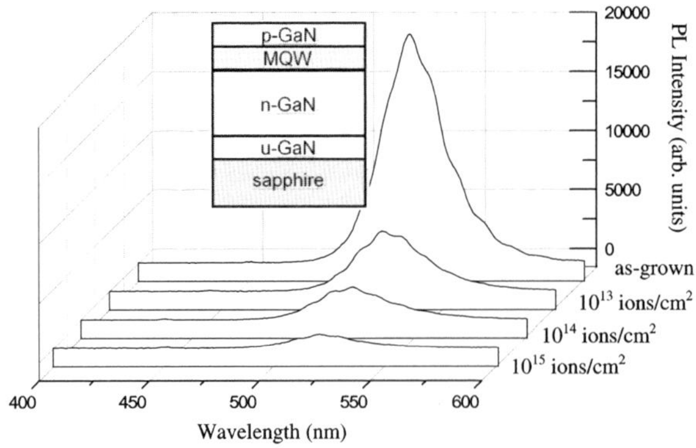

Fig. 16. Photoluminescence spectra of the epi-LED wafer samples with different implantation concentrations.

3.2.2. *Effects of different H-Ion implantation concentrations on Epi-LED wafers*

Figure 16 shows the PL spectra of the epi-LED wafers with different implantation concentrations. All emission peak wavelengths are located at 523 nm. The PL intensity of as-grown epi-LED wafer samples is relatively strongest as compare with that the other implanted epi-LED wafer samples.

The PL intensity weakens to accompany the implantation concentration raise, which is attributed to the low electron-hole recombination probability and the photons absorbed by the surface dislocations. It should be noted that 10^{15} ions/cm^2 epi-LED wafer samples exhibited weakest PL intensity due to the lattice deformation by implantation-induced damage.[21] The result confirms to the condition of the carrier mobility decreases while the sheet resistivity increases by the Hall measurement.

HRXRD measurements of the as-grown and ion-implanted epi-LED wafer samples were also performed. The HRXRD spectra of the as-grown and ion-implanted epi-LED wafer samples in the (0002) phase are shown in Fig. 17. The as-grown epi-LED wafer sample shows a sharp GaN (0002) peak with a narrow FWHM of 0.16° located at around 34.6°, and several satellite peaks from InGaN/GaN MQW structures as shown in the inset of this figure. The phenomenon exhibits the high crystalline quality and sharp GaN/InGaN interface of the as-grown epi-LED wafer. In contrast, the ion-implanted epi-LED wafer samples of the 10^{13}, 10^{14} and 10^{15} ions/cm^2 exhibits broadening reflection spectra with FWHM values of 0.19°, 0.25° and 0.29° in the angle region of 34.55–34.8°, respectively. This phenomenon indicates that the partial *p*-GaN regions was bombarded by hydrogen ion implantation which resulted in the winding grain and non-uniform

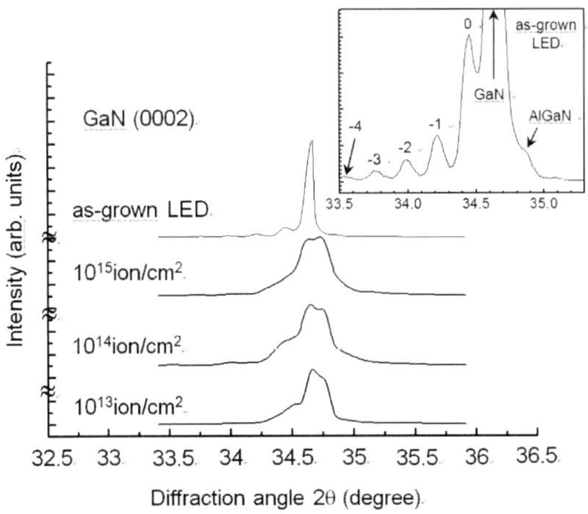

Fig. 17. XRD spectra of the as-grown, 10^{13}, 10^{14} and 10^{15} ions/cm^2 epi-LED wafer samples in the (0002) phase. An enlarged XRD spectrum of the as-grown epi-LED wafer samples is also shown in the inset. Reprinted with permission from Ref. 31. Copyright (2010) Semiconductor Science and Technology.

strain existing between the crystal lattice of p-GaN and MQW epi-LED layer in the normal direction of GaN (0002).[22] The peak on the left side of the (0002) peak shifts towards smaller angles and gradually shrinks until disappearance with increasing implantation concentration. It could be attributed to the excessive lattice expansion results in the collapse of the crystal lattice into amorphous clusters. The peak on the right side of the (0002) peak at around 34.75° is presumably originated from the AlGaN layer beneath the top p-GaN layer.[23] Because of the severe damage and the defect accumulation in the top p-GaN layer,[22] XRD diffracted intensity of the GaN (0002) peak from p-GaN layer was badly weakened, leading to the relative enhancement of the XRD signals from the underneath p-AlGaN layer. The XRD spectra of the ion-implanted epi-LED wafer samples fully exhibit that the lattice damage by hydrogen ion implantation, which could degrade the optical properties. The demonstration conforms to the results of Fig. 16, where the PL intensity weakens to accompany the implantation concentration raise.

In order to investigate the effect of the hydrogen ion implantation on the ohmic contacts of the p-GaN and metal layer, the as-grown and 10^{13}, 10^{14} and 10^{15} ions/cm^2 epi-LED wafer samples are deposited with the ITO (200 nm) films by E-beam evaporation system and then thermal annealed at 550°C in N_2 ambient for 5 minutes. Figure 18 shows the specific

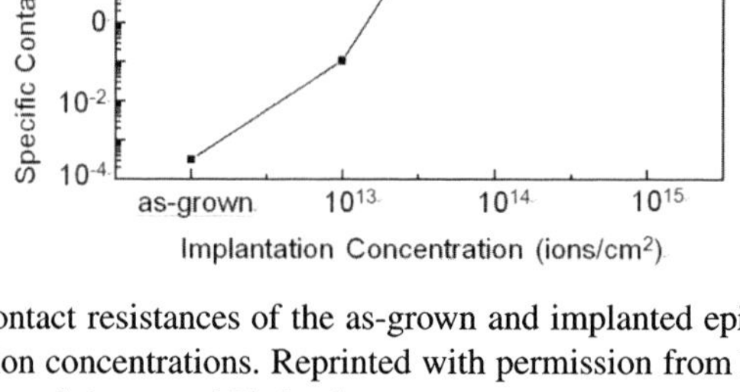

Fig. 18. Specific contact resistances of the as-grown and implanted epi-LED wafer under different implantation concentrations. Reprinted with permission from Ref. 31. Copyright (2010) Semiconductor Science and Technology.

contact resistance of the as-grown and ion-implanted epi-LED wafers under different implantation concentrations. The epi-LED wafer sample with 10^{13} ions/cm^2 has three orders increase of resistivity from $3.2 \times 10^{-4} \Omega \cdot cm^2$ to $1.1 \times 10^{-1} \Omega \cdot cm^2$, compared with the as-grown epi-LED wafer sample. As the implantation concentration increases up to 10^{14} ions/cm^2, the specific contact resistance value sharply increases to $1.4 \times 10^5 \Omega \cdot cm^2$. These phenomena indicate that the implantation concentration of 10^{14} ions/cm^2 is sufficient for an insulator of the RCLED devices which facilitated the formation of high-resistivity ohmic contacts to *p*-GaN.

3.2.3. *Characteristics of RCLEDs with Different Implantation Concentrations*

Figure 19 shows the EL spectra from the nonencapsulated RCLEDs with different implantation concentrations at a forward operation current of 20 mA, which the emission peak wavelength is located at 525 nm. An EL spectrum from a SiO_2 insulated (i.e. as-grown) RCLED is also included in the figure for a comparison. The optical properties of these devices are observed in the normal direction from the top light emission of the device. It is observed that the EL signals from the SiO_2 insulated RCLED are much weaker than all H^+-implanted RCLED. It is also seen clearly that the emission spectra from all RCLEDs are periodically modulated with

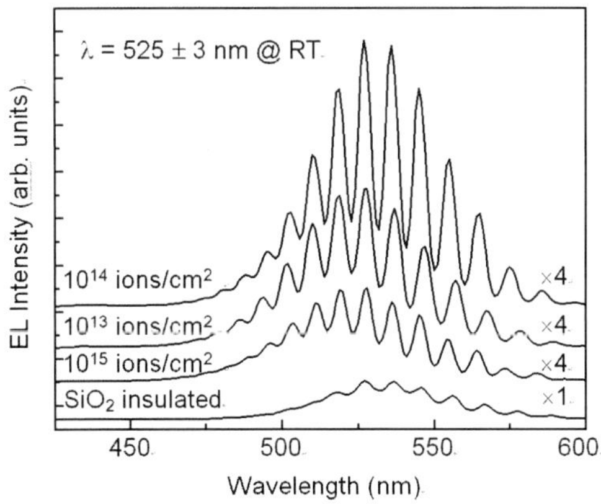

Fig. 19. EL intensity of the as-grown and ion-implanted RCLEDs under different implantation concentrations. Reprinted with permission from Ref. 31. Copyright (2010) Semiconductor Science and Technology.

several narrow resonant wavelength peaks due to the Fabry-Pérot cavity effect. As the H^+-implantation concentration increases from 10^{13} ions/cm^2 to 10^{14} ions/cm^2, the EL intensities increase greatly. This is because of the high carrier quantity injected efficiently into the light aperture region by the increased hydrogen ion implantation. It produced the electrons combining with holes effectively and resulted in more spontaneous emission light from the light aperture region.[24] However, as the H^+-implantation concentration increases further to 10^{15} ions/cm^2, the EL signals from the implanted RCLED decrease severely, which are lower than that of 10^{13} and 10^{14} ion/cm^2 implanted RCLED. This is due to the too heavy implantation concentration resulting in the effect of the large lattice expansion and severe distorting on the light aperture region (non-implanted region). The above results indicate that the the 10^{14} ion/cm^2 implantation concentration into the p-GaN is the most optimum H-ion implantation condition for fabricating RCLEDs and that RCLEDs with 10^{14} ion/cm^2 implantation concentration provide the best and superior optical property.

Figure 20 shows the 3D far-field light emission patterns of (a) the SiO_2-insulated and (b) the 10^{14} ions/cm^2 implanted RCLED at 5 mA. The light distributions of the SiO_2-insulated and the 10^{14} ions/cm^2 implanted RCLED are all showing a sketch of the chip and the only light emission aperture in a diameter of 40 μm, respectively. The sideward light emission of the SiO_2-insulated RCLED is more than that of the 10^{14} ions/cm^2 implanted RCLED. The result demonstrates that the 10^{14} ions/cm^2 implanted RCLEDs have superior directionality as compare with the SiO_2-insulated RCLED. The current path of implanted RCLEDs was effectively confined into the light aperture, which results in an enhancement of the photon confinements laterally. In order to avoid the light emission intensity saturation in the charge coupled device (CCD) a polarizer in front of the light emission beam of the RCLEDs was used. The light emission patterns of the SiO_2-insulated and the 10^{14} ions/cm^2 implanted RCLED with a 0.1% filter at 5 mA are shown in Fig. 20(c) and (d), respectively. It indicates that the intensity of 10^{14} ions/cm^2 implanted RCLED is higher than that of the SiO_2-insulated RCLED at the same filter and current injection. Obviously, the light output from the 10^{14} ions/cm^2 implanted RCLED, i.e. its performance, is superior to that of the SiO_2-insulated RCLEDs at the normal direction.

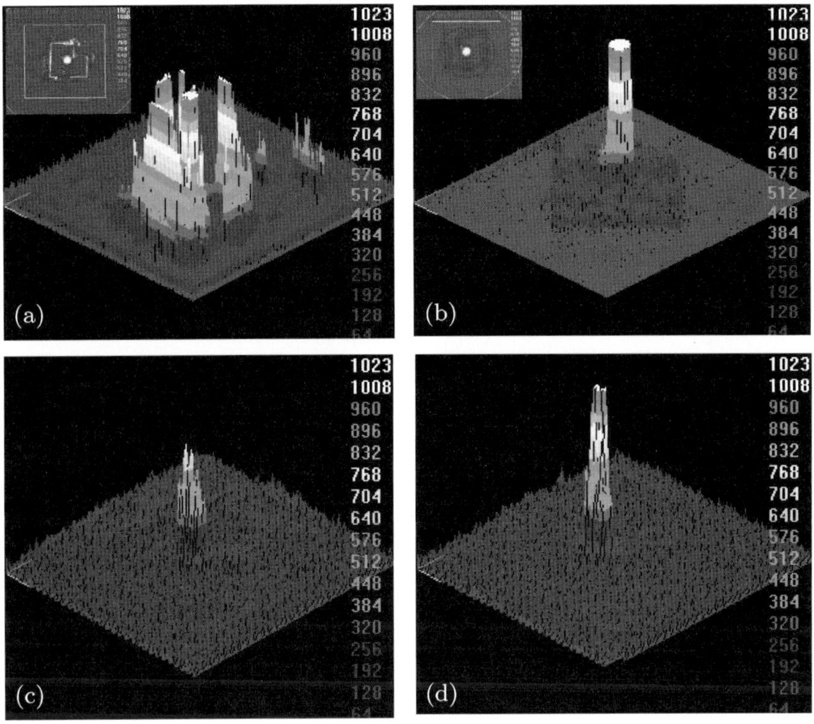

Fig. 20. Light emission patterns of (a) SiO_2-insulated (b) 10^{14} ions/cm^2 implanted RCLEDs at 5 mA (c) SiO_2-insulated (d) 10^{14} ions/cm^2 implanted RCLEDs at 5 mA with a 0.1% filter. Reprinted with permission from Ref. 31. Copyright (2010) Semiconductor Science and Technology.

3.2.4. *Comparison of green RCLEDs fabrication by hydrogen ion-implantation and SiO_2 insulation layers*

Figure 21 shows the typical secondary ion mass spectrometry (SIMS) depth distribution profiles of H^+ implantation (dose: 1×10^{14} ions/cm^2). The results indicate that the H^+ concentration is higher than the intrinsic H^+ concentration of epi-LED wafer after the ion-implanted processes. Notably, H^+ has a region of very high concentration extending to $\sim$0.05 μm from the epi-wafer surface. Furthermore H^+ concentration rapidly decreased when it was relatively distant from the surface of *p*-GaN layer. This indicates that H^+ in the *p*-GaN layer follows an in-diffusion profile where H^+ is trapped by Mg or defects. As concerning the distribution of H^+ implantation, the

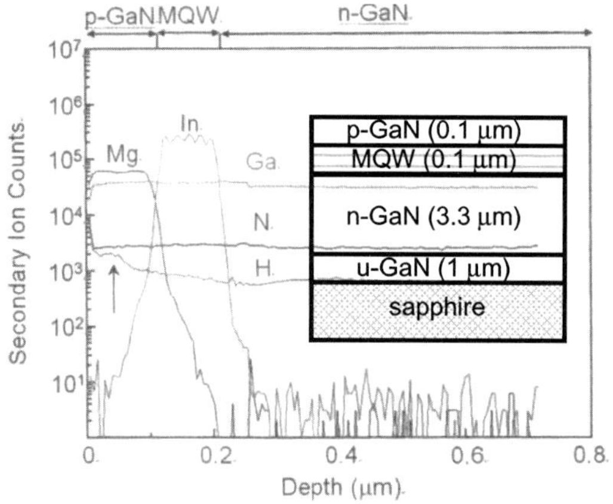

Fig. 21. SIMS depth profiles of H^+ implantation with energy of 70 KeV and dose of 1×10^{14} ions/cm^2 in p-GaN layers. The inset shows a cross-sectional diagram of the epi-LED wafer structure. Reprinted with permission from Ref. 34. Copyright (2010) IEEE Photonics Technology Letters.

epi-LED wafer with and without H^+ implanted *p*-GaN layer were also measured by cathodoluminescence (CL). Figure 22(a) shows a plane-view SEM micrograph of the epi-LED wafer with H^+ implanted *p*-GaN layer. The electron beam energy and current are 15 KeV and 0.3 nA, respectively. Figure 22(b) indicates that the CL spectra of the epi-LED wafer with and without H^+ implanted *p*-GaN layer. The spectra consist of one band centered at 371-nm peak and yellow band. They are attributed to the GaN band to band emission and epitaxial defects, respectively. However, it is worthy to note that there exists an extra spectrum centered at ~512 nm peak in the epi-LED wafer with H^+ implanted *p*-GaN layer. In order to evaluate the spectrum, the CL image obtained for a detection energy 2.4 eV corresponding to ~512 nm spectrum was shown in Fig. 22(c). Most of the area is bright and corresponding to the implantation area. On the contrary, the area without H^+ implantation presents dark. The results suggest that the 512-nm extra peak arising from the H^+ implantation. This suggests that H^+ implantation results in Mg trapping H^+ again or more complex vacancy defects formation.[20]

Following, characteristics of the green RCLEDs fabrication by hydrogen ion-implantation and SiO_2 insulation layers are reported. The RCLED with hydrogen ion-implantation is called RCLED-III. Another structures of

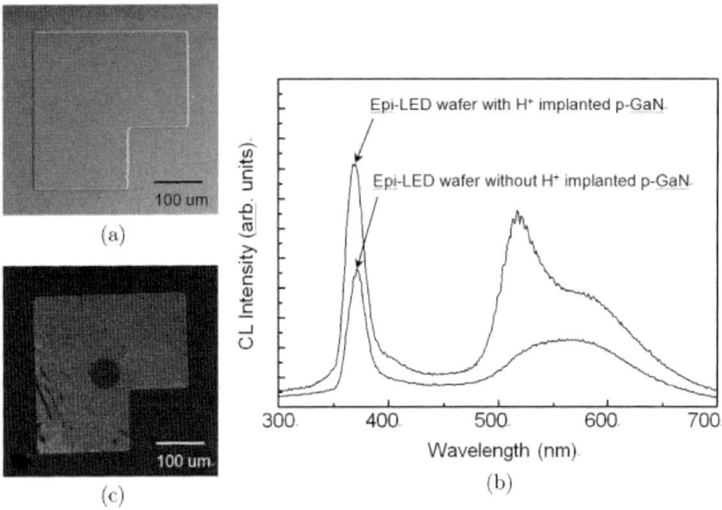

Fig. 22. (a) A plane-view SEM photograph of the epi-LED wafer with H^+ implanted p-GaN, (b) CL spectra of the epi-LED wafer with and without H^+ implanted p-GaN and (c) CL image obtained for detection energy 2.4 eV corresponding to ~512 nm spectrum. Reprinted with permission from Ref. 34. Copyright (2010) IEEE Photonics Technology Letters.

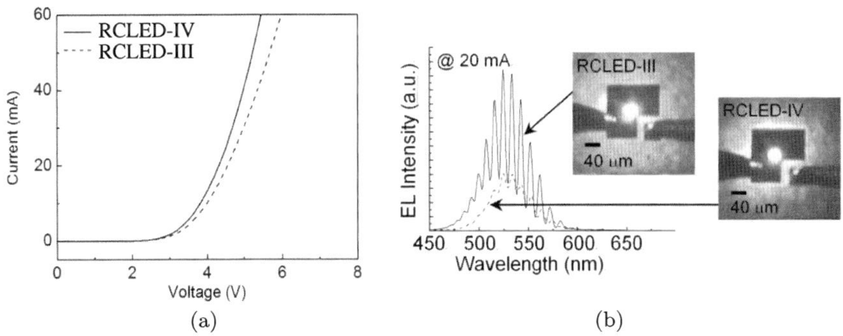

Fig. 23. (a) Forward $I-V$ characteristics of both types of RCLEDs at the same current injection. (b) The EL spectra and the light emission photographs of both types of RCLEDs at 20 mA. Figure (b) is reprinted with permission from Ref. 34. Copyright (2010) IEEE Photonics Technology Letters.

RCLED-IV denotes the type of RCLEDs with SiO_2 insulation layers. The forward $I-V$ curves of both types of RCLEDs are shown in Fig. 23(a). The corresponding forward turn-on voltage at 20 mA dc current injection was determined ~4.6 V and ~4.4 V, respectively. The result indicates that the

RCLED-III structure has similar $I-V$ as compared to that with the RCLED-IV. The EL spectra of both types of RCLEDs at 20 mA forward operation current injection are shown in Fig. 23(b). The spectra were measured from the normal direction of the bare chips without epoxy encapsulating. The emission peak wavelength is located at 525 nm. The emission spectra from both types of RCLEDs were periodically modulated with several narrow resonant wavelength peaks. This phenomenon indicates that the interference results from the multiple reflections at the mirrors on both sides of the Fabry-Pérot cavity. A detailed inspection of two emission spectra shows that the FWHM of the light emission of the RCLED-III and RCLED-IV are 20 nm and 36 nm, respectively. The results reveal a significant change in the spectral purity of the RCLED-III, suggesting that the more spontaneous emission light generates within the resonant-cavity. In addition, the EL intensity of the RCLED-III is higher than that of the RCLED-IV. This phenomenon indicates that the spontaneous emission light towards the escaping cone of the resonant-cavity is strongly redirected along the normal direction of the cavity due to the fact that the RCLED-III provides the high current density at the same current injections. The light emission photographs of both types of RCLEDs are also shown in the inset of Fig. 23(b). It is obvious that the lateral light emission confinement of the RCLED-III is superior to that of the RCLED-IV, as indicates in the micrographs. The result could be attributed to the H^+ implantation's ability to effectively control the carrier recombination at the center of the resonant-cavity. Moreover, the current route was confined to below the light aperture of devices, therefore attaining the superior directionality of light emission.

Figure 24 shows the far-field patterns of both types of RCLEDs at 20 mA current injection by the CCD. The far-field patterns were measured by placing a polarizer in front of the light emission beam of the RCLEDs. It was found that the light output patterns of both types of RCLEDs were a nearly Gaussian distribution. The RCLED-III has much higher light-emission intensity than that of the RCLED-IV. The far-field patterns of both types of RCLEDs without placing a polarizer for light distribution condition were also measured and are shown in the insets of Fig. 24. Obviously, the directionality of RCLED-III performs superior to that of the RCLED-IV. From the device structure as shown in Fig. 3, it can be found that

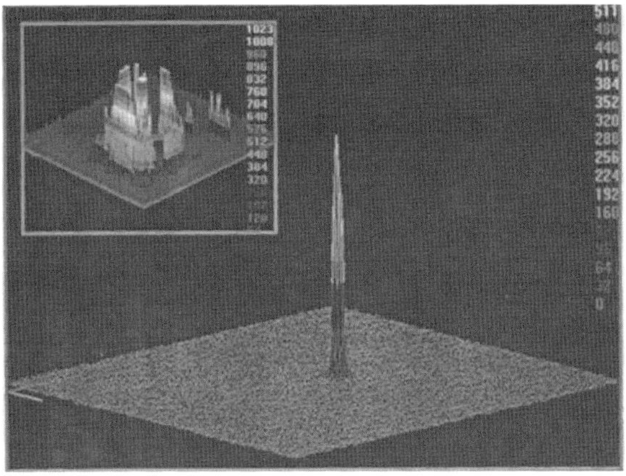

(a)

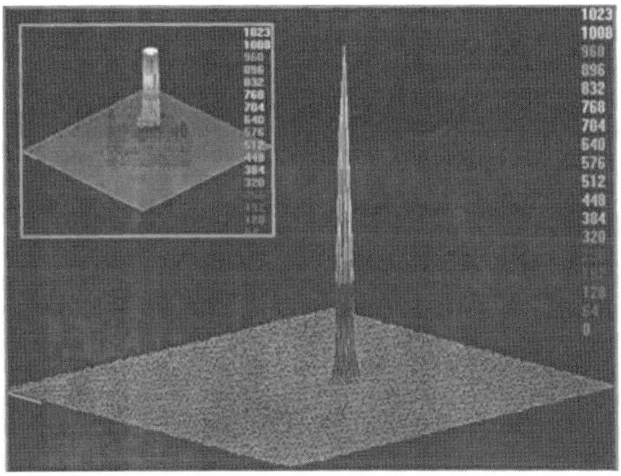

(b)

Fig. 24. Far-field patterns of the RCLEDs (a) without H^+ implantation and (b) with H^+ implantation by the CCD with placing a polarizer in front of light-emission beam of the RCLEDs. Far-field patterns of both types of RCLEDs without placing a polarizer are also shown in the inset. Reprinted with permission from Ref. 32. Copyright (2007) Journal of the Electrochemical Society.

the current can spread into the MQW for the RCLED-IV, but the current can be confined to below aperture for the RCLED-III. Thus, some light can escape from the edge for the RCLED-IV. However, H^+ implantation restricts current flows to the area below light-emission aperture. There is less light emission from the edge. These results indicated that the insulation

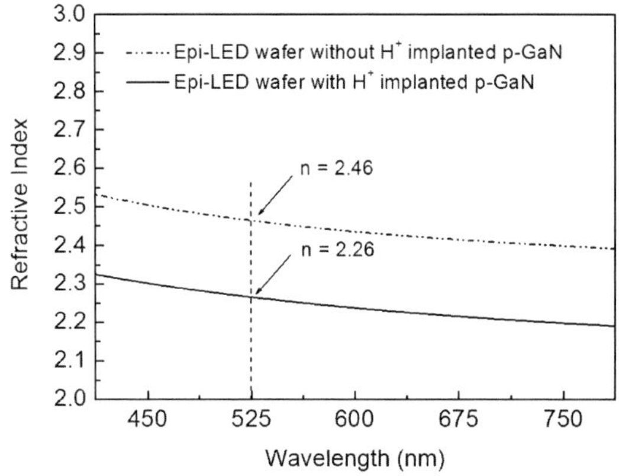

Fig. 25. Refractive index of the epi-LED wafer with and without H^+ implantation. Reprinted with permission from Ref. 34. Copyright (2010) IEEE Photonics Technology Letters.

layer can effectively control the light-emission route by the H^+ implantation manner.

Figure 25 shows the refractive index of the epi-LED wafer with and without H^+ implantation. The refractive indexes of the epi-LED wafer with and without H^+ implantation are 2.26 and 2.46, respectively. The refractive indexes of the epi-LED wafer with H^+ implantation indicate the lower than that of the epi-LED wafer without H^+ implantation. This could be because the partial p-type GaN region bombarded by H^+ results in the crystal structure transferring to the partially amorphous structure. This partially amorphous region presents a lower refractive index than the singe crystal due to the low pile density of crystalline grain. The refractive index change can provide superior optical confinement, and the probability of total reflection of the emission ray increases in the light aperture region. This can be supported by the EL measurement.

3.3. *Characteristics of green RCLEDs with vertical electrodes*

Figure 26 shows the shift in peak wavelength caused by increasing the mount temperatures for both types of RCLEDs at 20 mA. With increasing the mount temperature, the temperature-dependence of RCLED/Cu shifts to a long wavelength at a rate of 0.03 nm/°, which is lower than that of the RCLED/sapphire shift at a rate of 0.14 nm/°. The wavelength shift of the RCLED/Cu is approximately five times slower than that of the

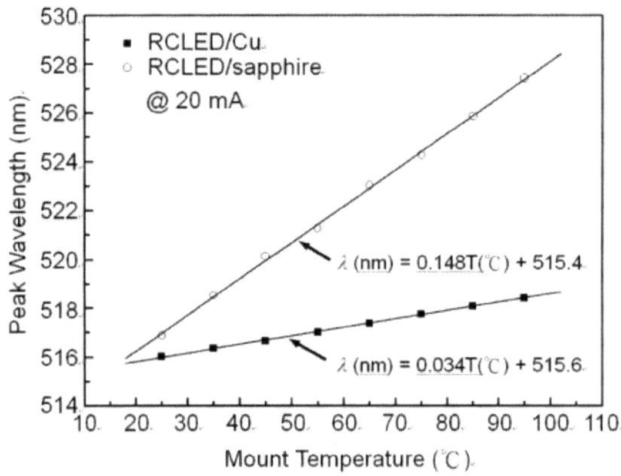

Fig. 26. Peak wavelength as a function of the mount temperatures for both types of RCLEDs at 20 mA. Reprinted with permission from Ref. 35. Copyright (2009) Journal of Lightwave Technology.

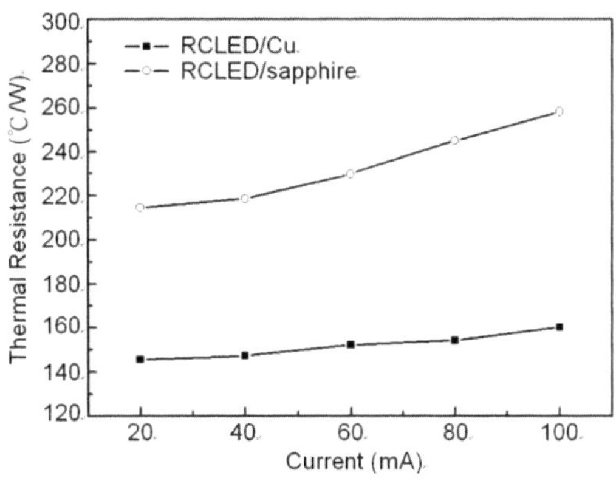

Fig. 27. Thermal resistances of both types of RCLEDs as a function of injection current. Reprinted with permission from Ref. 35. Copyright (2009) Journal of Lightwave Technology.

RCLED/sapphire. The peak wavelength shift is principally determined by the InGaN energy gap and the optical thicknesses of the epilayer with temperature. This result indicated that the RCLED/Cu emission wavelength stability is due to the superior thermal dissipation by a Cu substrate. The result is superior to the finding reported by Akhter *et al.*[25] The thermal resistance (R_{th}) of both types of RCLEDs at different injection current is shown in Fig. 27. The junction temperature (T_j) is related to the thermal resistance by[26]:

$$R_{th} = \frac{T_j - T_a}{P} \tag{4}$$

where T_a is the ambient temperature and P is the drive power of the device. The R_{th} was calculated to be 145.42°C/W and 214.55°C/W at 20 mA for the RCLED/Cu and RCLED/sapphire, respectively. The result indicates that the R_{th} of the RCLED/Cu is lower than that of the RCLED/sapphire. It suggests that a decrease in the Joule heating leads to a corresponding increase in the thermal dissipation due to the T_j decreasing.

The surface temperature distribution along the diagonal for both types of RCLEDs at 100 mA current injections is shown in Fig. 28. In this measurement, the emissivity of these two devices is assumed the same. The thermal energy emitted from the surface of devices by noncontact infrared thermal imaging systems. The surface temperature profile indicates that the RCLED/Cu is lower than that of the RCLED/sapphire. However, the dashed circle position present is contrary to the result of the above. The phenomenon could be attributed to the lateral current path by a sapphire substrate result in the current crowding effect; therefore the

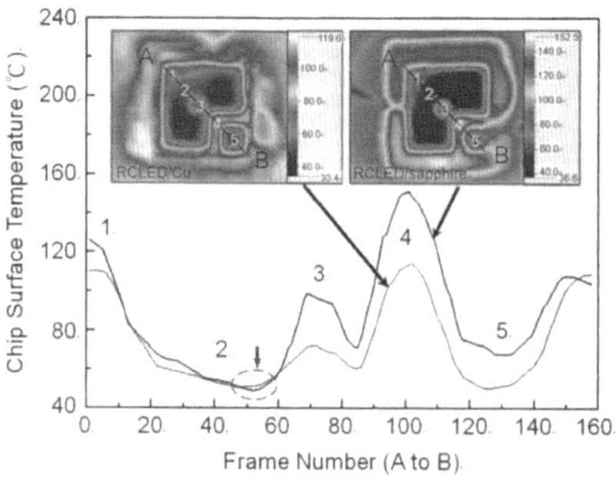

Fig. 28. Surface temperature distributions along the diagonal for both types of RCLEDs at 100 mA current injection. Relative to the surface temperature distribution, the infrared thermal images of both types of RCLEDs are also shown in the inset. The symbol of each chip is added for clarity. Reprinted with permission from Ref. 35. Copyright (2009) Journal of Lightwave Technology.

RCLED/sapphire has a lower temperature as compared with the RCLED/Cu in and around the dashed circle position. Relative to the surface temperature distribution, the infrared thermal images of both types of RCLEDs were also shown in the inset of Fig. 28. The symbol of each chip is added for clarity. The infrared thermal image of the RCLED/sapphire surface presents the thermal dissipation around the chip. In contrast, the thermal dissipation of the RCLED/Cu was transferred to the TO-can heat sink by a Cu substrate. The result indicates that the conductive heat transfer of RCLED/Cu is superior to that of the RCLED/sapphire. The light output power as a function of ambient temperature for both types of RCLEDs at 20 mA is shown in Fig. 29. The light output power of both types of RCLEDs decreases with the increasing ambient temperature. The light output power of RCLED/Cu and RCLED/sapphire decayed by 3% and 9% under the ambient temperature from 25 to 85°C, respectively. The RCLED/Cu has less degradation in the light output power than that of the RCLED/sapphire.

During modulation-speed measurement, the RF signal was injected into devices and a POF serve as the optical probe and collect the modulated optical power from the top of the chips without using lens and then fed into a low noise Si-based photoreceiver with a 125-MHz electrical bandwidth, which was connected with an RF spectrum analyzer. The

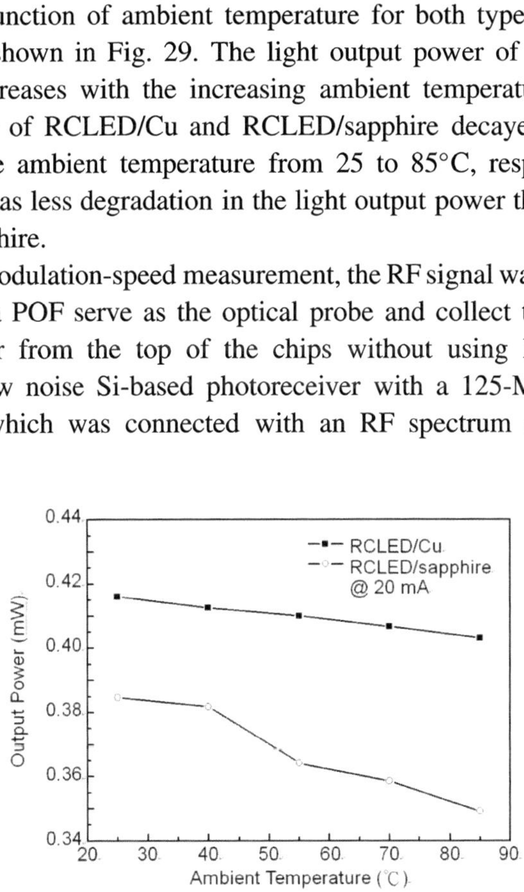

Fig. 29. Light output power as a function of ambient temperature for both types of RCLEDs at 20 mA. Reprinted with permission from Ref. 35. Copyright (2009) Journal of Lightwave Technology.

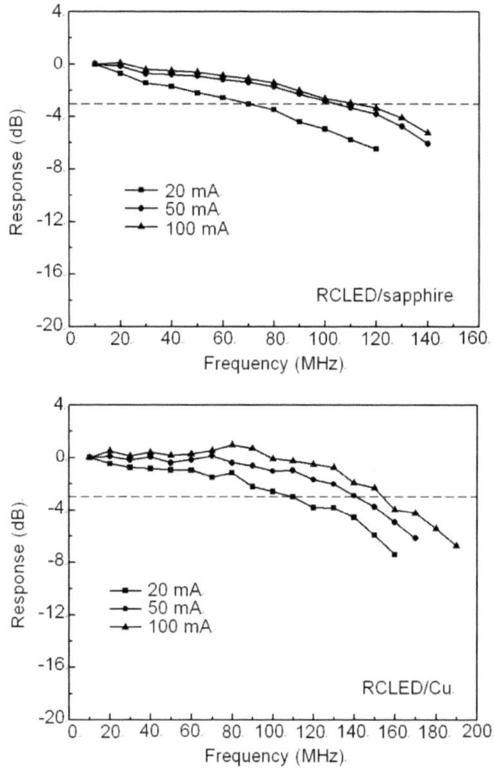

Fig. 30. Measured frequency responses of both types of RCLEDs under 20, 50, and 100 mA. Reprinted with permission from Ref. 35. Copyright (2009) Journal of Lightwave Technology.

frequency responses of both types of RCLEDs under different drive currents are shown in Fig. 30. The result indicates that the -3dB bandwidth ($f_{-3\text{dB}}$) of both types of RCLEDs rises with the increase in drive current. It could be a result from the fact that the recombination rate of carriers is proportional to the current injected into the MQW. It is worth noting that the frequency responses of the RCLED/Cu (~110 MHz and ~155 MHz) are superior to that of the RCLED/sapphire (~70 MHz and ~110 MHz) under the drive currents of 20 mA and 100 mA, respectively. The RCLED/Cu exhibits a significant improvement in this frequency response under a high injected current as compared with the RCLED/sapphire. The improvement is due to an improved electrical injection associated with the vertical current path.

The phenomenon can be evidenced from the following equation[27]:

$$f_{-3\text{dB}} = \frac{1}{2\pi}\left(\frac{1}{T_{nr}} + \frac{1}{T_r}\right) \tag{5}$$

where T_{nr} and T_r are the non-radiation lifetime and radiation lifetime, respectively. Based on the above equation, the T_{nr} decreasing results in the non-radiation recombination rate $(1/T_{nr})$ increasing and the high $f_{-3\text{dB}}$ being obtained. The result demonstrated that the RCLED/Cu under a high current operation could be suitable for the 155 MHz application for fiber communication systems. This modulation speed is comparable to the data reported by Shi *et al.*[28] Figure 31 shows the eye diagram for both types of RCLEDs at a transmission rate of 100 MHz for a 100-m long POF

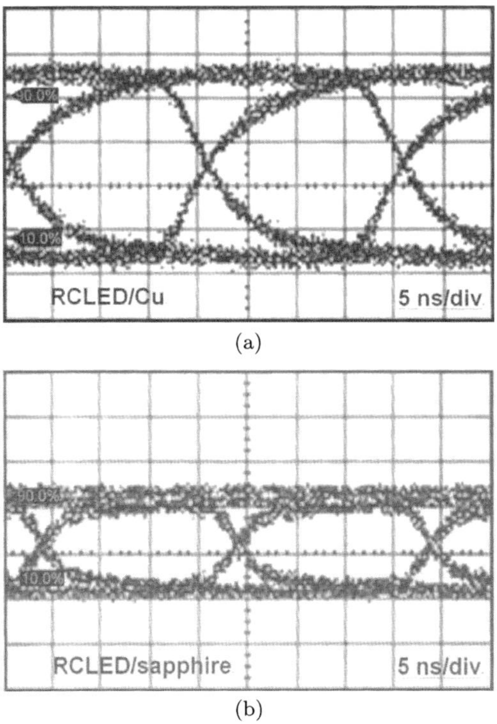

Fig. 31. Eye diagram of both types of RCLEDs at a transmission rate of 100 MHz for a 100-m long POF (NA = 0.5), under a 100-mA driving current. Reprinted with permission from Ref. 35. Copyright (2009) Journal of Lightwave Technology.

($NA = 0.5$) under a 100 mA driving current. The eye diagram indicates that the amplitude of RCLED/Cu is larger than that of the RCLED/sapphire. The result shows that the light source of the POF using the RCLED/Cu can tolerate more noise amount than the RCLED/sapphire.

4. Conclusions and Outlook

In this chapter, we have described the characteristics of the green RCLEDs fabricated using epitaxial lift-off and wafer transfer techniques. For the all dielectric DBRs RCLEDs with top DBRs, a narrow cavity modes of 5 nm for dominant mode at $\lambda = 525$ nm, which gives a quality factor of approximately 105. The stable EL emission peak and FWHM versus injection current were demonstrated. An improved directionality resulted from the resonant-cavity was also observed. The phenomena could be attributed to the superior Fabry-Pérot cavity effects. We also demonstrate the hybrid mirrors to fabricate the RCLEDs, as compared with the all dielectric DBRs RCLEDs, the hybrid mirrors RCLEDs is weak in the optical properties under neglected thermal effects. It could be attributed to the metal mirrors have high absorbency as compared with the dielectric DBRs result in the weakly light output. We also investigate the influence of different H-ion implantation concentrations on the GaN-based RCLEDs and try to find out the optimum implantation condition and revealed that the characteristics of H^+-implanted RCLED. An optimum concentration with 10^{14} ions/cm^2 dose acts as the current confinement layer was demonstrated from the sheet resistivity and carrier mobility as well as optical analyses of the p-GaN under different implantation concentrations (as-grown, 10^{13}, 10^{14} and 10^{15} ions/cm^2). The turn-on voltage of 10^{14} ions/cm^2 implanted RCLED is 4.6 V at 20 mA, which have similar to that of the SiO_2-insulated RCLEDs because the tailored implantation concentration have not create the damages seriously. The superior light emission directionality and high fiber-coupled power was achieved in the 10^{14} ions/cm^2 implanted RCLED devices. A phenomenon arose wherein a tailored implantation concentration can effectively control the lateral current flow path to improve the directionality and then increase the probability of the carrier recombination and optic characteristic changes result in light output advancement. Moreover, a vertical-conducting RCLED/Cu was fabricated by a combination of LLO and electroplating techniques. The temperature dependent on the rate of peak emission wavelength is only 0.03 nm/° under the high-temperature environment. The light output decayed by 3% under the ambient temperature from 25 up to 85°. Furthermore, the superior frequency responses was achieved a 155 MHz under the drive current of 100 mA. The eye diagrams have a clear and good open eye pattern. These superior performances can be attributed to the decreased thermal resistance and the enhanced thermal dissipation by vertical-conducting structures with a Cu substrate. Based on our research results, the green RCLEDs using wafer transfer techniques are suitable for POF-based multimedia networks and also have high potential for future mini projector display applications.

Recently, the green RCLEDs have obtained great attention due to it could be applied in multimedia data transmission of automobiles. Because of green wave band has a minimum attenuation loss at wavelength of $510 \sim 570$ nm. However, in order for the signal in POF can transmit farther and clearer. The characteristics of light sources, such as the light emission power, modulation speed, and spectral purity, which must be enhanced. Therefore, detune (thinned) the resonant cavity length is absolutely required. Generally, the cavity thickness of RCLEDs was detuned through using an epitaxial growth processes. It will result in the FWHM of the emission spectra and forward operating voltage increased with decreasing n-GaN thickness.[29,30] The photoelectrochemical (PEC) techniques could be applied in the thinned the cavity length to prevent these problems. Compared to the cavity thickness of RCLEDs was detuned through using an epitaxial growth processes, the advantage of the n-GaN surface after LLO was thinned using a PEC process as following:

(1) The compressive strain can decrease in epi-wafer LED with thicker n-GaN epilayers and result in the light emission intensity incrcasc duc to the enhancement of localized states in InGaN wells.
(2) The thick n-GaN epilayers can avoid the GaN film cracking when a sapphire substrate was removed after LLO processes.
(3) FWHM and forward voltage decrease with the thick n-GaN epilayers. Therefore, the future study is focused on the subject of detune the resonant cavity length for high-quality factor GaN-based RCLEDs.

References

1. E. F. Schubert, Y. H. Wang, A. Y. Cho, L. W. Tu, and G. J. Zydzik, *Appl. Phys. Lett.*, **60**, 921 (1992).
2. M. A. Mastro, J. D. Caldwell, R. T. Holm, R. L. Henry, and C. R. Eddy Jr., *Adv. Mater.*, **20**, 115 (2008).
3. R. H. Birkner, J. Kaiser, W. Elsäßer, and C. Jung, *Appl. Phys. B.*, **79**, 963 (2004).
4. M. Pessa, M. Guina, M. Dumitrescu, I. Hirvonen, M. Saarinen, L. Toikkanen, and N. Xiang, *Semicond. Sci. Technol.*, **17**, R1 (2002).
5. F. B. Naranjo, S. Fernández, M. A. Sánchez-García, F. Calle, and E. Calleja, *Appl. Phys. Lett.*, **80**, 2198 (2002).
6. A. M. Green, D. G. Gevaux, C. Roberts, and C. C. Phillips, *Phys. E*, **20**, 531 (2004).
7. L. Y. Ying, X. L. Hu, W. J. Liu, J. Y. Zhang, B. P. Zhang, and H. C. Kuo, *Superlattices Microstruct.*, **88**, 561 (2015).
8. T. Moudakir, F. Genty, M. Kunzer, P. Börner, T. Passow, S. Suresh, G. Patriarche, K. Köhler, W. Pletschen, J. Wagner, and A. Ougazzaden, *IEEE Photonics J.*, **5**, 8400709 (2013).
9. Y. C. Chu, Y. K. Su, C. H. Chao, and W. Y. Yeh, *Jpn. J. Appl. Phys.*, **52**, 01AG03 (2013).
10. F. Calle, F. B. Naranjo, S. Fernández, M. A. Sánchez-García, E. Calleja, and E. Muñoz, *Phys. Stat. Sol. (a)*, **192**, 277 (2002).
11. R. Wirth, C. Kamutsch, S. Kugler, and K. Streubel, *IEEE Photon Technol. Lett.*, **13**, 421 (2001).
12. Y. ShiuanLiu, A. F. M. Saniul, T. T. Kao, K. Mehta, S. C. Shen, T. Detchprohma, P. D. Yoder, R. D. Dupuis, H. Xie, and F. A. Ponce, *J. Cryst. Growth.*, **443**, 81 (2016).
13. S. Yoshida, K. Ikeyama, T. Yasuda, T. Furuta, T. Takeuchi, M. Iwaya, S. Kamiyama, and I. Akasaki, *Jpn. J. Appl. Phys.*, **55**, 05FD10 (2016).
14. C. Bergera, A. Lesnik, T. Zettler, G. Schmidt, P. Veit, A. Dadgar, J. Bläsing, J. Christen, and A. Strittmatter, *J. Cryst. Growth.*, **440**, 6 (2016).
15. J. H. Kim and S. H. Park, *J. Korean Phys. Soc.*, **63**, 1621 (2013).
16. J. Y. Zhang, L. E. Cai, B. P. Zhang, S. Q. Li, F. Lin, J. Z. Shang, D. X. Wang, K. C. Lin, J. Z. Yu, and Q. M. Wang, *J. Lightwave Technol.*, **27**, 55 (2009).
17. D. P. James, D. D. Michael, and B. G. Peter, *Silicon VLSI technology: fundamentals, practice, and modeling 1st*, Prentice-Hall Inc., Taipei, Taiwan, 86pp, 2002.
18. P. de Mierry, J. M. Bethoux, H. P. D. Schenk, M. Vaille, E. Feltin, B. Beaumont, M. Leroux, S. Dalmasso, and P. Gibart, *Phys. Stat. Sol. (a)*, **192**, 335 (2002).
19. I. Radu, R. Singh, R. Scholz, U. Gösele, S. Christiansen, G. Brüderl, C. Eichler, and V. Härle, *Appl. Phys. Lett.*, **89**, 031912 (2006).
20. M. A. Reshchikov and H. Morkoç, *J. Appl. Phys.*, **97**, 061301 (2005).
21. J. K. Sheu, M. L. Lee, C. J. Tun, C. J. Kao, L. S. Yeh, S. J. Chang, and G. C. Chi, *IEEE J. Sel. Top Quant. Electron.*, **8**, 767 (2002).
22. B. D. Cullity and S. R. Stock, *Elements of X-Ray Diffraction 3rd ed.*, Prentice-Hall Inc., New Jersey, USA, 174 (2001).
23. T. D. Moustakas, T. Lei, and R. J. Molnar, *Physica B*, **185**, 36 (1993).
24. M. M. Dumitrescu, M. J. Saarinen, M. D. Guina, and M. V. Pessa, *IEEE J. Sel. Top. Quantum Electron.*, **8**, 219 (2002).

25. M. Akhter, P. Maaskant, B. Roycroft, B. Corbett, P. de Mierry, B. Beaumont, and K. Panzer, *Electron. Lett.*, **38**, 1457 (2002).
26. J. Park and C. C. Lee, *IEEE Electron Device Letters*, **26**, 308 (2005).
27. S. J. Chang and C. S. Chang, *IEEE Photon. Technol. Lett.*, **10**, 772 (1998).
28. J. W. Shi, H. Y. Huang, J. K. Sheu, C. H. Chen, Y. S. Wu, and W. C. Lai, *IEEE Photon. Technol. Lett.*, **18**, 1636 (2006).
29. Y. C. Lee, C. E. Lee, B. S. Cheng, T. C. Le, H. C. Kuo, S. C. Wang, and S. W. Chiou, *Jpn. J. Appl. Phys.*, **46**, 2450 (2007).
30. J. F. Carlin, P. Royo, R. P. Stanley, R. Houdré, J. Spicher, U. Oesterle, and M. Llegems, *Semicond. Sci. Technol.*, **15**, 145 (2000).
31. S. Y. Huang, R. H. Horng, Y. J. Tsai, P. R. Lin, W. K. Wang, Z. C. Feng, and D. S. Wuu, *Semicond. Sci. Technol.*, **25**, 035013 (2010).
32. S. Y. Huang, R. H. Horng, H. C. Kuo, and D. S. Wuu, *J. Electrochem. Soc.*, **154**, H962 (2007).
33. R. H. Horng, W. K. Wang, S. Y. Huang, and D. S. Wuu, *IEEE Photon. Technol. Lett.*, **18**, 457 (2006).
34. S. Y. Huang, R. H. Horng, P. H. Tseng, J. H. Tu, L. W. Tu, and D. S. Wuu, *IEEE Photon. Technol. Lett.*, **22**, 404 (2010).
35. S. Y. Huang, R. H. Horng, J. W. Shi, H. C. Kuo, and D. S. Wuu, *J. Lightwave Technol.*, **27**, 4084 (2009).

CHAPTER 10

NANOTEXTURING EFFECTS IN GAN/INGAN MULTI-QUANTUM-WELLS LED PLANAR STRUCTURES

S. J. Xu

Department of Physics,
Shenzhen Institute of Research and Innovation (SIRI),
HKU-CAS Joint Laboratory on New Materials,
The University of Hong Kong,
Pokfulam Road, Hong Kong, P. R. China
sjxu@hku.hk

Conventional 2D GaN/InGaN multi-quantum-wells LED planar structures were nanotextured into quasi-1D nanopillars with two "top-down" approaches of self-assembled Ni nanoislands mask + dry etching and direct focused ion beam milling. Such nanotexturing effects on the optical properties of the nanopillars were investigated in detail with confocal micro-Raman microscopy and precise photoluminescence spectroscopy. Two main nanotexturing effects, namely greatly altered light scattering (i.e., appearance of new surface optical vibration mode) and light emission (i.e., strongly enhanced internal quantum efficiency), have been demonstrated. In this chapter, we aim to give a comprehensive review and summary of the nanotexturing effects in GaN/InGaN multi-quantum-wells LED planar structures. Further investigation suggestions and prospective are also briefly discussed.

1. Introduction

With GaN emerging as a key wide bandgap semiconductor in LED-based solid state lighting,[1,2] III-nitride nanostructures containing GaN/InGaN multi-quantum wells (MQW) have drawn considerable research interests in recent years.[3–5] Nanostructures can be formed either through atomic nucleation-starting growth (bottom-up approach)[3,4,6] or nano-patterning followed by etching (top-down approach).[5,7,8] Amongst established top-down approaches, metal nanoislands mask + following dry etching and focused ion beam (FIB) milling have been widely recognized as two promising techniques for the fine patterning of GaN based nanostructures.[8,9] In the past a few years, we have prepared the GaN/InGaN MQW nanopillars (NPs) from conventional 2D planar structures by using the above-mentioned two "top-down" approaches and investigated optical properties of these prepared nanostructures.[9–11] Some interesting findings have been obtained by us in these studies. For example, local surface optical (SO) phonon mode predicted by relevant theoretical calculations in 1970s[12,13] was unambiguously observed in the nanotextured GaN/InGaN MQW NPs with average diameter $<$200 nm.[9–11] In the author's view, the appearance of SO phonon mode shall be Raman signature of light scattering greatly altered in the quasi-1D nanostructures. Due to dimensionality change induced by nanotexturing, actually, the dielectric constant and thus refractive index of material system shall be largely modified, so that light propagating and scattering mechanisms in these nanoscale materials may be significantly altered.[14,15] Even in the highly disordered interface separating two semiconductors, Raman light scattering may show very different behaviours from predicted by the existing model.[16] On the other hand, the stress and electronic states in GaN/InGaN MQW planar structure may be also substantially reformed. For instance, deeper localization of carriers and significant improvement in quantum efficiency (e.g., from 4.76% of the planar MQW structure to 12.5% of the 160 nm MQW NPs) of light emission in the whole interested temperature range from 4 K to 300 K have been shown by us in a latest study.[11] Meanwhile, with the aid of localized-state ensemble (LSE) luminescence model developed by us,[17–20] the photoluminescence (PL) spectra of the GaN/InGaN MQW NPs are quantitatively interpreted in the entire temperature range. Moreover, a concept of "negative" thermal activation energy is tentatively proposed in terms of distinctive temperature dependence of PL of the nanowire samples.[11] These findings could lead to a deeper insight into the physical nature of localization and luminescence mechanism of excitons in InGaN/GaN nanopillars.

In this chapter, we want to provide a comprehensive review of aforementioned studies. In addition, we would also like to write some suggestions for further investigations on the relevant subjects.

2. Experimental

2.1. *Sample preparation*

As mentioned earlier, two different "top-down" techniques were employed to prepare GaN/InGaN MQW NPs. They are nanosized Ni islands mask + dry etching approach and direct focused ion beam milling method. The starting wafer was a MQW LED planar structure containing five periods of GaN/InGaN quantum wells epitaxially grown by metal-organic chemical vapor deposition (MOCVD) on a c-plane sapphire substrate. In the MOCVD epitaxy process, trimethylindium, trimethylgallium, and ammonia (NH_3) were used as indium, gallium, and nitrogen precursors, respectively. Bicyclopentadienyl magnesium (Cp_2Mg) and silane (SiH_4) were employed as the sources of p-type and n-type dopants, respectively. The details of the starting wafer include a 3.0 μm n-type GaN:Si layer, then five periods of GaN/InGaN (30 nm/3 nm) MQW as an active layer, and a 0.2 μm thick p-type GaN:Mg top contact layer. For preparation of GaN/InGaN MQW NPs, a 300 nm thick SiO_2 thin film was first deposited on the LED wafer by plasma-enhanced chemical-vapour deposition, followed by the electron-beam evaporation of a Ni metal film. Then a rapid thermal annealing at 850°C under nitrogen gas ambient was used to form self-assembled Ni nanoislands. Afterwards the wafer was etched down to the n-type GaN layer with inductively coupled plasma (ICP) etching technique and by using self-assembled Ni nanoislands as etching mask to transfer the island pattern on the nano-sized Ni metal to the wafer. Finally, the Ni nanoislands + SiO_2 layer was removed by a buffer oxide etchant (BOE) to expose the MQW NPs. Such preparation process was schematically shown in Fig. 1.

The second approach, namely the FIB direct milling technique, was schematically drawn in Fig. 2. Before the FIB milling, a thin layer of indium tin oxide (ITO) was deposited on top of the wafer to enhance surface conductivity. An excellent surface conductivity could not only prevent surface charging but also facilitate obtaining pillars of good uniformity and

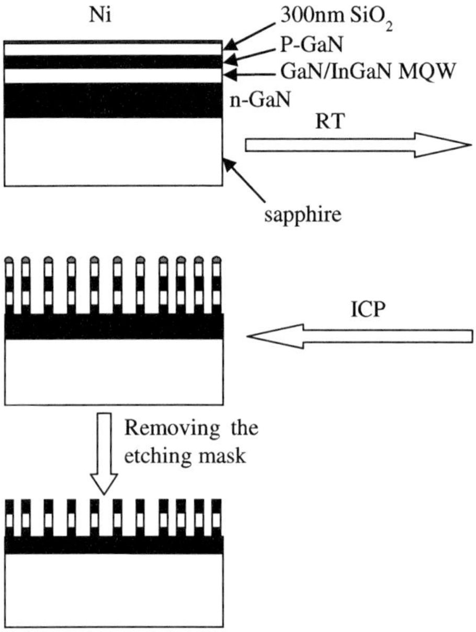

Fig. 1. A schematic diagram of the preparation procedures of GaN/InGaN MQW NPs with Ni nanoislands mask + dry etching technique.

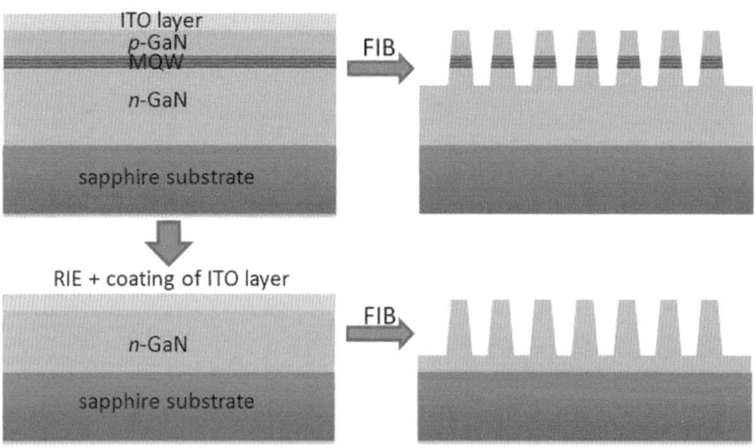

Fig. 2. A schematic diagram illustrating FIB fabrication of NPs on a GaN/InGaN MQW wafer. For comparison purpose, an identical array of NPs was also prepared from a similar wafer with the MQWs etched away by reactive ion etching.

high-aspect ratios during FIB milling.[9] The FIB nano-patterning was carried out on a Quanta 200 3D DualBeam (FIB/SEM) system. 30 keV gallium ions with a beam current of 0.1 nA/0.3 nA were used to mill the sample according to the pattern image uploaded to the software. The supreme advantage of FIB direct milling method could be nano-pattern uniformity, while time consuming may be its obvious disadvantage.

Figure 3 shows scanning electron microscopic (SEM) images of GaN/InGaN MQW NPs prepared with Ni nanoisalnds + ICP etching approach. Figure 3(a) depicts a top view while (b), (c) and (d) show cross-sectional views of GaN/InGaN MQW NPs with different heights. In fact, by adjusting the preparation conditions such as Ni film thickness, thermal annealing time durations and ICP etching times etc., one can control the average diameter, height and areal density of the nanopillars.[8,10] For

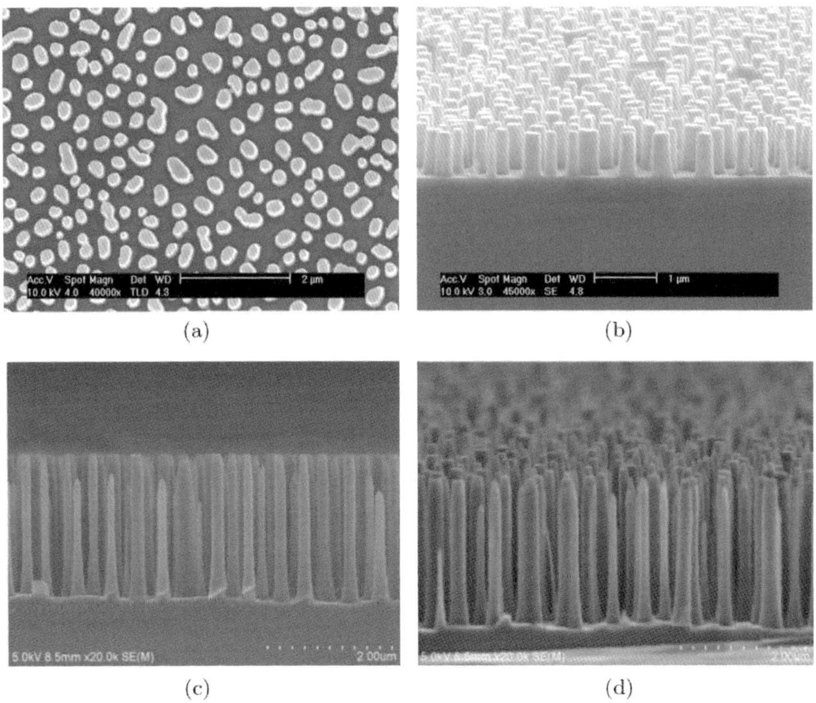

Fig. 3. SEM images of GaN/InGaN MQW nanopillars prepared with Ni nanoislands mask + ICP etching. (a) Top view; (b), (c), and (d) Cross-sectional views of the NPs with different heights.

example, the nanowire diameter can be varied from 250 nm to 100 nm, whereas the height can be changed from 50 nm to 2000 nm. It should be pointed out that large distribution in size is one major disadvantage of the Ni nanoislands + dry etching approach compared with the FIB direct milling method.

Shown in Fig. 4 are SEM images of the nanopillar array at different stages of FIB fabrication. Figure 4(a) shows the SEM image of the nanopillar array just after FIB milling, with base diameters of 300 nm. The SEM image clearly shows that an extra layer has been adhered onto the nanopillars and the sidewalls of the array boundary. This adhesive layer is attributed to the quick oxidization of the FIB damaged outer shell when the sample

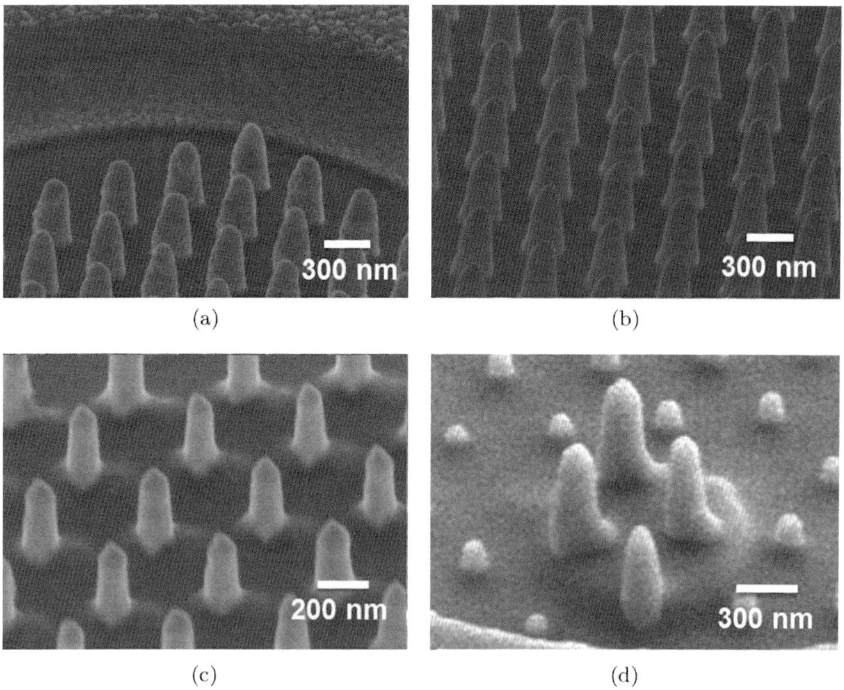

Fig. 4. SEM images of GaN/InGaN MQW nanopillar array at different stages prepared with FIB direct milling. (a) Immediately after FIB patterning; (b) Etched in acid to remove the ITO layer as well as the amorphous GaN layer due to ion bombardment damage, leaving a relatively smooth surface; (c) Further etched in KOH solution, shrinking the diameters of the NPs down to desirable diameter such as 150 nm; (d) Fine milling of the array by FIB to form a small cluster of NPs.

was exposed to air. The sample was then subjected to successive wet etch treatments for removal of ion damage. For example, the ITO layer and the adhesive layer were etched simultaneously in an acid solution of $1HNO_3$:$2H_2O$:4HCl. The field-emission SEM image in Fig. 4(b) shows nanopillars with clean and smooth surfaces after the chemical treatment. The sample was further treated with KOH of 20 wt. % dissolved in water, heated to 50°C, to remove the amorphous shell leaving the straight pillars and to reduce the base diameters of the pillars down to 180 nm, as shown in Fig. 4(c). For comparison, another set of nanopillar arrays were fabricated by the same process, except that the p-GaN and MQW layers have been etched away by reactive ion etching (RIE) prior to FIB patterning, also shown in the schematic diagram of Fig. 2.

2.2. *Optical characterization*

The optical properties of the GaN/InGaN MQW NPs prepared in the two different "top-down" approaches were characterized with room-temperature micro-Raman scattering and variable-temperature PL techniques. The micro-Raman light scattering and imaging measurements were carried out on a WITec-Alpha confocal scanning micro-Raman system under the back scattering geometric configuration. The excitation source was 514.5 nm line of an Ar^+ laser with an output power of 30 mW. The laser beam was focused onto the sample with a 60× Nikon objective lens. Scattered light signal was collected by the same objective lens and guided by a multiple mode optical fiber with 50 μm diameter into an Acton SP2300i monochromator. Before the light signal was transported by the optical fiber, the strong Rayleigh scattered light was blocked by a long-pass filter. Finally, the Raman scattering light was detected by an Andor charge coupling device (CCD) detector which was thermoelectrically cooled and maintained at −61°C. For Raman imaging, the sample was mounted on a piezoelectric stage where two dimensional scanning was manoeuvrable. Variable-temperature PL spectroscopic measurements were carried out on a home-made PL system using a Kimmon 325 nm He-Cd laser as the excitation light source. The maximum output power of the He-Cd laser was 38 mW. In the variable-temperature PL measurements, the samples were mounted on the cold figure of a Janis closed cycle cryostat providing a varying temperature range of 3.6–300 K. The PL signal was dispersed by a Spex750 monochromator before it was detected by a Hamamatsu R928 photomultiplier detector. The dispersive element was a 1200 lines/mm reflection grating.

3. Results and discussion

3.1. *Raman light scattering*

As mentioned in the section of introduction, a surface optical phonon mode has been theoretically predicted to exist in small (e.g. nanoscaled) crystals since 1970s.[12,13] For such SO phonon mode, the first experimental observation could be the work done by Gupta *et al.* in 2003.[21] In this study of Raman light scattering from the crystalline GaP nanowires, they observed a strong additional peak in the first-order Raman spectrum that can be clearly assigned to the light scattering by SO phonons. Three years later, Adu *et al.* observed the SO mode in ZnS nanowires.[22] The nanowires which show the SO vibration modes in these studies were grown with "bottom-up" approach. In the GaN/InGaN MQW NPs fabricated with "top-down" methods, the existence of the SO modes was demonstrated by us in 2011.[9,10] Figure 5 depicts the recorded micro-Raman scattering spectra of four GaN/InGaN MQW NPs with different average diameters and a planar control sample. The average diameters of the nanopillar samples were 120, 160, 200, and 250 nm, respectively, as seen in the SEM images in Fig. 6.

As seen in Fig. 5, the two distinct peaks at about 569 and 734 cm^{-1} were well observed for all the samples, which are assigned to the E_2 (high) and the A_1 (LO) phonon mode of wurtzite GaN, respectively.[23] In addition, a weak peak located around 417 cm^{-1} was also observed, which has been already identified to be a vibration mode of N-rich Ga-N bond configuration in GaN based nanostructures.[14] Besides these known modes, an outstanding spectral structure appears between the E_2 (high) and the A_1 (LO) phonon modes. More interestingly, the location (frequency) of this spectral structure exhibits a clear dependence on the nanopillar size. For instance, it shows a well resolved redshift with decreasing the nanopillar diameter. From the location of this additional spectral structure, especially from the observed unique dependence of its location (frequency) on the nanopillar size, one can speculate that this additional structure shall stem from light scattering by the SO mode. According to the well-established theoretical mode, the SO phonon frequency is dependent on the dielectric

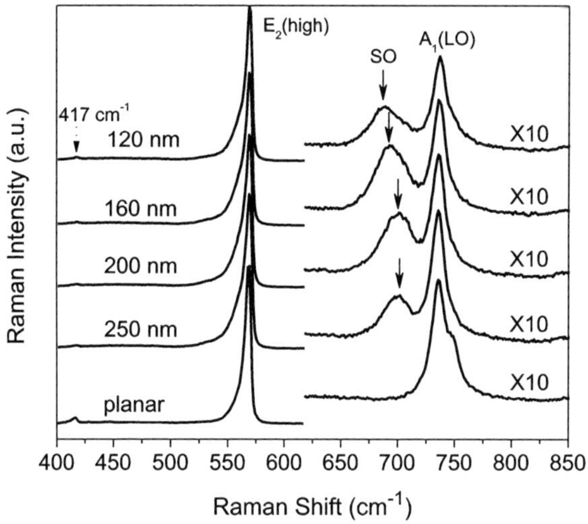

Fig. 5. The recorded room-temperature micro-Raman spectra of the GaN/InGaN MQW NPs with average diameters varying from 120 to 250 nm. The bottom one is the micro-Raman spectrum of the planar control sample. An additional peak marked by vertical arrows appears in all the NPs samples and shows a redshift as the average diameter decreases, which is assigned to the SO mode.

constant of the surrounding medium, as well as the nanostructure shape and size.[12,13] In our case (the nanopillars + atmosphere), the key parameter determining the SO phonon frequency shall be the diameter of the NPs. The expression for the SO phonon frequency in a 1-D nanostructure is given by[12]

$$\omega_{SO}^2 = \omega_{TO}^2 \frac{\varepsilon_0 - \rho_{nx}\varepsilon_m}{\varepsilon_\infty - \rho_{nx}\varepsilon_m}, \tag{1}$$

where ω_{TO} is the transverse optical (TO) phonon mode frequency, ε_0 and ε_∞ the static and high-frequency dielectric constants of the material, respectively, and ε_m the dielectric constant of the surrounding medium which is air in our case. And ρ_{nx} is given by,[12]

$$\rho_{nx} = \frac{K_0'(x)I_0(x)}{K_0(x)I_0'(x)} = \frac{K_1(x)I_0(x)}{K_0(x)I_1(x)}, \tag{2}$$

where $I_j(x)$ and $K_j(x)$ $(j = 0, 1)$ are the modified Bessel functions and $x = q$r (r is the radius of the nanopillars and q is the wave vector). The dielectric constant of the air ε_m was taken as 1. Values of $\varepsilon_0 = 10.3$

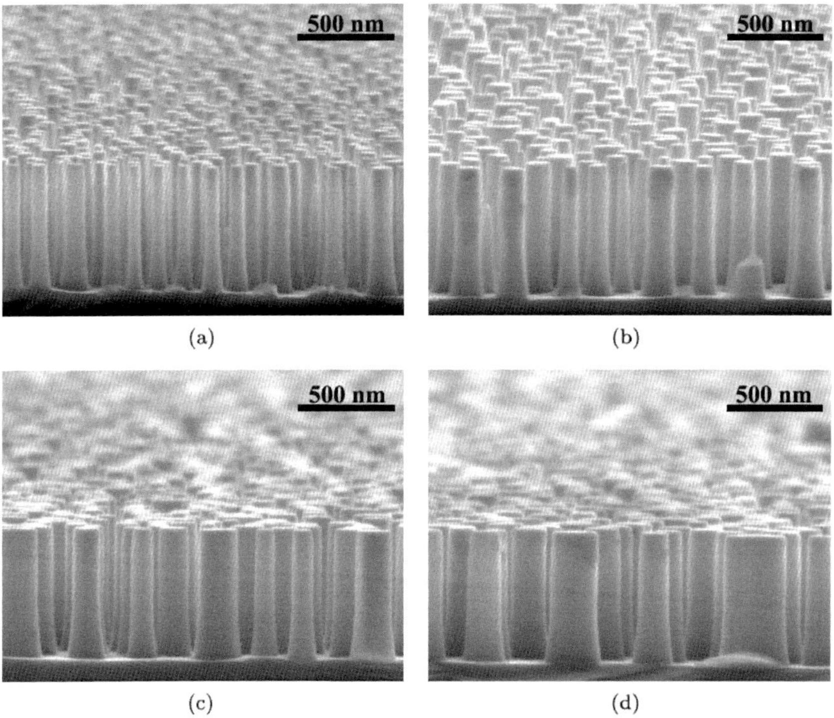

Fig. 6. SEM images of the GaN/InGaN MQW nanopillars with various average diameters: (a) 120 nm, (b) 160 nm, (c) 200 nm, (d) 250 nm.

and $\varepsilon_\infty = 6.7$ were used for InN and $\varepsilon_0 = 10.4$ and $\varepsilon_\infty = 5.8$ for GaN, respectively.[24,25] Assuming that the In composition is 0.15 for our InGaN/GaN MQWs sample, the linear interpolate values of ε_0 and ε_∞ were calculated to be 10.4 and 5.9 for $In_{0.15}Ga_{0.85}N$, respectively. It is known that there are two TO phonons, A_1(TO) and E_1(TO), in the wurtzite structure. In principle, two SO phonon modes should be expected. If A_1 (TO) at 534 cm^{-1} and E_1 (TO) at 561 cm^{-1} were used for GaN and A_1 (TO) at 447 cm^{-1} and E_1 (TO) at 478 cm^{-1} for InN,[24,25] the A_1(TO) and E_1(TO) in $In_{0.15}Ga_{0.85}N$ alloy were calculated to be 521 cm^{-1} and 549 cm^{-1}, respectively. It has been known that only SO phonon mode pertaining to E_1(TO) character can be distinctly observed for InGaN by Raman scattering.[24] So the observed SO phonon structure in our experiment shall be associated with E_1(TO) mode.

Figure 7(a) shows the E_1 related ω_{SO} as a function of qr with Eq. (1) and (2). A horizontal line was drawn to indicate the SO phonon

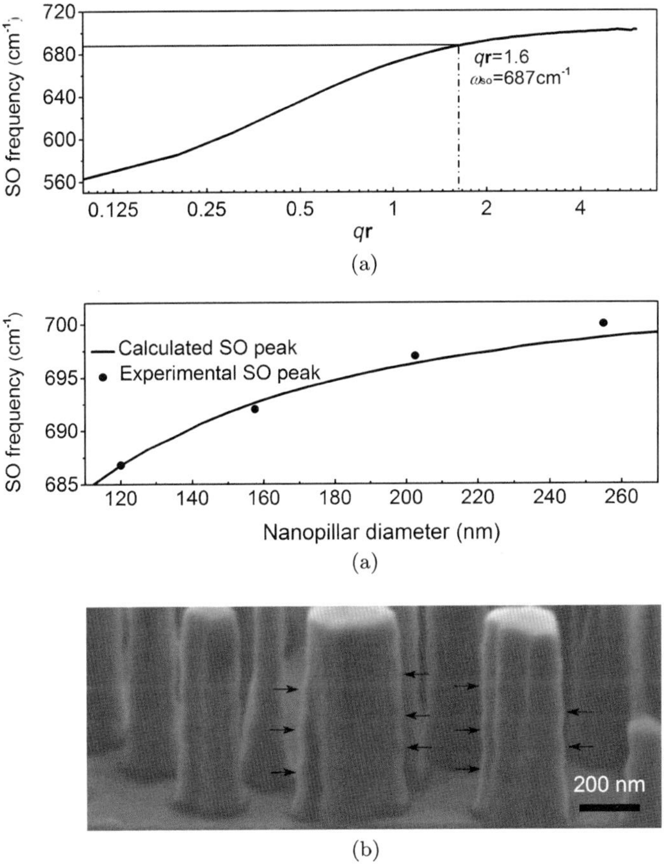

Fig. 7. (a) Calculated SO phonon modes of GaN/InGaN MQW NPs as a function of qr. (b) The measured and calculated Raman SO modes for NPs with different diameters. (c) SEM image showing surface roughness in typical GaN/InGaN MQW NPs. Arrows show the modulation in the sidewall surface of NPs.

frequency ($\omega_{SO} = 687\,\text{cm}^{-1}$) determined by Raman scattering for the NPs with diameter of 120 nm. Then, the corresponding wave vector q can be calculated as shown in Fig. 7(a) (for r $=$ d$/2$ $=$ 60 nm, qr $=$ $687\,\text{cm}^{-1}$, qr $=$ 1.6, q $\approx$ $0.027\,\text{nm}^{-1}$). Here, q $\approx$ $0.027\,\text{nm}^{-1}$ was taken for all the NPs. Then the values of qr were calculated to be 2.13, 2.67, and 3.33 for the GaN/InGaN MQW NPs with diameters of 160 nm, 200 nm, and 250 nm, respectively. The experimental (solid circles) and calculated SO phonon mode peaks (solid curve) are plotted against different diameters in

Fig. 7(b). Clearly, it can be seen that good agreement between experiment and calculation has been achieved. Figure 7(c) show a SEM image of the several NPs where the surface roughness may be estimated.

Regarding the SO phonon intensity, it has been argued to be determined by the magnitude of surface roughness due to the breakdown of translation symmetry of the surface potential.[22] Therefore, the wavelength of the surface potential which is responsible for breaking the symmetry and activating the SO phonon was then evaluated by k = 2p/q and calculated to be 236 nm when $q = 0.027\,nm^{-1}$ for the GaN/InGaN MQW NPs. As seen in Fig. 7(c), surface roughness with an undulation period of ~120 nm was observed in the NPs. Modulation period with any integer multiple of λ may be sufficient to initiate breakdown of translational symmetry for the surface potential, leading to the SO phonons. In our case, $2 \times 120\,nm = 240\,nm$ which is approximately equal to λ can make the intensity of the surface modes comparable to those of other observed phonons modes, as observed indeed in Fig. 5.

Due to surface localization nature of the SO phonon mode,[13] principally, the SO phonon mode shall be resolved in high-resolution Raman images. We have measured the Raman images of the GaN/InGaN MQW NPs and unetched planar structure for an area of $3\,\mu m \times 3\,\mu m$. The obtained Raman images were shown in Fig. 8: (a) for GaN E_2(high) mode integrated intensity of the planar sample; (b) for GaN E_2(high) mode of the 160 nm NPs; and (c) for the SO phonon mode integrated intensity of the 160 nm NPs. In sharp

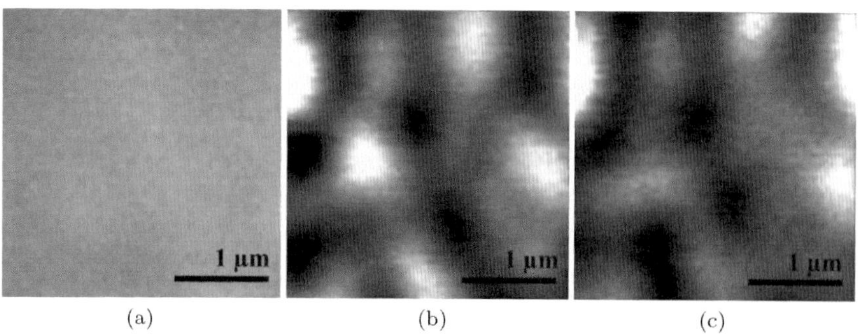

Fig. 8. The integrated intensity Raman images: (a) for the GaN E_2(high) mode of the planar sample; (b) for the GaN E_2(high) mode of the 160 nm NPs; (c) for the SO phonon mode of the 160 nm NPs.

contrast to the homogeneous distribution of the GaN E_2(high) mode in the planar sample, the Raman images of the NPs clearly exhibit the pillar distribution with bright spots. More interestingly, the Raman images of the GaN E_2(high) mode and the SO mode in the NPs sample show a distinct spatial correlation, indicating that the SO phonon modes are indeed localized on the nanopillar's surface.

For the SO phonon intensity in Raman spectrum of the GaN/InGaN MQW NPs prepared in FIB method, it may exceed the intensity of A_1(LO) mode, as shown in Fig. 9 where the micro-Raman spectra of the FIB patterned nanopillar arrays with different diameters of 150, 180 and 190 nm. To further investigate the SO phonon with regard to nanopillar surface roughness, two nanopillar arrays were prepared with identical diameters and geometry fabricated with the same FIB settings but different KOH treatment durations as the KOH treatment time was deliberately controlled to obtain different surface roughness. The pillars were examined by the low-voltage field-emission SEM and the obtained images were shown in Fig. 10. It is well known that low-energy electrons penetrate shallowly into the structure, thereby producing a much better surface profile. The

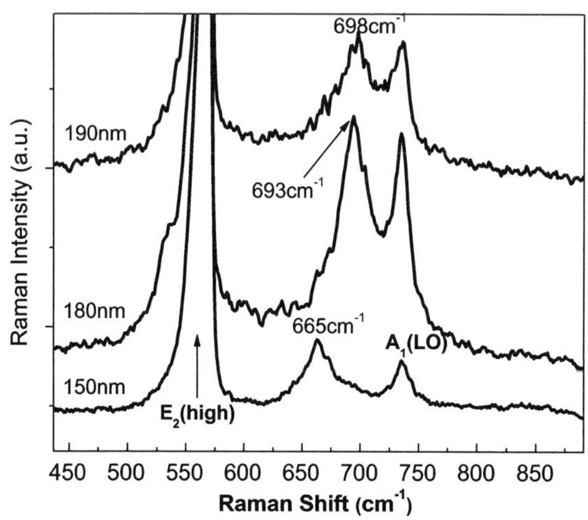

Fig. 9. Micro-Raman spectra of three FIB patterned GaN/InGaN MQW nanopillar arrays with different diameters of 150, 180, and 190 nm. The intensity of SO mode is even stronger than that of symmetry-allowed A_1(LO) mode.

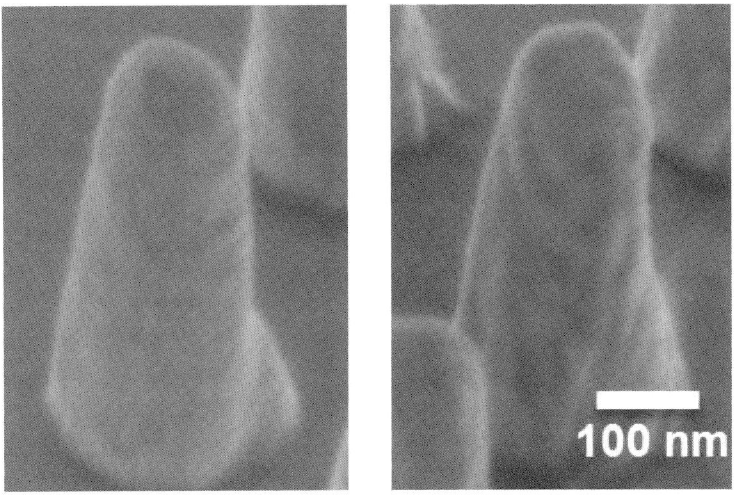

(a) (b)

Fig. 10. Low-voltage field-emission SEM pictures showing surface profile of the nanopillar from two arrays which were prepared with identical FIB parameters and settings, but with different KOH etch duration.

SEM images show an obvious distinction in surface roughness between the two samples. The micro-Raman spectra of these two nanopillar arrays with different roughness surfaces were measured under the same conditions and illustrated in Fig. 11. The SO phonon corresponding to the smoother surface (i.e., as shown in Fig. 11(a)) has a smaller frequency than that of the rougher surface, and this shift of the SO phonon peak is consistent with our calculations: the smoother surface brought in a smaller disturbance to the surface potential, hence leads to a smaller k along the measuring surface. Thus the SO phonon peak in the Raman spectrum shifts further away from the A_1(LO) mode according to the dispersion relation.[9]

In a short summary, GaN/InGaN MQW NPs have been successfully fabricated by using two different "top-down" approaches. The SO phonon was unambiguously observed in the Raman spectra of the NPs with diameters less than ~200 nm. The frequency of SO mode shows a distinct dependence on the feature size of NPs. Its intensity and even frequency exhibit dependence on the surface roughness, reflecting the surface localization nature of the mode.

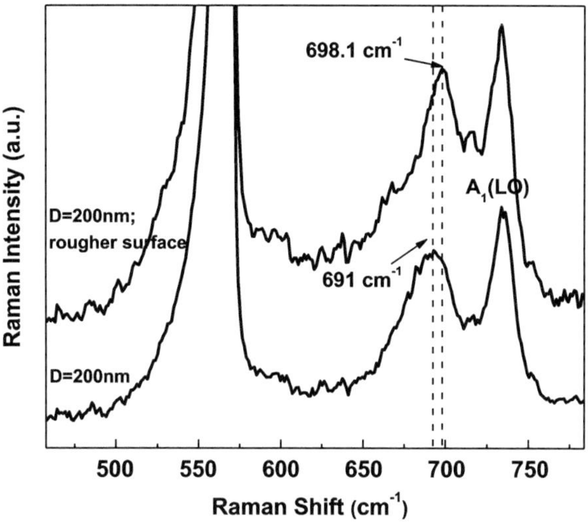

Fig. 11. Micro-Raman spectra of the nanopillar arrays with different surface roughness as shown in Fig. 10. With a smoother surface, the peak position of the SO mode downshifts $8\,cm^{-1}$.

3.2. *Spontaneous light emission*

In the preceding section, we have shown that nanotexturing can lead to the appearance of the SO phonon mode in the prepared NPs. We might say: The light inelastic scattering mechanism could be greatly changed by nanotexturing. In this section, we want to discuss the spontaneous light emission mechanism in the NPs. Before we have a detailed discussion on our results, it could be better for us to give a brief review on the luminescence mechanism in conventional GaN/InGaN MQW LED planar structures. It has been well recognized that localized electronic states induced by indium content fluctuations and In-rich nanostructures *etc.* structural imperfections play a key role in the light emission in traditional GaN/InGaN MQW LED planar structures.[26,27] Carrier localization is commonly believed to be beneficial to luminescence quantum efficiency of the GaN/InGaN MQW LED. To date, however, luminescence efficiency of GaN/InGaN MQW LED is still insufficient to support the emerging lighting technique revolution.[2] Meanwhile, efficiency droop problem in the operation of the blue LED under large injection current is another severe hinder.[28,29] GaN-based nanostructures such as nanowires and nanopillars are hence being explored

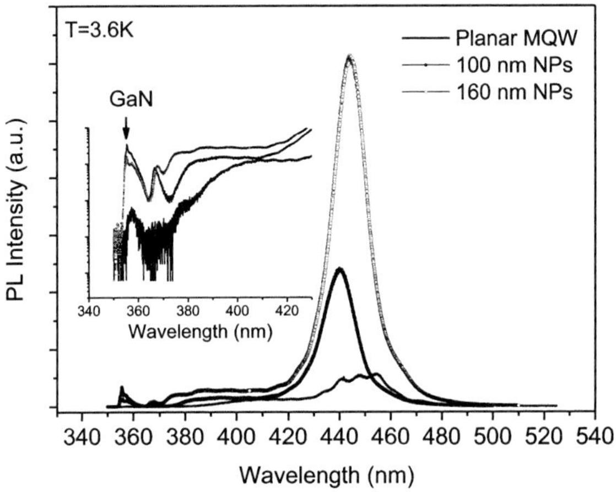

Fig. 12. 3.6 K PL spectra of the three samples measured under the same conditions. The inset shows close-up spectra near the band-edge emission of GaN. Compared with the planar MQW sample, both NPs samples show significantly improved emission intensity at low temperatures, especially the 160 nm NPs sample.

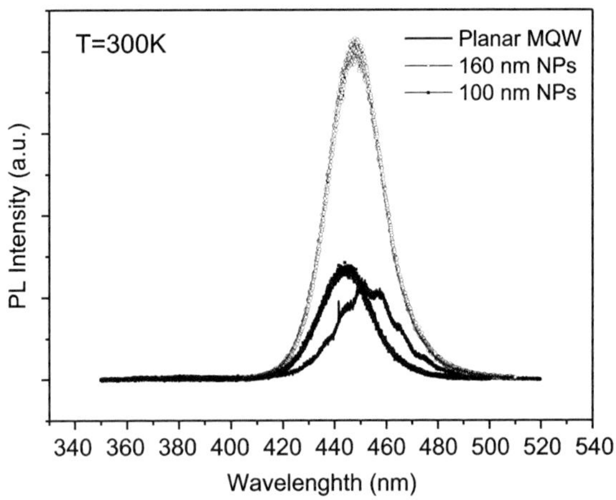

Fig. 13. Room-temperature PL spectra of the three samples. Again, the NPs samples exhibit improved emission intensity at room temperature. In particular, the 160 nm NPs sample shows a substantial improvement in PL intensity at room temperature.

as a promising way for solving aforesaid problems, at least enhancing luminescence efficiency.[5] It is also one of our main objectives.

Figures 12 and 13 show 3.6 K and room-temperature PL spectra of three samples measured under the same conditions except temperature, respectively. Compared with the planar control sample, both NPs samples exhibit substantially enhanced PL emission, indicating significant improvement in luminescence efficiency in nanotextured structures. In particular, the 160 nm NPs sample shows a remarkable improvement in PL intensity at room temperature. The two NPs samples were fabricated by Ni nanoislands mask + ICP etching approach already described in the experimental section. Their SEM images are shown in Fig. 14.

From the top-view ((a) and (c)) and the tilted-view ((b) and (e)) SEM images in Fig. 14, it can be seen that the NPs are about 160 nm ((a) and (b)) and 100 nm ((c) and (d)) in diameter on average, respectively, and both are about 700 nm in height. Nanopillar sheet densities of the 160 nm and 100 nm NPs samples are estimated to be 1.1×10^7 and 2.6×10^7 cm^{-2}, respectively. The two NPs samples also show the distinct SO phonon scattering structures in their micro-Raman spectra.[11] Here, we concentrate on their luminescence

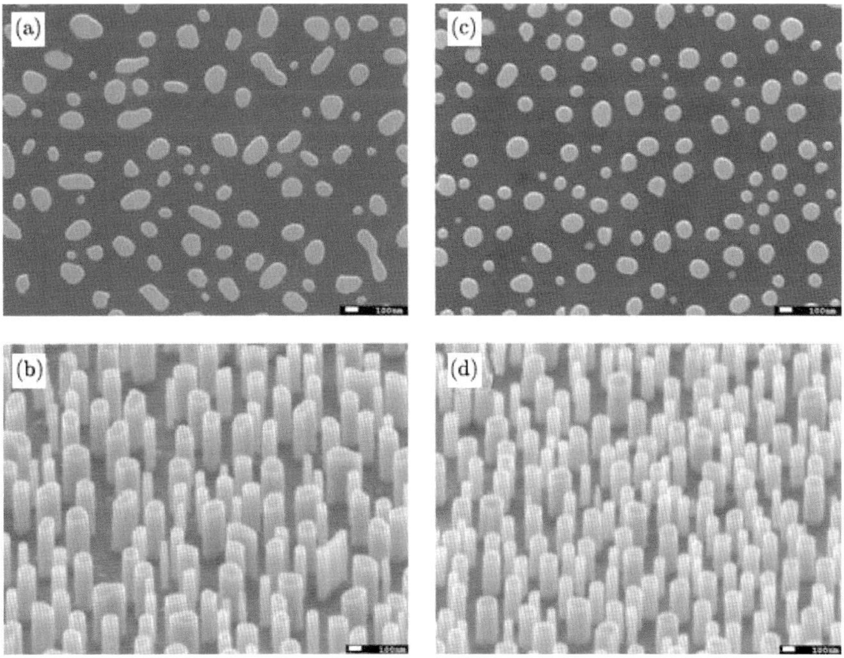

Fig. 14. Top-view and tilted-view SEM images of the two NPs samples. (a) and (b) for the 160 nm NPs, while (c) and (d) for the 100 nm NPs.

mechanisms. To the end, we have actually measured their PL spectra at various temperatures. To save space, we only show 3.6 K (Fig. 12) and 300 K PL spectra (Fig. 13) of the planar control sample (solid line) and the two NPs samples (solid lines + symbols). We now extract all the three chief spectral parameters, namely peak position (spectral center), integrated intensity and linewidth, from the measured variable-temperature PL spectra of the three samples. The obtained data were depicted in various solid symbols in Figs. 15, 16 and 17 for the three samples. For the planar sample, we use a single Gaussian lineshape function to do a best fit to the measured PL spectra for getting the chief spectral parameters because the spectra exhibit a thin-film interference oscillatory structure.

As seen in Figs. 16 and 17, S-shaped temperature dependence of the PL peak positions is observed for the two NPs samples, which is a typical temperature behavior of collective luminescence of localized excitons.[17,18] Taking into account generation, thermal activation (e.g., non-radiative

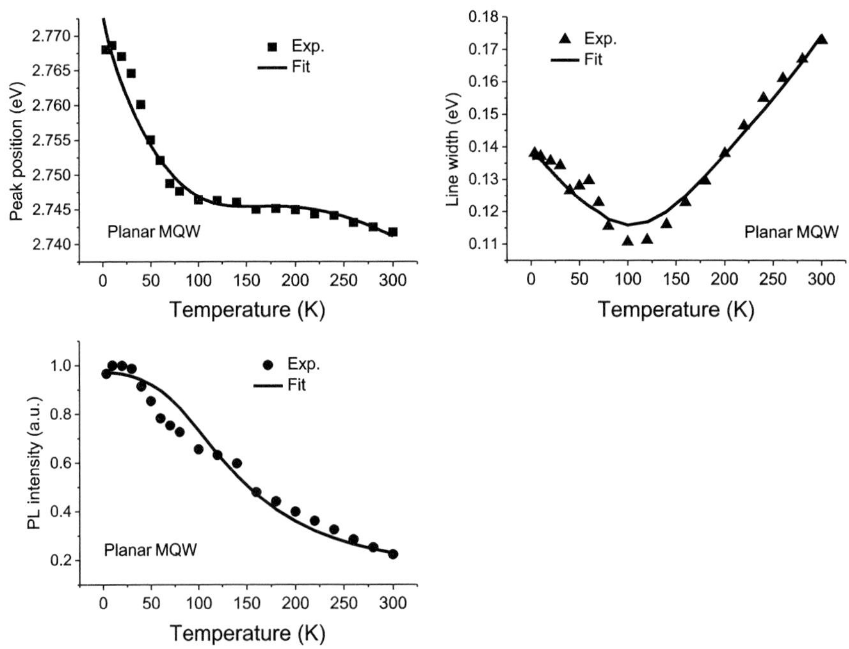

Fig. 15. Extracted peak positions, linewidth and integrated intensities of the planar MQW sample at different temperatures. The solid lines are the fitting curves with LSE model.

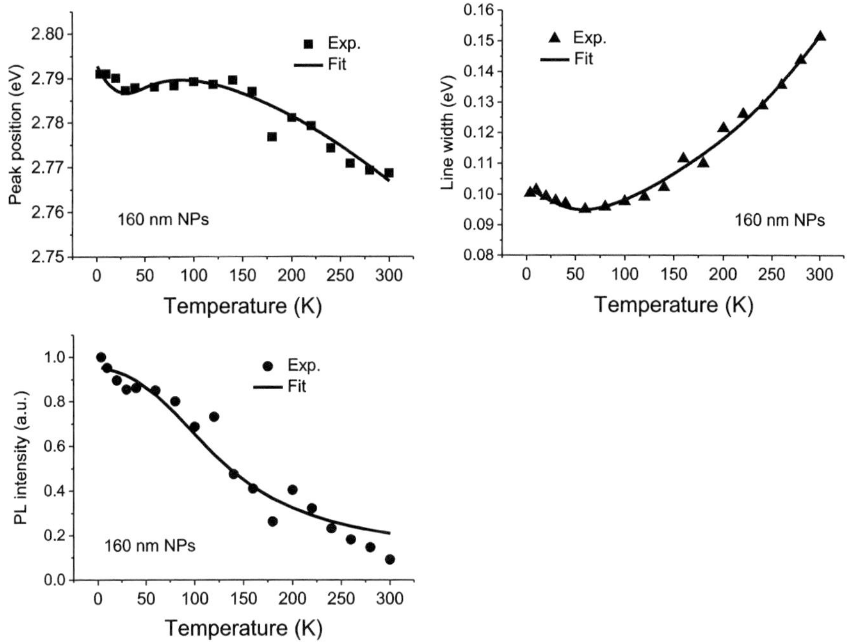

Fig. 16. Extracted peak positions, linewidth and integrated intensities of the 160 nm NPs sample at different temperatures. The solid lines are the fitting curves with LSE model.

recombination), radiative recombination, and re-capturing of excitons, as well as assuming a Gaussian-type density of states for localized-state ensemble, a generalized luminescence model (e.g., LSE model) for localized-state ensemble was developed by Li and Xu *et al.*[17–20] According to this model, temperature-dependent PL peak positions can be written as

$$E(T) = E_0 - \frac{\alpha T^2}{T + \Theta} - \beta(T) \cdot k_B T, \tag{3}$$

where α is Varshni's parameter, Θ Debye temperature, τ_{nr} time constant related to thermal activation (e.g., non-radiative recombination), τ_r time constant of radiative recombination, E_0 and σ energetic center and width of the Gaussian-type density of states for localized-state ensemble distribution, respectively, and k_B the Boltzmann constant. Here, $\beta(T)$ is a temperature-dependent dimensionless parameter whose value can be

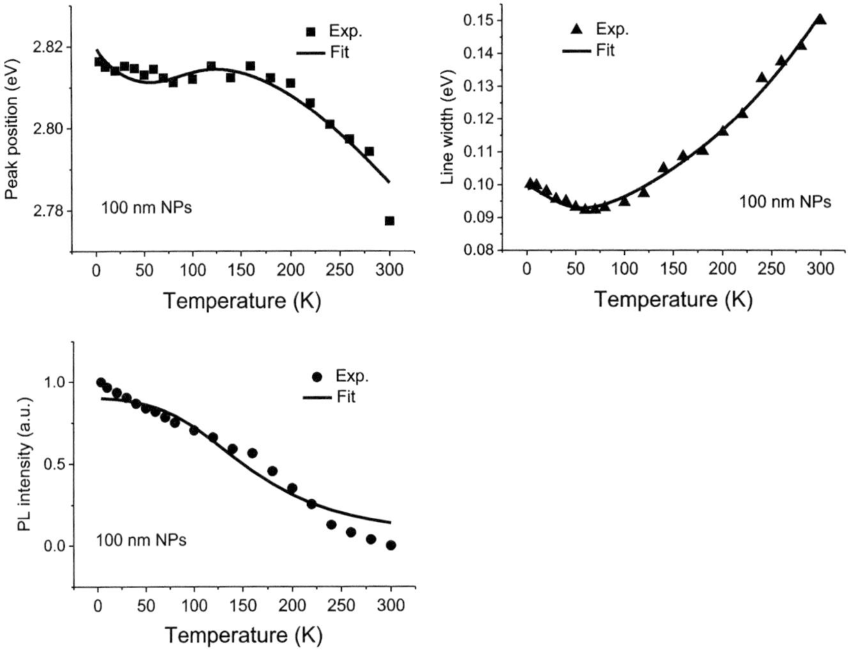

Fig. 17. Extracted peak positions, linewidth and integrated intensities of the 100 nm NPs sample at different temperatures. The solid lines are the fitting curves with LSE model.

obtained by numerically solving the following equation:[11]

$$\left[\frac{1}{\beta}\left(\frac{\sigma}{k_B T}\right)^2 - 1\right] e^{-\beta} = \frac{\tau_{\mathrm{nr}}}{\tau_{\mathrm{r}}} \times e^{-(E_0 - E_a)/k_B T}. \tag{4}$$

From the model itself and its application practice, we can justify that $\Delta E = E_0 - E_a$ actually acts as a more important role than both parameters themselves in determining temperature behavior of luminescence.[19] Depending on material and experimental conditions, ΔE can take positive (+) or negative (−) sign.[19] For the GaN/InGaN material system investigated in the present work, ΔE usually takes a positive sign.[17,18] If luminescence quantum efficiency η is simply defined as $\eta = \frac{\tau_{\mathrm{nr}}}{\tau_{\mathrm{r}} + \tau_{\mathrm{nr}}}$, Eq. (4) can be rewritten as

$$\left[\frac{1}{\beta}\left(\frac{\sigma}{k_B T}\right)^2 - 1\right] e^{-\beta} = \left(\frac{\eta}{1-\eta}\right) \times e^{-\Delta E/k_B T}. \tag{5}$$

Table 1. Parameters for the three samples.

Sample	σ (meV)	E_0 (eV)	ΔE (meV)	η	γ_c	α (meV/K)
160 nm NPs	11.10	2.8065	14.3	12.5%	0.21	0.451
100 nm NPs	16.30	2.8472	27.2	7.69%	0.36	0.643
Planar	19.80	2.7850	11.2	4.76%	0.34	0.245

By using Eqs. (3) and (5), one may get the values of several chief parameters characterizing luminescence of localized excitons via making a best fit to the temperature dependence of PL peak position. Parameters adopted in the fitting are tabulated in Table 1. A unified Debye temperature $\Theta = 630\,\mathrm{K}$ was taken in the fitting for all the three samples. As mentioned earlier, noticeable blueshift in peak position and large increase in emission intensity are the two major effects observed after the nanotexturing. The former is evidenced by the increase of parameter E_0. For example, it increases from 2.7850 eV of the planar MQW structure to 2.8065 eV of the 160 nm NPs sample, further to 2.8472 eV of the 100 nm NPs sample. This phenomenon has been well documented and interpreted as reverse quantum confined Stark effect (QCSE) largely due to the strain relaxation induced by nanotexturing.[5] However, for the significant luminescence intensity increase after the nanotexturing, its origin has not yet well understood. In the present study, the luminescence quantum efficiency is obtained for the three samples. From Table 1, one can see that the luminescence quantum efficiency of the two NPs samples is much higher than that of the planar MQW sample. In particular, the luminescence efficiency of the 160 nm NPs sample is about 2.6 times stronger than that of the planar sample. Such result is very encouraging because enhancing luminescence quantum efficiency is our central objective. But, when the nanopillars' average diameter is further reduced to about 100 nm, their luminescence quantum efficiency decreases, e.g., from 12.5% of the 160 nm NPs to 7.69% of 100 nm NPs. The efficiency decrease could be attributed to the enhanced surface state effect leading to greater nonradiative recombination in small size nanopillars.

It has been known for a long time that S-shaped temperature dependence of luminescence peak position is a characteristic behaviour of localized carriers in semiconductors. On the basis of LSE luminescence model, S-shaped temperature dependence of PL peak positions of both 160 nm and 100 nm NPs samples can be well reproduced, as shown in Figs. 16 and 17. As argued previously, $\Delta E = E_0 - E_a$ rather than E_a itself plays a more significant role in determining abnormal temperature dependence of luminescence of localized carriers.[19,20] For the unetched MQW LED planar structure, its $\Delta E(=11.2\,\text{meV})$ is much smaller than parameter $\sigma = 19.80\,\text{meV}$ and the luminescence linewidth. This could be a main reason why the PL peak of the planar sample doesn't show obvious S-shaped temperature dependence. Instead of, its PL peak positions show a "chair" shaped temperature dependence due to large parameter σ characterizing the distribution of localized states, as seen in Fig. 15. When the planar structure was nanotextured into 160 nm NPs, however, ΔE increases to 14.3 meV. For the resulting 100 nm NPs, ΔE further increases to 27.2 meV. It should be noted that in usual cases, ΔE may take negative sign,[19] which implies that E_a is above E_0. In these so-called normal cases, localized carriers are thermally activated to a delocalized state (or sites in real space) with higher energy, as schematically shown by a dashed horizontal line in Fig. 18. In such cases, ΔE taking a negative value is usually assigned to be a "positive" thermal activation energy. However, in the GaN/InGaN system, especially in their nanostructures, ΔE takes a positive value. We thus argue that thermal quenching of luminescence in GaN/InGaN NPs may be characterized by a "negative" thermal activation energy in terms of their abnormal temperature dependence of luminescence peak and positive ΔE.[11] Figure 18 schematically illustrates such cases, e.g., $E_a < E_0$, then $\Delta E(=E_0 - E_a)$ takes a positive value, but shall be referred to as a "negative" thermal activation energy. In the figure, the wavy arrows stand for the thermal activation processes, while the straight downward arrow represents radiative transition. The bottom curve represents a luminescence spectrum of localized-state ensemble. In the studied NPs samples, as argued earlier, strongly reduced QCSE effect takes place due to the strain relaxation and hence piezoelectric electric field reduction. As a result, E_0 shall increase and the absolute value of ΔE may increase. Indeed, our experimental results demonstrate such tendency.

In order to further investigate the QCSE effect in the NPs samples, we conducted the variable-temperature PL measurements on both the NPs samples under different excitation intensities. Shown in Fig. 19 are the peak positions (symbols) of measured PL spectra of the 160 nm and 100 nm

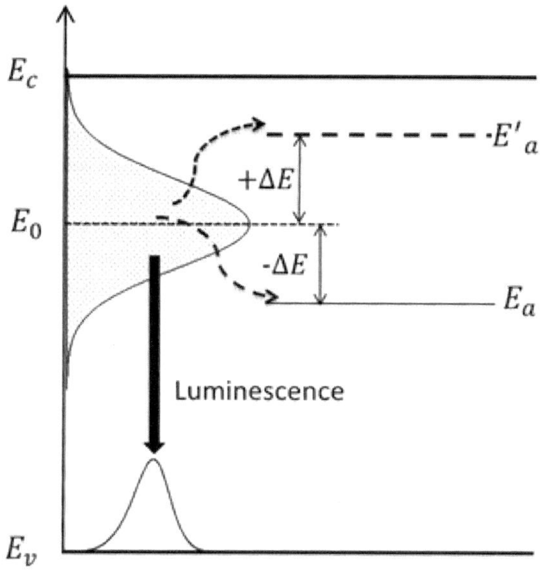

Fig. 18. A physical picture showing the luminescence mechanism of localized-state ensemble. Several key energy parameters are denoted in the figure, especially $+\Delta E$ and $-\Delta E$ which represent positive and negative thermal activation energy, respectively, corresponding to the normal thermal quenching and abnormal thermal quenching of luminescence.

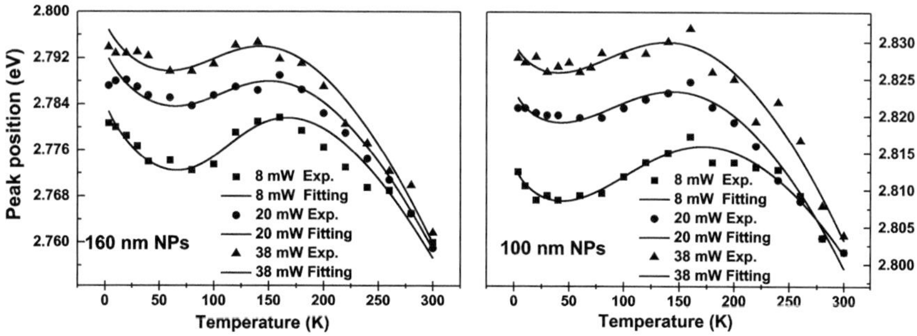

Fig. 19. Peak positions (various symbols) of the PL spectra vs. temperature for the 160 nm NPs and 100 nm NPs samples under three different excitation powers. The solid lines are the fitting curves with LSE model.

NPs samples against temperature for three different excitation powers. As expected, the PL peak shifts toward higher energy direction as the excitation intensity increases, especially at low-medium temperatures. Again, we employ Eqs. (3) and (4) to make a fit to the experimental data, and good agreement between theory and experiment is achieved. In the figure the solid lines represent the fitting curves. At a fixed temperature, blueshift of PL peak with the excitation intensity and hence photogenerated carrier density can be ascribed to the reduced QCSE effect because of screening effect of photogenerated carriers on the piezoelectric field.[30] The screening effect in the GaN/InGaN MQW NPs is efficient at low temperatures, which suggests that strong piezoelectric field still exists in the NPs samples, especially at low temperatures. From the results in Fig. 19, one can see that piezoelectric field shows an important influence on carrier localization.[11,30]

In addition to the peak position, full width at half maximum height or simply linewidth is another key parameter which also carries important information on localization mechanism of carriers.[19,20] Reduction in linewidth with temperature in a low temperature range is another PL signature of carrier localization, which may be principally understood as thermal redistribution of localized excitons among different localized states.[17,19] According to LSE model, the thermal distribution of localized excitons is jointly determined by Gaussian-type density of states and a newly-derived distribution function. In mathematic form, the thermal distribution of localized excitons can be described by $n(E, T) = \rho(E) \cdot f(E, T)$. Here $\rho(E) = \rho_0 \exp[-(E - E_0)/2\sigma^2]$ is a standard Gaussian function and $f(E, T) = [\exp((E - E_a)/k_B T) + \tau_{nr}/\tau_r]^{-1}$ is a newly derived distribution function for localized carriers.[19,20] It has been previously proven that $n(E, T)$ essentially represents the spectral profile of localized-state ensemble luminescence.[19,20] The linewidth $\Gamma_c(T)$ of $n(E, T)$ is thus obtained by numerically solving $n(E, T) = n(E_{pk}, T)/2$. Besides, phonon scattering induced broadening shall be taken into account. By taking into account these two effects, we may write down an expression for the actual linewidth of localized-state ensemble luminescence[19]:

$$\Gamma = \Gamma_0 + \Gamma_c(T) + \sigma_A T + \frac{\gamma_{LO}}{e^{\hbar\omega_{LO}/k_B T} - 1}, \tag{6}$$

where Γ_0 is a temperature-independent constant accounting for scattering by unavoidable impurity and imperfection,[31] $\sigma_A T$ represents the broadening

Table 2. Parameters for the fitting to linewidth data of the three samples.

Sample	σ_A (meV/K)	γ_{LO} (eV)	$\hbar\omega_{LO}$ (meV)	Γ_0 (eV)
160 nm NPs	0.0885	0.8721	91.3	0.0561
100 nm NPs	0.0916	0.9149	91.3	0.0539
Planar	0.0389	1.2909	91.3	0.0657

term induced by acoustic phonon scattering which is linearly dependent on temperature,[31] and the last term on the right side of Eq. (6) stems from optical phonon scattering. In polar semiconductors like GaN, the dominant scattering mechanism arises from LO phonon mode.[31,32] Here γ_{LO} represents exciton-LO phonon coupling strength, while $\hbar\omega_{LO}$ is the characteristic energy of LO phonon. By using Eq. (6), we have made a fit to the experimental linewidth data. Parameters adopted in the fitting are tabulated in Table 2. It can be seen from Figs. 15, 16 and 17 that good agreement between theory and experiment is achieved in the entire interested temperature range for all the three samples. Compared with the as-grown planar MQW sample, both the NPs samples show weaker exciton-LO-phonon coupling, e.g., $\gamma_{LO} = 0.9149\,\text{eV}$ for the 100 nm NPs sample, while $\gamma_{LO} = 1.2909\,\text{eV}$ for the planar MQW sample. Our results seem to be consistent with Renwick *et al.*'s data.[33] In another study done by Chen *et al.*,[34] significant enhancement of electron-LO-phonon coupling was observed in highly strained GaN/InGaN quantum well structures in terms of the relative intensity between the first-order LO phonon replica and its principal emission line. They attribute such enhancement to the interface features such as interface optical phonon mode and the high residual strain. In our case, noticeable release of the residual lattice-mismatch strain and new surface optical phonon mode located at the lower frequency side of A_1(LO) mode (See, Fig. 5) were observed in the NPs samples. Therefore, weaker exciton-LO-phonon coupling is expected and indeed observed in the GaN/InGaN NPs.

4. Conclusions and Outlook

In conclusion, GaN/InGaN MQW nanopillars were prepared with two different "top-down" approaches, and their optical properties were investigated with confocal micro-Raman scattering and temperature-dependent photoluminescence techniques. Localized surface optical phonon mode and distinct Raman light scattering have been unambiguously demonstrated in the fabricated GaN/InGaN nanopillars with average diameter $< \sim 200$ nm. Meanwhile, localized electronic states and piezoelectric electric field have been firmly shown to play an important role in the spontaneous light emission mechanism in the nanopillars. Greatly enhanced internal luminescence efficiency has been also detected in the nanopillars. In one sentence, nanotexturing has a significant impact on the electronic states, lattice vibration states and optical properties of GaN/InGaN MQW LED planar structure.

For further studies, we would like to suggest these points: influence of nanostructure shape on the localized surface optical mode; electron (exciton)-SO phonon coupling; further investigations of physical meaning of "negative" thermal activation energy; and time-resolved luminescence of localized excitons in nanostructures.

Acknowledgments

The study was supported by HK-RGC-GRF Grants (No. HKU 705812P), NSFC Grants (No. 11374247), and HKU SRT on New Materials, as well as in part by HK-UGC AoE Grants (No. AoE/P-03/08).

Taking this opportunity, I would like to thank all my research collaborators, students and friends in the semiconductor luminescence community for their helpful contributions and stimulating discussions over the years. My special thanks are given to Dr. Q. Li, Dr. J. H. Zhu, Mr. Z. C. Su, Dr. X. H. Wang, Prof. J. Q. Ning, Mr. W. Bao, and Dr. C. C. Zheng of my group at The University of Hong Kong for their hard work and outstanding contributions to the study of this issue. I am indebted to Prof. D. G. Zhao, Prof. H. W. Choi, Prof. L. X. Zheng, Prof. M. H. Xie, Prof. H. Yang, Prof. C. M. Che and Prof. S. Y. Tong for their extensive collaboration and friendship over the last almost 20 years. Last but not least, I want to thank Prof. Z. C. Feng, the editor of this special book, for his kind invitation and long-term friendship.

References

1. E. F. Schubert and J. K. Kim, *Science*, **308**, 1274 (2005).
2. S. Pimputkar, J. S. Speck, S. P. DenBaars, and S. Nakamura, *Nat. Photon.*, **3**, 180 (2009).

3. H. M. Kim, Y. H. Cho, H. Lee, S. I. Kim, S. R. Ryu, D. Y. Kim, T. W. Kang, and K. S. Chung, *Nano Lett.*, **6**, 1059 (2004).
4. Fang Qian, Yat Li, S. G. Cak, H.-G. Park, Yajie Dong, Yong Ding, Z. L. Wang, and C. M. Lieber, *Nat. Mater.*, **7**, 701 (2008).
5. Shunfeng Li and A. Waag, *J. Appl. Phys.*, **111**, 071101 (2012), and references therein.
6. Xiangfeng Duan, and C. M. Lieber, *Adv. Mater.*, **12**, 298 (2000).
7. See for example, a good review article, D. Mijatovic, J. C. T. Eijkel and A. van den Berg, *Lab Chip*, **5**, 492 (2005).
8. J. H. Zhu, S. M. Zhang, X. Sun, D. G. Zhao, J. J. Zhu, Z. S. Liu, D. S. Jiang, L. H. Duan, H. Wang, Y. S. Shi, S. Y. Liu, and H. Yang, *Chin. Phys. Lett.*, **25**, 3485 (2008).
9. X. H. Wang, J. Q. Ning, S. J. Xu, and H. W. Choi, *J. Appl. Phys.*, **110**, 093111 (2011).
10. J. H. Zhu, J. Q. Ning, C. C. Zheng, S. J. Xu, S. M. Zhang, and Hui Yang, *Appl. Phys. Lett.*, **99**, 113115 (2011).
11. Wei Bao, Zhicheng Su, Changcheng Zheng, Jiqiang Ning, and Shijie Xu, *Sci. Rep.*, **6**, 34545 (2016).
12. R. Ruppin and R. Englman, *Rep. Prog. Phys.*, **33**, 149 (1970).
13. B. E. Sernelius, Surface Modes in Physics (Wiley, New York, 2001).
14. J. Q. Ning, S. J. Xu, D. P. Yu, Y. Y. Shan, and S. T. Lee, *Appl. Phys. Lett.*, **91**, 103117 (2007).
15. J. Q. Ning, S. J. Xu, P. W. Wang, Y. P. Song, D. P. Yu, Y. Y. Shan, S. T. Lee, and H. Yang, *Mater. Character.*, **73**, 153 (2012).
16. J. Q. Ning, C. C. Zheng, L. X. Zheng, and S. J. Xu, *J. Appl. Phys.*, **118**, 073101 (2015).
17. Q. Li, S. J. Xu, W. C. Cheng, M. H. Xie, S. Y. Tong, C. M. Che, and H. Yang, *Appl. Phys. Lett.*, **79**, 1810 (2001).
18. Q. Li, S. J. Xu, M. H. Xie, and S. Y. Tong, *J. Phys.: Conden. Matters*, **17**, 4853 (2005).
19. Q. Li, S. J. Xu, M. H. Xie, and S. Y. Tong, *Europhys. Lett.*, **71**, 994 (2005).
20. Q. Li and S. J. Xu, *Physics (Wu Li)*, **35**, 659 (2006) (in Chinese).
21. R. Gupta, Q. Xiong, G. D. Mahan, and P. C. Eklund, *Nano Lett.*, **3**, 1745 (2003).
22. K. W. Adu, Q. Xiong, H. R. Gutierrez, G. Chen, and P. C. Eklund, *Appl. Phys. A*, **85**, 287 (2006).
23. D. G. Zhao, S. J. Xu, M. H. Xie, S. Y. Tong, and H. Yang, *Appl. Phys. Lett.*, **83**, 677 (2003).
24. C. L. Hsiao, L. W. Tu, T. W. Chi, M. Chen, T. F. Young, C. T. Chia, and Y. M. Chang, *Appl. Phys. Lett.*, **90**, 043102 (2007).
25. S. E. Wu, S. Dhara, T. H. Hsueh, Y. F. Lai, C. Y. Wang, and C. P. Liu, *J. Raman Spectro.*, **40**, 2044 (2009).
26. S. Nakamura, *Science*, **281**, 956 (1998).
27. K. O'Donnell, R. Martin, and P. Middleton, *Phys. Rev. Lett.*, **82**, 237 (1999).
28. J. Iveland, L. Martinelli, J. Peretti, J. S. Speck, and C. Weisbuch, *Phys. Rev. Lett.*, **110**, 177406 (2013).
29. R. Sarkissian, S. T. Roberts, T.-W. Yeh, S. Das, S. E. Bradforth, J. O'Brien, and P. D. Dapkus, *Appl. Phys. Lett.*, **103**, 041123 (2013).
30. Q. Li, S. J. Xu, M. H. Xie, S. Y. Tong, X. H. Zhang, W. Liu, and S. J. Chua, *Jpn. J. Appl. Phys.*, **41**, L1093 (2002).
31. S. Rudin, T. Reinecke, and B. Segall, *Phys. Rev. B*, **42**, 11218 (1990).

32. S. J. Xu, L. X. Zheng, S. H. Cheung, M. H. Xie, S. Y. Tong, and H. Yang, *Appl. Phys. Lett.*, **81**, 4389 (2002).
33. P. Renwick, H. Tang, J. Bai, and T. Wang, *Appl. Phys. Lett.*, **100**, 182105 (2012).
34. D. Chen, Y. Luo, L. Wang, H. T. Li, G. Y. Xi, Y. Jiang, Z. B Hao, C. Z. Sun, and Y. J. Han. *J. Appl. Phys.*, **101**, 053712 (2007).

CHAPTER 11

GROUP III-NITRIDE NANOSTRUCTURES FOR LIGHT-EMITTING DEVICES AND BEYOND

Je-Hyung Kim, Young-Ho Ko and Yong-Hoon Cho*

Department of Physics,
Korea Advanced Institute of Science and Technology (KAIST),
Daejeon 34141, Republic of Korea

Beyond epitaxially grown planar-thin film structures, nanostructure-based group III-nitride materials have attracted intensive interest for light-emitting device applications. Nanostructures have distinctive optical and structural properties such as improved light extraction, fast spontaneous emission rate, large surface area, high crystal quality, and high indium incorporation efficiency. Therefore, nanostructure-based light-emitting diodes may overcome many problems of planar light-emitting diodes, including high dislocation density, limited light extraction, efficiency droop, large built-in electric field, and poor efficiency at long visible wavelengths. In the past decade, various types of three-dimensional nanostructures and numerous fabrication approaches have been suggested to realize nanostructure-based high-performance light-emitting diodes. As a result, these nanostructures allow efficient phosphor-free white light sources. Now beyond light-emitting diodes, the group III-nitride nanostructures are expanding their field of application such as cavity-polariton devices, photonic diodes, and single photon sources.

1. Introduction

Over the past few decades, there have been radical developments in the field of light-emitting devices since the appearance of efficient

*yhc@kaist.ac.kr

blue-colored, solid-state light-emitting diodes (LEDs),[1] and LEDs have rapidly replaced existing light sources such as bulbs and cathode lamps. Group III-nitride based LEDs are leading this revolution, and have made all many new developments possible after researchers' extensive efforts to improve the internal and external quantum efficiencies of LEDs. However, group III-nitride based LEDs still suffer from many problems, including low crystal quality,[2] low light extraction,[3] long radiative recombination time, efficiency droop at high power level,[4] low indium incorporation efficiency,[5] and so on. Although the majority of work to date has been done with a GaN film on a sapphire substrate, this *c*-plane planar structure is becoming a main factor limiting the quality and performance of LEDs. A GaN film on a sapphire substrate receives a compressive strain, resulting in high density of dislocations. Light extraction from the GaN film is limited by the total internal reflection. The lack of inversion symmetry along the growing direction (*c*-axis) in Wurtzite structures yields a huge internal polarization field in the group III-nitride materials.

Recently, a number of works have reported various types of nanostructured LEDs that can overcome the above limitations of the planar LEDs, and these results have suggested strong potential to high-performance LEDs.[6–9] Therefore, three-dimensional nanostructures are taking an important position in the area of next generation of LEDs, and more reliable and more economical fabrication methods for the various types of nanostructures are continually being developed these days.

Here we overview the group III-nitride nanostructures formed using several different approaches. In Sec. 2 we deal with the advantages of the use of nanostructures from various points of view and consider why nanostructures are important for LED applications. In Secs. 3 and 4, we introduce fabrication methods for high-quality nanostructures and optical properties of nanostructures. Sec. 3 focuses on the nanostructures formed by top-down techniques, and Sec. 4 introduces the nanostructures grown by bottom-up techniques. In Sec. 4, we feature both vertical rod-shaped and pyramidal-shaped nanostructures. Finally, in Sec. 5, we look at the strong potential of group III-nitride nanostructures for new applications beyond LEDs.

2. Nanostructures for LED Applications

Nanostructures with a high surface to volume ratio and multi-crystalline facets show optical and structural properties distinctive from those of planar films and thus have received great attention for use in various optoelectronic applications such as waveguides, field emitters, sensors, and detectors.[10–14] Nanostructures are also becoming increasingly important for nano-photonic devices, in particular high-performance LED applications. In the early stages of research in this field, the nanostructures formed by reactive ion etching (RIE)[15,16] were studied in the group III-nitride system; although these nanostructures showed enhanced light extraction, mechanical etching caused severe etching damage on the nanostructures. Recently, instead of simple RIE techniques, several approaches for the fabrication of high-quality nanostructures have been developed.

We can classify nanostructures according to their fabricating approaches, namely top-down etching and bottom-up growth. More specifically, dry etching using reactive ions and chemical etching using liquid- or vapor-phase etchants are the top-down etching approaches. Direct crystal growth methods using molecular beam epitaxy (MBE),[17–19] hydride vapor phase epitaxy (HVPE),[20,21] or metal-organic chemical vapor deposition (MOCVD)[22–24] are the bottom-up growth approaches. It is also possible to combine both top-down etching and bottom-up regrowth techniques.[9]

For LED applications, active regions, *i.e.*, quantum wells (QWs) need to be inserted in the host material having a higher bandgap energy. Based on the position of the active regions in the nanostructure, we can divide the cases of nanostructures once more (Fig. 1). For axial-type nanostructures, we form the QWs along the vertical direction (polar direction), and the QWs are inserted as disks in the middle of the nanostructures, while radial-type nanostructures consist of several over-layers of QWs and barriers from the inner to outer direction (non-polar direction), resulting in core-shell structures. Top-down etching methods mostly lead to axial-type nanostructures because they etch pre-grown planar QW wafers, which have QWs along a growth direction.[25,26] In the case of bottom-up approaches, the growing equipment mostly determines the type of the nanostructures. MBE strongly drives direct vertical crystal growth, and thus axial-type nanostructures are formed by the subsequent

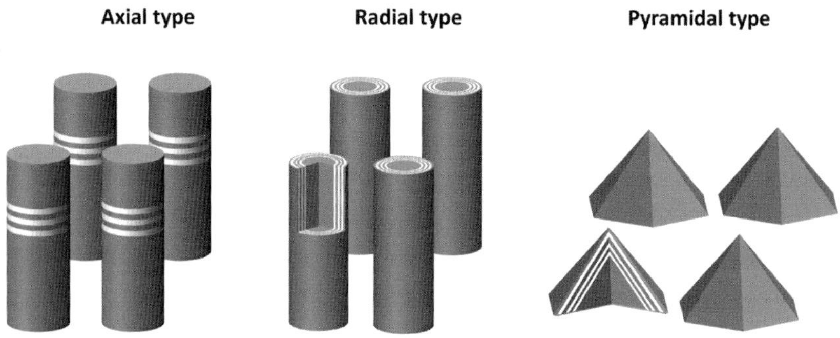

Fig. 1. Schematic illustrations for various types of nanostructures.

change of growing materials.[27,28] With MOCVD, a lateral growth mode is predominant, so MOCVD is useful for radial-type nanostructures. However, the lateral growth in MOCVD generally leads to a thin film, instead of the formation of nanostructures, and therefore, MOCVD requires additional local growth techniques to form nanostructures. The local growth can be accomplished by using a catalyst,[29] Si dopant,[22,30] or patterned substrates.[19,31] In addition, a change of polarity,[32] strain,[23] and growth conditions[24] modifies the surface energy between crystalline facets, resulting in local vertical growth of GaN. Once vertical GaN nanostructures are formed, GaN/InGaN core-shell structures are easily grown by using MOCVD.[30,31,33] In addition, it is possible to grow QWs on semi-polar facets of GaN that results in pyramidal structures shown in Fig. 1. Pyramidal structures are easily formed on a patterned substrate using the conventional MOCVD process.

In this review, we focus on the core-shell and pyramidal nanostructures because these nanostructures have more advantages than axial-type nanostructures, and MOCVD technique can lead to better industrial productivity than MBE technique for LED applications.

We summarize the important merits of core-shell and pyramidal nanostructures that have QWs on their non/semi-polar facets as follows.

- Improvement in light extraction efficiency

Planar GaN film has a poor light extraction efficiency of around 4 % due to total internal reflection. The formation of nanostructures effectively reduces the total internal reflection and efficiently extract photons into free space.[34] It is also possible to manipulate far-field patterns by controlling the size and position of nanostructures.[35]

- Large surface area

Core-shell and pyramidal nanostructures have a large active surface area due to their enlarged surface to volume ratio. This increases a brightness of LEDs of the same chip size.[6] Furthermore, the large surface area can reduce the current density in LEDs, reducing the efficiency droop problem at high current level.

- Improvement in crystal quality

Planar GaN film generally contains many dislocations that originate from mismatches in the lattice constants and thermal expansion coefficients between the GaN epilayer and the substrate. The use of nanostructures effectively minimizes the strain from the substrate,[23] and the lateral growth in nanostructures yields additional strain relaxation and changes the direction of dislocations.[36] Therefore, we can obtain high-quality nanostructures with a reduced number of dislocations.[24,37,38]

- Reduction of internal electric fields

The lack of inversion symmetry in the Wurzite structure of group III-nitride materials causes a huge built-in electric field along the *c*-axis. This electric field suppresses the radiative recombination process of excitons by separating electrons and holes. In the core-shell and pyramidal structures, QWs are placed on the non/semi-polar surface of nanostructures, where the influence of electric fields is minimized.[39] Therefore, we can expect improved internal quantum efficiency with fast radiative recombination of excitons in QWs.[9]

- Improvement of indium incorporation efficiency

In planar InGaN QW structures, high indium composition is difficult to achieve due to the low indium incorporation efficiency of InGaN QWs.[5] In addition, as we increase the indium composition in the QWs, the enlarged strain induces dislocations and large built-in electric fields, leading to high non-radiative and low radiative recombination rates of excitons. This limits the internal quantum efficiency of the LEDs at the long visible wavelengths of green and red. Nanostructures show great potential for the development of high indium composition in QWs. Indium atoms have a longer diffusion length than gallium atoms, resulting in high indium composition in three-dimensional nanostructures.[9,40] The relaxed strain also improves the indium incorporation efficiency in the nanostructures.[41] This improved indium incorporation efficiency enables the formation of efficient green- and red-emitting LEDs, and also of phosphor-free white LEDs.[9,41]

- Increase of QW thickness

The reduced strain in nanostructures enables the growth of thick InGaN QWs[9] or InGaN/GaN heterostructures.[41] without the generation of dislocations. The thick QWs improve indium incorporation efficiency by reducing the compositional pulling effect.[42] Furthermore, the thick QWs can reduce the carrier density, and thus be helpful for the efficiency droop problem.[43]

As shown above, the nanostructures possess unique features that solve the existing problems in planar LEDs. Therefore, these nanostructures may provide a promising platform for high-performance LEDs with high quality, high brightness, high power, and high indium composition, and fast recombination.

3. Top Down Etched Nanostructure

Top-down etching is an intuitive approach to form vertical nanostructures. In order to fabricate nanostructures, top-down etching normally requires masks, which can be formed by optical/electron-beam lithography,[26,44] nanoimprint,[45] and nanomaterial deposition.[46,47] After employing a mask, the RIE process with Cl_2 and Ar flows can be used to etch GaN and its relevant materials.[15]

Although the RIE easily forms vertical nanostructures and controls their position and shape, the attendant high energy ion bombardment forms surface defects on the etched side walls, causing severe non-radiative surface recombination and large leakage current.[16,48,49] Especially, the strong atomic bonding of the group III-nitride materials requires strong etching conditions including high bias voltage and therefore induces profound effects on the optical and electrical properties of the etched nanostructures. To improve the crystal quality of the etched surface, several surface passivation methods such as subsequent thermal annealing,[50] gas treatment,[51] and chemical treatments[25,52] have been suggested. These treatments can alleviate the etching damage by reducing defect-induced surface recombination and also change the etched surface from irregular facets to crystalline facets.

Chemical etching is an alternative way to form top-down nanostructures, and is widely used for other semiconductor materials because it maintains high crystal quality of the etched surface. However, chemical etching of GaN films on a sapphire substrate is very difficult because they have a top surface with Ga-polarity (0001), which has a very low surface energy.[53]

Comparing to the Ga-polarity (0001) facet, N-polar or non/semi-polar facets are easier for chemical etching. A GaN film with a N-polar top surface can be obtained from the backside of a normal Ga-polar film by removing a sapphire substrate[54] or by introducing an AlN buffer layer before growing GaN on a sapphire substrate.[55] The surface energy of the N-polar surface is higher than that of the Ga-polar surface, i.e., the N-polar surface is more reactive to the chemical etching process. Pyramidal-shaped nanostructures have been reported by chemically etching N-polar GaN films.[54]

A more effective approach for chemical etching is to increase the chemical etching energy itself. This can be accomplished by using vapor-phase chemical etchant at a higher temperature. Chemical wet etching experiments are typically performed at relatively low temperatures below 400°C,[53,56] but chemical vapor-phase etching can increase the temperature even above 1000°C, which is the range of growth temperature of GaN materials. At this high temperature, the chemical decomposition process actively occurs. The chemical vapor-phase etching effectively eliminates dislocations and forms high-quality nanostructures on a *c*-plane GaN layer. The successful demonstration of the chemically etched nanostructures on a Ga-polar GaN film has been reported using HCl gases at about 1000°C that showed high-quality vertical GaN nanostructures.[38]

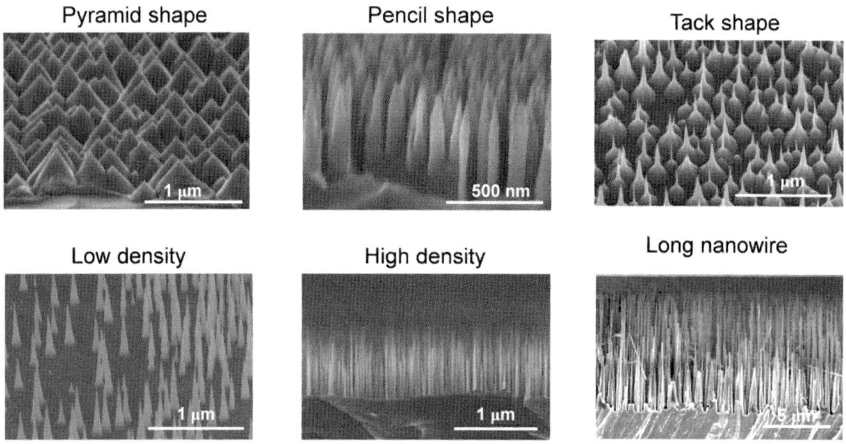

Fig. 2. SEM image of chemically etched GaN nanostructures with various shape, length, and density. Reprinted with permission from Ref. 38. Copyright (2012) American Chemical Society.

The chemical vapor-phase etching forms nanostructures uniformly on a wafer without any masks as shown in scanning electron microscopy (SEM) images in Fig. 2 The etching conditions such as etching temperature, gas flow rate, and doping level strongly affect the shape, size, and density of the nanostructures. The formation of self-etched nanostructures begins with the region of dislocations because these are the high energetic sites compared to other regions. Although chemical wet etching also forms hexagonal-shaped etch pits on dislocation regions, the pit density is much lower than the dislocation density in GaN, and so the chemical wet etching does not reveal all dislocations, but it is known to form the etch pits on nanopipe-types dislocations.[57] On the other hand, the chemical vapor-phase etching, with its higher etching energy, exposes most dislocations. Atomic scale terrace-shaped step edges on a GaN surface[58] can also become another type of initial etching point during the chemical vapor-phase etching process (Fig. 3).

The chemically etched, dislocation-free nanostructures show improved optical characteristics compared with those of the GaN film before etching. Figure 4 provides a comparison of the photoluminescence (PL) spectra of the GaN film (before etching) and the GaN nanostructures (after etching) at 10 K. In this comparison, the nanostructures show a large increase of PL intensity (by a factor of 10) and a red shift (12 meV) of the exciton peak

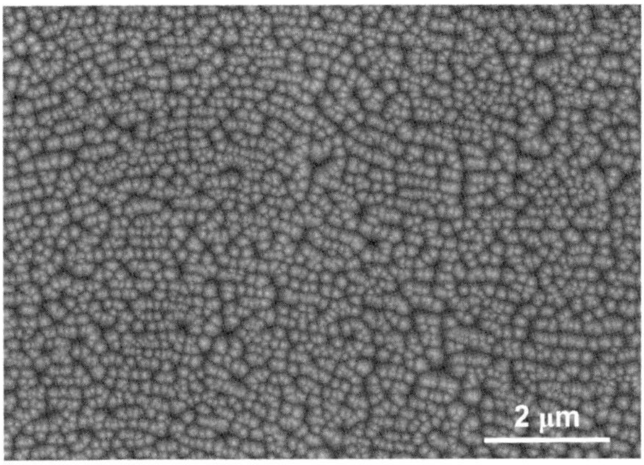

Fig. 3. SEM image of GaN surface at early stage of etching.

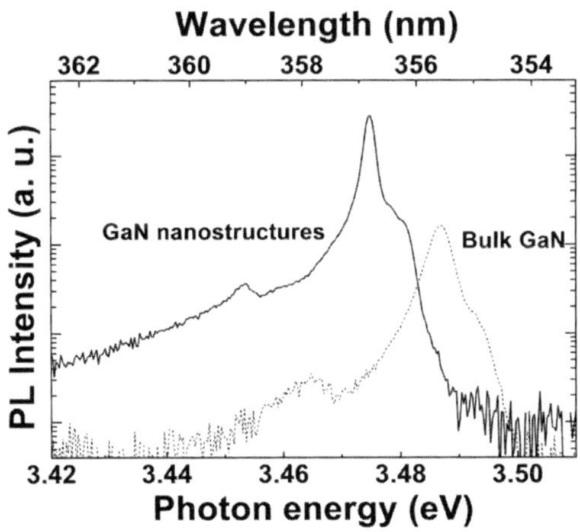

Fig. 4. Low temperature (10 K) PL spectra of as-etched GaN nanostructures (solid line) and a bulk GaN wafer (dotted line).

due to the improved light extraction and reduced residual strain in the GaN material.[59] The reduction of defects in the material improves the crystal quality of the chemically etched nanostructures, leading to a narrowing of the exciton peak from 2.5 meV for the GaN film to 1.9 meV for the etched nanostructures.

Regrowing InGaN QWs on the etched GaN nanostructures using MOCVD forms core-shell nanostructures [Fig. 5(a)].[9] Scanning

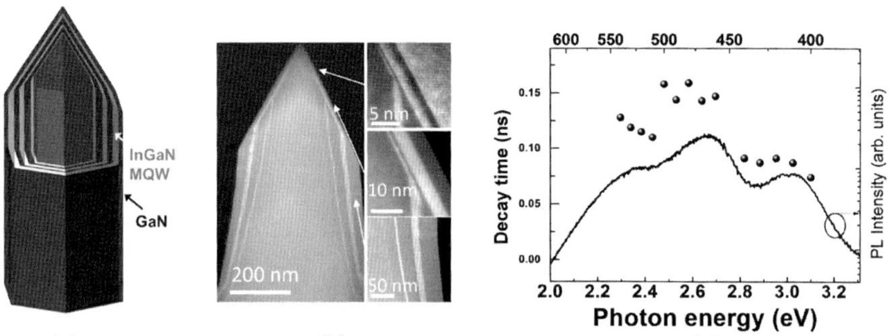

(a) (b) (c)

Fig. 5. (a) Schematic image of an InGaN/GaN core-shell nanostructure and (b) its cross-sectional STEM images. (c) PL spectrum of InGaN QWs in core-shell nanostructures and their decay times at different wavelengths. Reprinted with permission from Ref. 9. Copyright (2014) Royal Society of Chemistry.

transmission electron microscopy (STEM) images in Fig. 5(b) show that the nanostructure does not include any dislocations inside and that the multi-QWs cover the entire surface of the core-shell nanostructure. The formed core-shell nanostructure has an obelisk-like shape consisting of a vertical column part and a pyramidal tip part at the top of the nanostructure. The InGaN QWs grown on semi/non-polar facets of the core-shell nanostructure show optical properties distinctive from those of planar QWs grown on the *c*-plane. For instance, the large built-in field in planar InGaN QWs causes long radiative recombination times on the order of a few ns, and these values rapidly increase up to a few tens of ns at long wavelengths due to the increased quantum confined Stark effect.[60] Whereas the QWs on semi/non-polar facets can have a fast recombination time due to a reduced build-in electric field. Figure 5(c) shows the lifetime of InGaN multi-QWs in the core-shell structure at different wavelengths, which show very fast radiative recombination times less than 200 ps even at long wavelengths. Moreover the core-shell nanostructures emit a broad spectrum, covering the entire visible wavelengths [Fig. 5(c)]. This originates from the large variation in the QW thickness and in the indium composition. As can be seen in the TEM images, the core-shell structures contain multi-QWs with thicknesses in a range from 1 nm to 15 nm. The degree of the quantum confinement effect varies between ultrathin QWs and thick QWs. The indium composition also depends on the position of the QWs because indium atoms have a

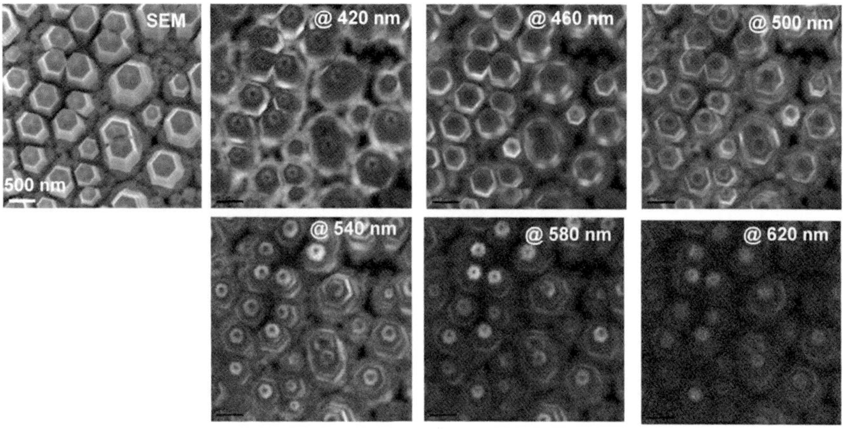

Fig. 6. SEM and CL mapping images at several different emission wavelengths. Reprinted with permission from Ref. 9. Copyright (2014) Royal Society of Chemistry.

longer diffusion length than that of gallium atoms, resulting in higher indium composition in the higher position of nanostructures where a longer diffusion length is required. Cathodoluminescence (CL) is a good tool to investigate the optical properties of nanostructures and provides spatially- and spectrally-resolved images. Figure 6 shows top-view SEM and monochromatic CL images at several wavelengths, showing different emission colors depending on the position.

These results show that top-down chemical etching method can produce high quality GaN nanostructures, and the core-shell nanostructures have strong potential for use in efficient LEDs over the planar film structures.

4. Bottom-up Crystal Growth Process

In this section, we introduce direct crystal growth of GaN nanostructures, having vertical rod and pyramidal structures. Both structures can be obtained by MOCVD process. To grow vertical structures, it is important to suppress lateral growth and enhance vertical growth, while the pyramidal structures can be obtained by conventional MOCVD process on patterned substrates. The sizes and positions of the pyramidal structures can be controlled using the hole-pattern design. Also, more complex patterns such as rings can be applied to form multi-faceted structures.

4.1. *Vertical rod structure*

Several approaches have been suggested to form vertical rod-shaped nanostructures using MOCVD. One common method is catalyst-assisted growth that strongly drives vertical growth at local positions. Vapor-liquid-solid (VLS) growth steps are the well-known mechanism of the catalyst-assisted growth.[29,61] However, these catalysts, as impurities, contaminate and/or form additional defects inside the nanostructures.[62]

It is possible to grow vertical nanostructures without the use of a catalyst, i.e. self-organized growth. Self-organized nanostructures are often found in several semiconductor systems such as Si[63] and ZnO,[64] but achieving self-organized nanostructures from GaN on a sapphire substrate is a challenging task due to a strong lateral growth in MOCVD. To drive vertical growth of GaN, control of the surface energy and of the adatom kinetics is very important.[23,65] These factors can be modified by implementing changes to the substrate materials or to the growth conditions, such as the pressure, temperature, and gas ratios of the composite atoms.

The thermodynamic equilibrium shape, which determines the shape of the grown nanostructures, has been investigated in the first-principle studies of the surface energy of each crystal plane,[66,67] and these studies predict that the vertical nanorods consisting of {1-100} side facets and a (000-1) top surface can be obtained from N-polar GaN. Figure 7(a) shows catalyst-free self-organized nanorods grown on a Si (111) substrate. They are vertically well aligned and have hexagonal sidewalls {1-100} and a

Fig. 7. SEM images of self-organized GaN nanostructures on (a) Si (111) and (b) Si (100) substrates. Reprinted with permission from Refs. 23 and 33. Copyright (2012 & 2014) American Chemical Society.

N-polar top surface. The growing direction of the nanorods can also be modified by changing the crystal orientation of the substrate. Figure 7(b) shows GaN nanorods grown along the tilted direction on a Si (100) substrate.

The modification of biaxial strain can also change the surface energy of crystal facets.[23] Figure 8 plots the surface energies of these nanorod facets as a function of biaxial strain, and it shows that a high aspect ratio requires high tensile strain. The Si (111) substrate imparts a strong tensile strain of 16.36% to the (000-1) facet, explaining the formation of the long nanorods in Fig. 7(a).

The material growth is not only determined by the surface energy, but also strongly influenced by kinetic factors such as adatom diffusion. Various growth conditions of temperature, pressure, and gas ratio affect the kinetic factors. For instance, Figure 9 shows how the growth rate of the nanorods changes depending on the reactor pressure. The growth rate (Vg) of the materials has the following relationship with the reactor pressure (P).[23]

$$Vg \propto \frac{1}{1/k_0 P exp(-E_a/k_b T) + \sqrt{\mathrm{p}}}, \tag{1}$$

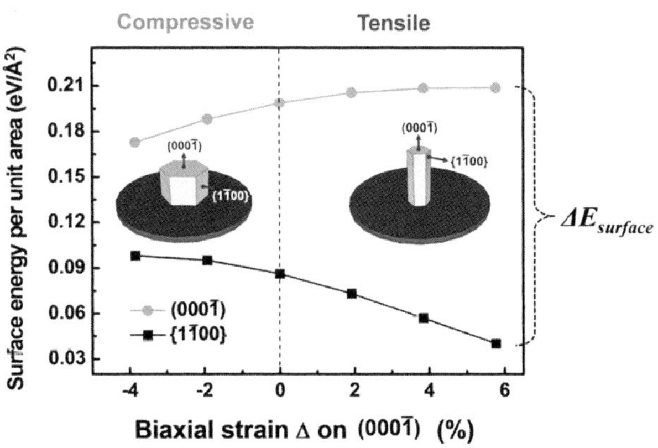

Fig. 8. Density functional theory based calculation of surface energy per unit area as a function of strain. Reprinted with permission from Ref. 23. Copyright (2012) American Chemical Society.

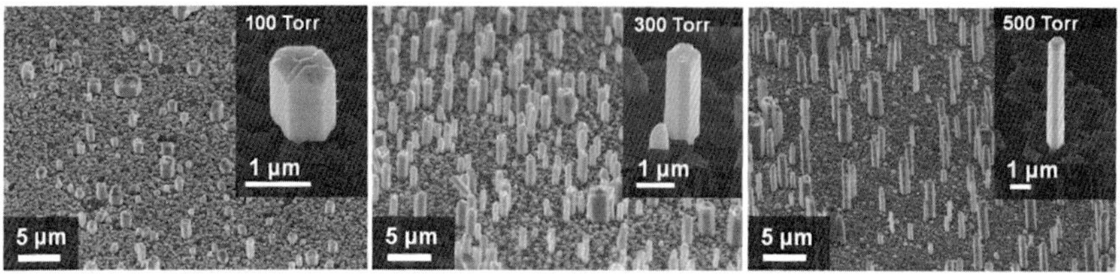

Fig. 9. SEM images of GaN nanorods grown on different reactor pressures of 100 Torr (left), 300 Torr (middle), and 500 Torr (right). Insets show the magnified images. Reprinted with permission from Ref. 23. Copyright (2012) American Chemical Society.

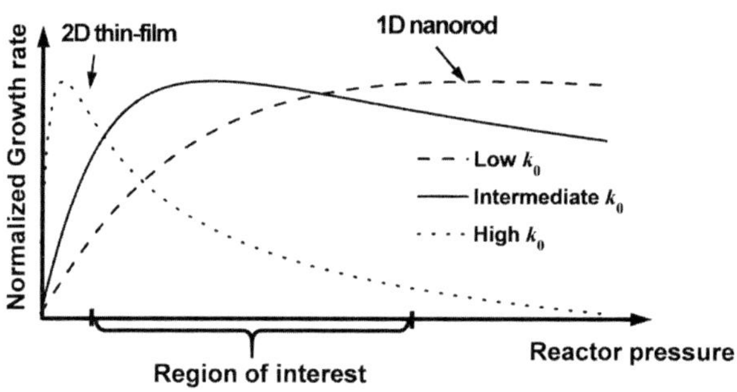

Fig. 10. Estimated curves of the growth rate as a function of the reactor pressure with the various pre-exponential factors k_0 in the reaction rate constant k. Reprinted with permission from Ref. 23. Copyright (2012) American Chemical Society.

where k_0 is the pre-exponential factor, related to the adsorption (k_{ads}) and desorption rate (k_{des}) of adatoms ($k_{ads} - k_{des} = k_0 P\exp(-E_a/k_b T)$), E_a is activation energy, k_b is the Boltzmann constant, and T is temperature. Equation (1) indicates that depending on the value of k_0, the growth rate of materials can either increase or decrease with the reactor pressure as shown in Fig. 10. In the case of two dimensional thin film growth with high k_0, the growth rate decreases as we increase the pressure, whereas the growth rate of one dimensional nanorods with a small k_0 increases with the pressure.

Not only Si substrates, but also sapphire substrates can be used for GaN nanorod growth. N-polar GaN nanorods have been obtained by nitridation of the Al-face of *c*-plane sapphire,[32] and Ga-polar nanorods have been demonstrated by using either pulsed or continuous flows in MOCVD.[22,24,30]

The use of a self-catalyst or Si dopant gas can play important roles to drive the VLS mode in Ga-polar nanorod growth on sapphire substrates.

Based on these techniques for vertical GaN nanostructures, GaN/InGaN core-shell nanostructures have been successfully obtained,[30,31,33] and they commonly show a large PL enhancement, reduced polarization fields, and improved indium incorporation efficiency similar to the core-shell structures shown in Sec. 3.

4.2. *Pyramidal structure*

Together with the rod-shaped vertical nanostructures, group III-nitride materials have another important nanostructure, a hexagonal pyramidal structure consisting of {11–22} or {10–11} semi-polar facets.[68] These facets are one of the stable crystal facets in Wurtzite GaN crystal and are frequently observed during growth or etching processes. The pyramidal structure can be obtained by selective-area growth technique with conventional MOCVD growth on a sapphire substrate.[68] Especially, a hole-patterned mask forms uniform hexagonal pyramid arrays with highly crystalline facets [Fig. 11(a)]. Compared to the vertical core-shell structure, the pyramidal structures do not require the use of catalysts or buffer layers. Also the pyramidal structure has the growth and device fabrication processes that are quite similar to those of the planar structure.

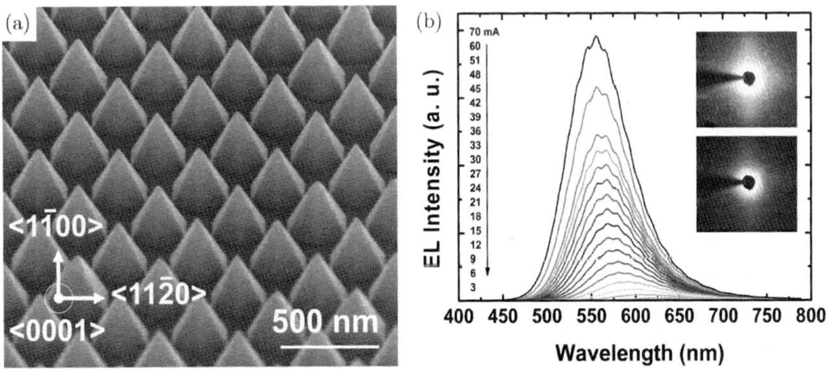

Fig. 11. (a) SEM images of GaN pyramidal structure array. (b) EL spectrum of InGaN/GaN nano-pyramids as a function of injection currents. Insets show photo images of green-emitting pyramid LEDs at current 100 mA (upper) and 50 mA (lower). Reprinted with permission from Ref. 68. Copyright (2011) John Wiley and Sons.

The pyramidal structure enables the InGaN QWs to be grown on the semi-polar facets of pyramidal structures, where the built-in electric fields are minimized.[39] Therefore, these QWs on the pyramids have a fast exciton recombination time less than 1 ns.[41] Another important advantage of the pyramidal structures is their ability to grow QWs with high indium composition and thick layer, leading to efficient long-wavelength emission.[40,41,69] Figure 11(b) shows electroluminescence (EL) spectrum and optical images of the InGaN/GaN pyramid structure-based LEDs, showing a bright green emission. Recently, efficient red-color emission from InGaN/GaN double heterostructures on GaN nanopyramid structures has been demonstrated.[41]

In the case of planar films, it is difficult to obtain a high-quality indium-rich, thick InGaN layer because as the indium content and thickness of the InGaN layer increase, the increased strain forms dislocations to reduce the strain, and the strain also induces compositional pulling effect that extracts indium atoms from an InGaN layer to reduce the deformation energy.[42] Meanwhile, the pyramidal structures effectively relax the strain and form an indium-rich, thick InGaN layer. Figure 12 presents a schematic and TEM images of the InGaN/GaN double heterostructures, showing a high-quality 88-nm thick InGaN layer on a GaN nanopyramid. Efficient lateral/vertical diffusion of indium atoms in the three-dimensional pyramidal structures on the patterned substrate enables the indium atoms to move from the

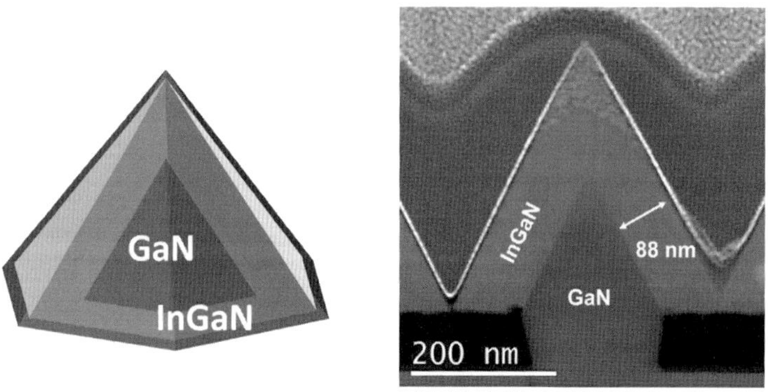

Fig. 12. Schematic illustration of a GaN/InGaN double heterostructure on a GaN pyramid (left). Cross-sectional scanning TEM image (right). Reprinted with permission from Ref. 41. Copyright (2015) American Chemical Society.

mask to the pyramids and from the bottoms to the tops of the pyramids. Therefore, the indium composition in the InGaN layer increases along the vertical direction of the pyramid. In addition, the three-dimensional growth mode and the thick InGaN layer relax the strain that reduces the compositional pulling effect resulting in the higher indium composition along the growth direction. Combination of the long indium diffusion and the reduced compositional pulling effects in the InGaN/GaN pyramidal heterostructure allows the InGaN layer to have high indium content of up to 50%. Figures 13(a, b) shows the indium composition variation along the vertical and growth directions. The results show high indium composition (∼50%) at the top pyramid region and relatively low indium composition

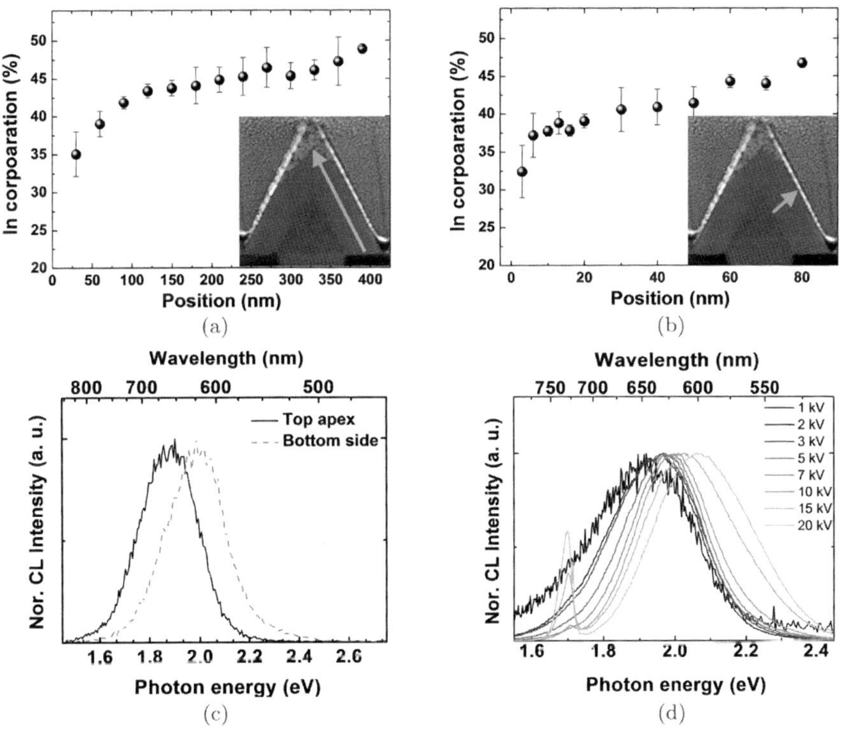

Fig. 13. (a, b) Variation in indium composition in InGaN layer measured along the vertical direction (a) and the growth direction (b). Insets display measured directions, respectively. (c) CL spectra measured at top apex and bottom side region of nanopyramids. (d) Depth-resolved CL spectra measured with various acceleration voltages. Reprinted with permission from Ref. 41. Copyright (2015) American Chemical Society.

($\sim$30%) at the bottom pyramid region. Along the growth direction, low indium composition ($\sim$30%) at the InGaN layer near the inner GaN (the strained region) and high indium composition ($\sim$45%) at the InGaN layer away from the inner GaN (the strain relaxed region) are observed.

Spatial- and depth-resolved CL techniques are useful to investigate the composition variations depending on the position. Figure 13(c) shows that the top apex of the pyramid emits longer wavelength than the bottom side region, as expected from Fig. 13(a). The change of electron beam energy in CL provides the depth information of the sample. Figure 13(d) shows the longer (shorter) emission wavelength at the surface (inner) region, which agree with the result in Fig. 13(b).

The formation of a thick InGaN layer has another important benefit. The increased volume decreases a carrier density and thus reduces the efficiency droop behavior, occurring at high carrier density. Figure 14 shows a comparison of the PL efficiencies between the 3 nm-thick thin InGaN QW and the 90-nm thick InGaN layer grown on GaN nanopyramids. Compared to the thin InGaN QW on GaN pyramids, the thick InGaN heterostructure on GaN pyramids shows a delayed beginning of droop and relatively lower droop slope at high excitation power. Therefore, the combination of pyramidal structures and a thick InGaN layer shows strong potential for efficient LEDs working at long visible wavelengths.[41]

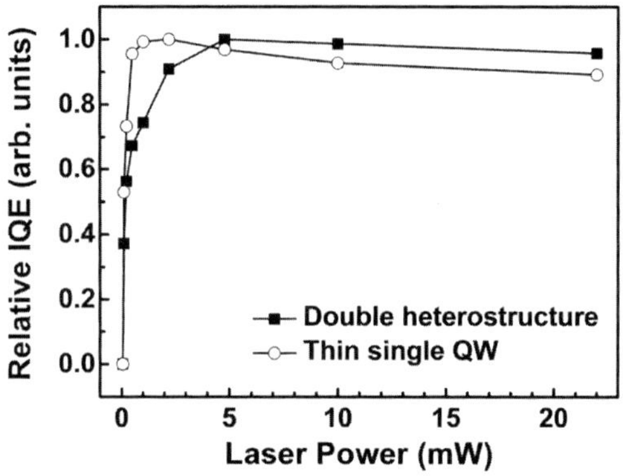

Fig. 14. Comparison of efficiency droop phenomenon between the InGaN/GaN double heterostructure and the thin single QW on nanopyramid. Relative internal quantum efficiency (Y axis) is defined as PL intensity divided by excitation powers. Reprinted with permission from Ref. 41. Copyright (2015) American Chemical Society.

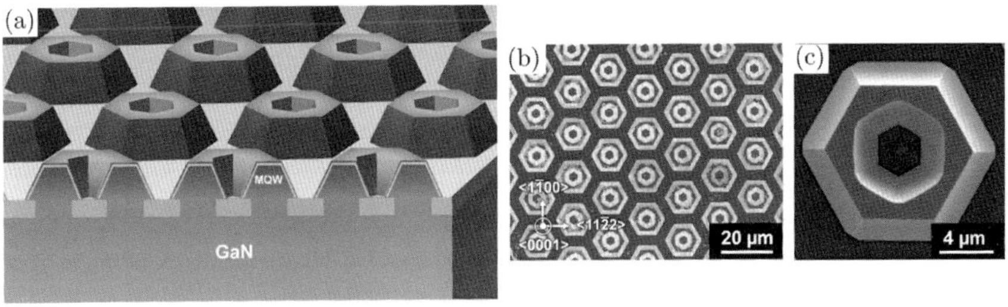

Fig. 15. (a) Schematic illustration of the hexagonal annular structure with InGaN/GaN MQWs. (b) Top-view SEM image of the hexagonal annular structure, and (c) the magnified image of a single structure. Reprinted with permission from Ref. 71. Copyright (2014) Nature Publishing Group.

To broaden the emission spectrum further, the use of ring patterns instead of hole patterns have also been suggested.[70,71] Figure 15 shows a schematic and SEM images of the GaN/InGaN multi-faceted pyramidal structures. These annular structures show both {11–22} and {10–11} semi-polar facets and also a (0001) top surface. The strain and surface energy strongly depend on their grown facets and cause a large variation in the thickness and indium compositions of the InGaN QWs. The multi-faceted, annular structure emits a broad spectrum, covering the entire visible wavelength.[70,71] Therefore, these structures show strong potential for phosphor-free white LEDs.

Another interesting aspect in the pyramids is that indium atoms tend to be localized at the apex and the edge of the nanostructures. This results in the formation of QW/quantum wire/quantum dot (QD) structures at the sidewalls, edges, and apexes of the InGaN/GaN pyramidal structures.[68] Figure 16 is CL images at several wavelengths, showing the different emission wavelengths at sidewalls, edges, and apex. The localized QD at the top apex can be used as site-controlled single photon sources. Next section, we introduce these applications in more detail.

5. Other Applications

As shown in the previous sections, group III-nitride nanostructures can provide a promising platform for high-performance LED applications. Beyond LED applications, nitride nanostructures are also attracting great attention for use in other nanophotonics applications such as waveguides,

Fig. 16. Top SEM of a single micro-pyramid and its CL images at different emission wavelengths of 510, 570, and 610 nm, respectively. Reprinted with permission from Ref. 68. Copyright (2011) John Wiley and Sons.

lasers, and single photon sources. Here we introduce recent important demonstrations of nitride-based nanostructures for new applications.

5.1. *Photonic diode*

Photonic waveguides are an important one-dimensional nanostructure for integrated photonic circuits.[10,72] It would be more useful if the waveguide allows photons to move in one direction like photonic diodes. The concept of photonic diodes has been demonstrated using GaN/InGaN core-shell nanorods, which have the spatially-graded bandgap energy along the growth axis.[33]

Figures 17(a–f) show STEM images of the GaN/InGaN nanorod grown by MOCVD. The thickness and the indium composition of the QWs show gradual increases along the growth direction. Both factors reduce the size of the bandgap from the bottom to the top of the nanorods. Figure 18

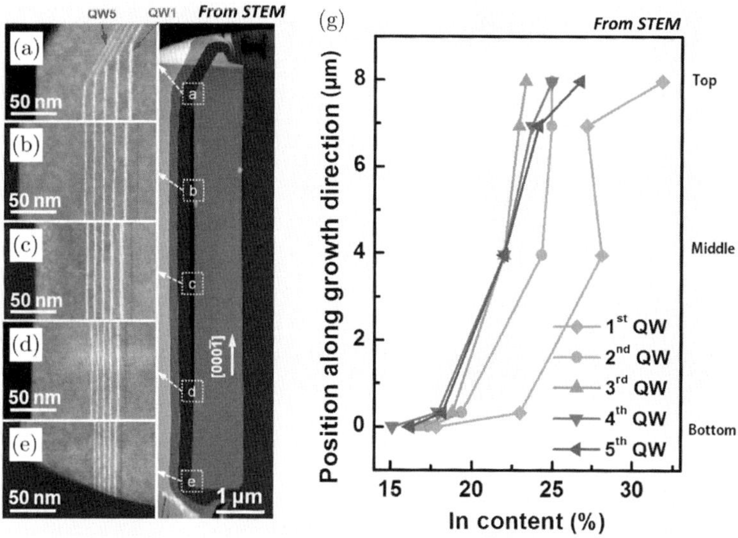

Fig. 17. STEM images of a GaN/InGaN core-shell nanorod: (a–e) Magnified images of QWs taken at various positions along a single rod. (f) Overall image of a GaN/InGaN core-shell rod. (g) Variation in the indium content inside a QW along the distance from bottom of the rod. Reprinted with permission from Ref. 33. Copyright (2014) American Chemical Society.

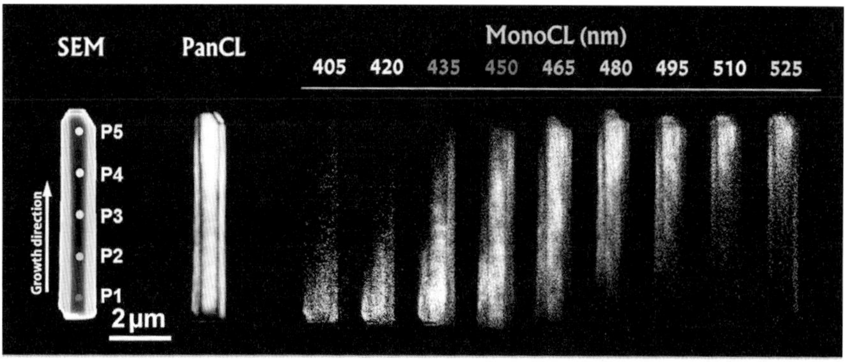

Fig. 18. SEM and Panchromatic CL images of a single GaN/InGaN core-shell rod and monochromatic CL images of the single rod. Reprinted with permission from Ref. 33. Copyright (2014) American Chemical Society.

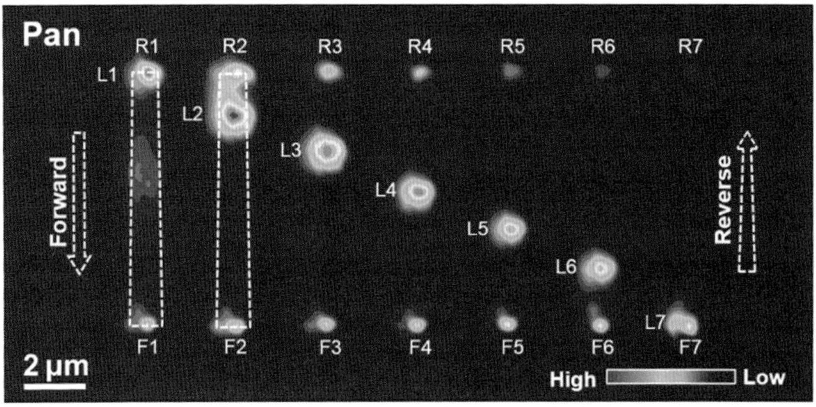

Fig. 19. Panchromatic PL image of Eg-graded rod with various local excitations (L1–L7) of a 405 nm laser. Reprinted with permission from Ref. 33. Copyright (2014) American Chemical Society.

shows spectrally-resolved CL images representing a continuous emission wavelength shift along the axis. This gradual change of the bandgap in the core-shell nanorod results in asymmetric light propagation along the axis. Figure 19 shows the PL images of the bandgap-graded nanorod excited at various local laser spots. At the bottom end of the nanorod, the output intensity is almost constant regardless of the laser excitation positions, while at the opposite top end, the output light intensity shows a significant decrease as the laser spot moves to the bottom side. This photonic diode behavior is due to higher absorption along one direction and smaller absorption along the other direction in the bandgap-graded core-shell nanostructures. The photonic diodes can provide an important component for optical on-chip devices.

5.2. *Cavity polariton*

The hexagonal cylinder shape of group III-nitride micro- or nano-structures provides a good optical cavity.[73] Total internal reflection occurs at the side facets of micro/nanorods and induces whispering gallery cavity modes without any complex cavity fabrication process [Figs. 20(a,b)]. The clean crystalline facets induce a high quality factor and lead to strong light-matter coupling, and thus enable us to study cavity quantum electrodynamics in

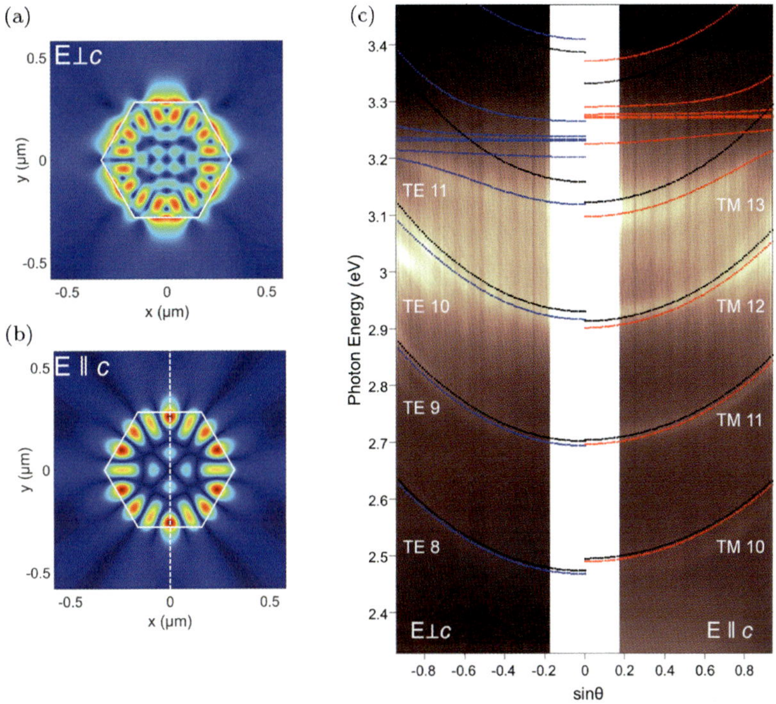

Fig. 20. Electric field distributions of whispering gallery cavity modes for (a) TE and (b) TM, respectively. (c) Polarization-resolved dispersion curve with high momentum. The black dots represent dispersion of pure whispering gallery cavity modes in the weak coupling regime, while the gray curves represent cavity polariton dispersion curves as a function of momentum (angle) for TE (left) and TM (right) polarized polaritons, respectively. Reprinted with permission from Ref. 77. Copyright (2015) American Chemical Society.

solid-state. The strong coupling between cavity photons and excitons in the semiconductors forms polariton states, which behave like a light boson.[74] Polariton condensation behavior has been reported in a semiconductor microcavity at 5 K.[75]

Wide-bandgap semiconductors such as ZnO and GaN have attracted great attention in the area of polariton physics because their large exciton binding energy preserves the polaritons up to room temperature.[76] In addition, GaN/InGaN core-shell nanorods locally confine the exciton in the QWs where the whispering gallery modes exist. Together with good spatial overlap between the photons and the excitons, the fast radiative recombination process in the QWs grown on non-polar facets results in a giant Rabi-splitting value of 180 meV.[77] Figure 20(c) shows the measured polariton dispersion curves in a GaN/InGaN core-shell nanorod.

As well as their academic interest, cavity-polaritons have an interesting application: a polariton laser. Conventional photon lasing requires population inversion while polaritons can accumulate in a single quantum state without population inversion. This enables the polariton laser to emit coherent light at a very low threshold.[78] Therefore, a polariton laser based on one dimensional micro/nano cavities can be very useful for an on-chip small light source operating at very low power, and this technology can provide a new source of coherent light.

5.3. *Single photon sources*

Previously we have introduced the possible applications of group III-nitride nanostructures for efficient LEDs and polariton lasers, which correspond to classical and coherent light sources, respectively. The group III-nitride nanostructures are also useful to generate non-classical light, *i.e.* single photons. Three-dimensionally confined QDs behave like atoms with their discrete energy levels and generate single photons at a time.

Single photon sources are an important element of quantum cryptography and also provide optical quantum bits for quantum information processing.[79] Semiconductor QDs with their high stability, fast generation rate, wavelength tunability, and the possibility of electrical operation provide a promising solid-state platform for single photon sources.[80–82] Especially, group III-nitride QDs have a large exciton binding energy[83] and large band offset, allowing single photon generation at high temperature. Single photon emission from GaN QDs has been reported at 200 K[84] and recently even at room temperature.[85] The important figure of merit of single photon sources is the high collection efficiency and fast generation rate of single photons. However, the conventional self-assembled group III-nitride QDs have a poor out-coupling efficiency of single photons into external optics system because they are buried in a planar structure. Huge built-in electric fields in the QDs also limit their recombination rate.[86] To overcome these problems several approaches that couple the QDs to photonic nanostructures have been suggested.[87–89]

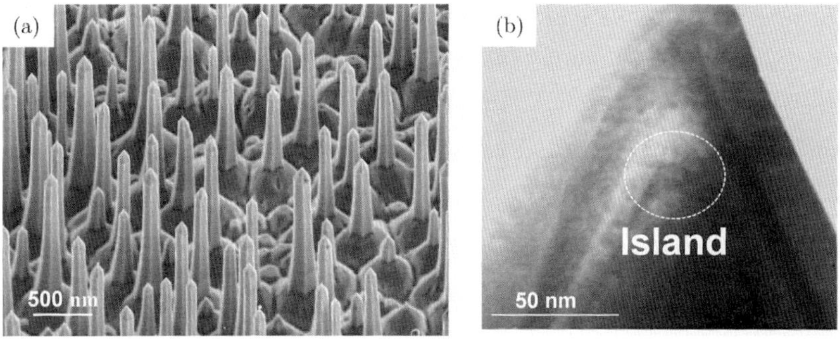

Fig. 21. (a) SEM image of obelisk-shaped nanostructures. (b) Cross-sectional TEM image of a pyramidal tip of nanostructure showing the formation of dot at the apex. Reprinted with permission from Ref. 90. Copyright (2013) Nature Publishing Group.

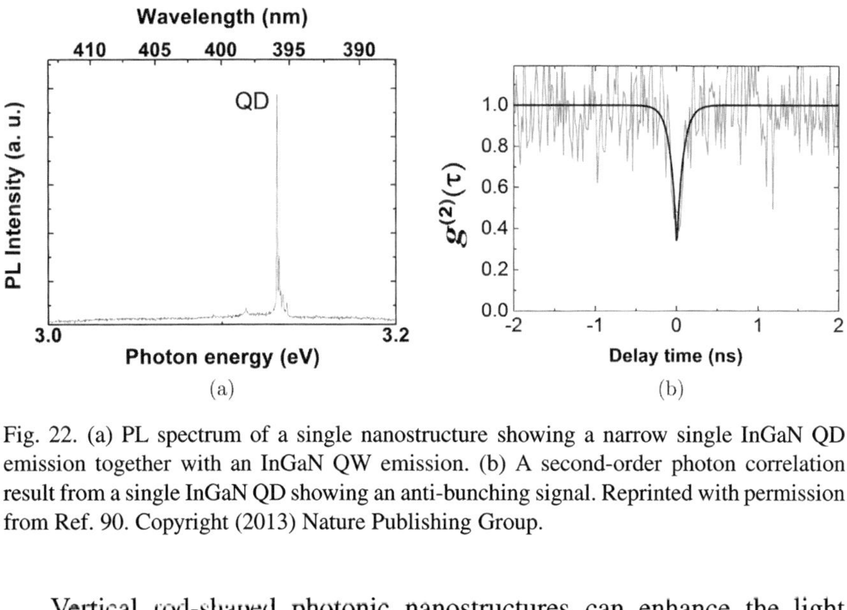

Fig. 22. (a) PL spectrum of a single nanostructure showing a narrow single InGaN QD emission together with an InGaN QW emission. (b) A second-order photon correlation result from a single InGaN QD showing an anti-bunching signal. Reprinted with permission from Ref. 90. Copyright (2013) Nature Publishing Group.

Vertical rod-shaped photonic nanostructures can enhance the light extraction efficiency and radiative recombination rate of coupled single QDs. Figure 21(a) shows a SEM image of nanophotonic structures formed by chemical vapor-phase etching and MOCVD regrowth process.[90] The InGaN QD is formed at the top apex of the nanostructure shown in the TEM image in Fig. 21(b). The single QD coupled to the nanostructure emits a narrow and bright single exciton line [Fig. 22(a)]. Figure 22(b) shows the second-order correlation of the singe QD emission, indicating non-classical light generation (*i.e.*, single photon state) as a clear anti-bunching dip at zero delay. The sharp anti-bunching signal with a time constant of 100 ps is strongly associated with the fast recombination time of the single QDs in the coupled nanostructures.[90]

It is also possible to employ plasmonic cavities to enhance spontaneous recombination rate of QDs. The site-selective formation of QDs at the apex of nanostructures allows deterministic coupling between the QD and the plasmonic cavity. This coupling has been demonstrated by simple deposition of a thin silver layer on site-controlled pyramid QDs [Fig. 23(a)].[91] In this configuration, three-dimensional plasmonic nanofocusing occurs at the top apex of a pyramid where the QD exists [Fig. 23(b)]. Therefore, deterministic coupling between the QD and the strong plasmonic mode occurs without any complex alignment procedure. This plasmonic coupling induces very large enhancement of the spontaneous emission rate of the QDs more than 20 [Fig. 23(c)].[91]

Group III-nitride QDs coupled to nanostructures can overcome many of the drawbacks of the conventional self-assembled QDs and also provide site-controllability. Therefore, they offer a promising platform for the realization of efficient on-demand single photon sources.

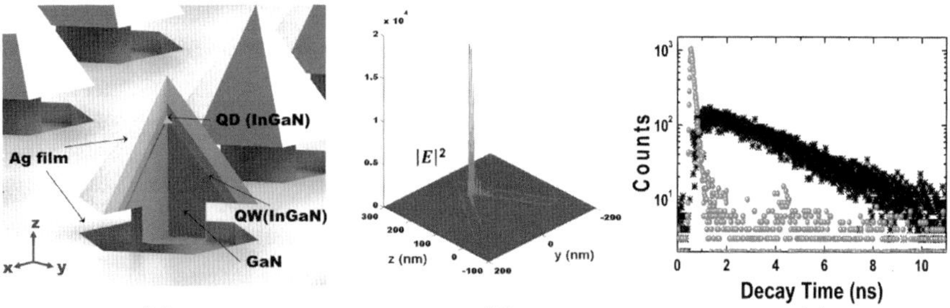

(a) (b) (c)

Fig. 23. (a) Schematic illustration of the site-controlled single QD array deterministically coupled to a plasmonic system. (b) Numerical simulation of the pyramid-plasmonic system showing the electric field intensity ($|E|^2$) showing the sharply enhanced local electric fields near the apex of the pyramid structure. (c) Decay dynamics of single QD with (gray dots) and without (black stars) the metal structure with their decay time constants of 0.14 ns and 4.4 ns, respectively. Reprinted with permission from Ref. 91. Copyright (2015) National Academy of Sciences, USA.

6. Conclusion

This chapter has reviewed recent works on group III-nitride nanostructures and their applications in LEDs and in other areas. GaN and its relevant materials are very useful systems for light-emitting applications in the range from ultraviolet to near infrared wavelengths, and from classical to non-classical lights. However, group III-nitride materials suffer from several problems; the absence of practical substrates for homoepitaxy creates a high density of dislocations, and the use of thin QWs on a polar surface leads to efficiency droop at high current. Nanostructures can increase light extraction, enlarge the active area, reduce the strain, and grow thick QWs on non/semi-polar surfaces. Therefore, the group III-nitride-based nanostructures can probably deflect the problems in existing planar structures. Here, we have introduced several approaches to the formation of high-quality nanostructures, and their good optical properties. With the advances in group III-nitride nanostructure technology, and given many advantages of these system over planar structures, they are becoming a promising alternative for high performance LEDs of the near future and continue to be of great interest with possible applications such as waveguides, lasers, and single photon sources. Finally, although nanostructures have shown promising results for various applications, further study for the practical, reliable mass-production of these nanostructures will be needed to win the competition against the well-established planar structures.

Acknowledgments

This work was supported by the National Research Foundation (NRF-2013R1A2A1A01016914) of the Ministry of Education, and the Climate Change Research Hub of KAIST (Grant No. N01150041).

References

1. S. Nakamura, T. Mukai, and M. Senoh, *Appl. Phys. Lett.*, **64**, 1687 (1994).
2. M. S. Ferdous, X. Wang, M. N. Fairchild, and S. D. Hersee, *Appl. Phys. Lett.*, **91**, 231107 (2007).
3. M. Boroditsky and E. Yablonovitch, *in* "Photonics West'97", Vol. 3002, p. 119. (ISOP,1997).
4. J. Piprek, *Phys. Status Solidi A*, **207**, 2217 (2010).
5. M. Bosi and R. Fornari, *J. Cryst. Growth*, **265**, 434 (2004).

6. A. Waag, X. Wang, S. Fündling, J. Ledig, M. Erenburg, R. Neumann, M. Al Suleiman, S. Merzsch, J. Wei, and S. Li, *Phys. Status Solidi C*, **8**, 2296 (2011).
7. S. Li and A. Waag, *J. Appl. Phys.*, **111**, 071101 (2012).
8. H. P. T. Nguyen, S. Zhang, A. T. Connie, M. G. Kibria, Q. Wang, I. Shih, and Z. Mi, *Nano Lett.*, **13**, 5437 (2013).
9. J.-H. Kim, Y.-H. Ko, J.-H. Cho, S.-H. Gong, S.-M. Ko, and Y.-H. Cho, *Nanoscale*, **6**, 14213 (2014).
10. M. Law, D. J. Sirbuly, J. C. Johnson, J. Goldberger, R. J. Saykally, and P. Yang, *Science*, **305**, 1269 (2004).
11. C. Lee, T. Lee, S. Lyu, Y. Zhang, H. Ruh, and H. Lee, *Appl. Phys. Lett.*, **81**, 3648 (2002).
12. Q. Wan, Q. Li, Y. Chen, T.-H. Wang, X. He, J. Li, and C. Lin, *Appl. Phys. Lett.*, **84**, 3654 (2004).
13. H. Kind, H. Yan, B. Messer, M. Law, and P. Yang, *Adv. Mater.*, **14**, 158 (2002).
14. Y. Li, F. Qian, J. Xiang, and C. M. Lieber, *Mater. Today*, **9**, 18 (2006).
15. H.-W. Huang, C.-C. Kao, T.-H. Hsueh, C.-C. Yu, C.-F. Lin, J.-T. Chu, H.-C. Kuo, and S.-C. Wang, *Mater. Sci. Eng., B*, **113**, 125 (2004).
16. F. Demangeot, J. Gleize, and J. Frandon, *J. Appl. Phys.*, **91**, 6520 (2002).
17. L. Tu, C. Hsiao, T. Chi, I. Lo, and K. Hsieh, *Appl. Phys. Lett.*, **82**, 1601 (2003).
18. H. Sekiguchi, K. Kishino, and A. Kikuchi, *Appl. Phys. Lett.*, **96**, 231104 (2010).
19. K. Kishino, H. Sekiguchi, and A. Kikuchi, *J. Cryst. Growth*, **311**, 2063 (2009).
20. H.-M. Kim, D. Kim, D. Kim, T. Kang, Y.-H. Cho, and K. Chung, *Appl. Phys. Lett.*, **81**, 2193 (2002).
21. G. Avit, K. Lekhal, Y. André, C. Bougerol, F. o. Réveret, J. Leymarie, E. Gil, G. Monier, D. Castelluci, and A. Trassoudaine, *Nano Lett.*, **14**, 559 (2014).
22. C. Tessarek, M. Heilmann, E. Butzen, A. Haab, H. Hardtdegen, C. Dieker, E. Spiecker, and S. Christiansen, *Cryst. Growth Des.*, **14**, 1486 (2014).
23. S.-M. Ko, J.-H. Kim, Y.-H. Ko, Y. H. Chang, Y.-H. Kim, J. Yoon, J. Y. Lee, and Y.-H. Cho, *Cryst. Growth Des.*, **12**, 3838 (2012).
24. S. D. Hersee, X. Sun, and X. Wang, *Nano Lett.*, **6**, 1808 (2006).
25. Q. Li, K. R. Westlake, M. H. Crawford, and S. R. Lee, *Opt. Express*, **19**, 25528 (2011).
26. J. B. Wright, S. Liu, G. T. Wang, Q. Li, A. Benz, D. D. Koleske, P. Lu, H. Xu, L. Lester, T. S. Luk, I. Brener, and G. Subramania, *Sci. Rep.*, **3**, 2982 (2013).
27. H. P. T. Nguyen, M. Djavid, S. Y. Woo, X. Liu, A. T. Connie, S. Sadaf, Q. Wang, G. A. Botton, I. Shih, and Z. Mi, *Sci. Rep.*, **5**, 7744 (2015).
28. Y. Park, S. Jahangir, Y. Park, P. Bhattacharya, and J. Heo, *Opt. Express*, **23**, A650 (2015).
29. L. Geelhaar, C. Chèze, W. M. Weber, R. Averbeck, H. Riechert, T. Kehagias, P. Komninou, G. P. Dimitrakopulos, and T. Karakostas, *Appl. Phys. Lett.*, **91**, 093113 (2007).
30. X. Wang, S. Li, M. S. Mohajerani, J. Ledig, H.-H. Wehmann, M. Mandl, M. Strassburg, U. Steegmüller, U. Jahn, and J. Lähnemann, *Cryst. Growth Des.*, **13**, 3475 (2013).
31. T.-W. Yeh, Y.-T. Lin, L. S. Stewart, P. D. Dapkus, R. Sarkissian, J. D. O'Brien, B. Ahn, and S. R. Nutt, *Nano Lett.*, **12**, 3257 (2012).

32. W. Bergbauer, M. Strassburg, K. Ch, N. Linder, C. Roder, J. Lähnemann, A. Trampert, S. Fündling, S. F. Li, H. H. Wehmann, and A. Waag, *Nanotechnology*, **21**, 305201 (2010).
33. S.-M. Ko, S.-H. Gong, and Y.-H. Cho, *Nano Lett.*, **14**, 4937 (2014).
34. J. H. Son, J. U. Kim, Y. H. Song, B. J. Kim, C. J. Ryu, and J.-L. Lee, *Adv. Mater.*, **24**, 2259 (2012).
35. C. Wiesmann, K. Bergenek, N. Linder, and U. T. Schwarz, *Laser Photonics Rev.*, **3**, 262 (2009).
36. R. Colby, Z. Liang, I. H. Wildeson, D. A. Ewoldt, T. D. Sands, R. E. García, and E. A. Stach, *Nano Lett.*, **10**, 1568 (2010).
37. S. D. Hersee, A. K. Rishinaramangalam, M. N. Fairchild, L. Zhang, and P. Varangis, *J. Mater. Res.*, **26**, 2293 (2011).
38. J.-H. Kim, C.-S. Oh, Y.-H. Ko, S.-M. Ko, K.-Y. Park, M. Jeong, J. Y. Lee, and Y.-H. Cho, *Cryst. Growth Des.*, **12**, 1292 (2012).
39. J. S. Speck and S. F. Chichibu, *MRS Bull.*, **34**, 304 (2009).
40. J. R. Riley, S. Padalkar, Q. Li, P. Lu, D. D. Koleske, J. J. Wierer, G. T. Wang, and L. J. Lauhon, *Nano Lett.*, **13**, 4317 (2013).
41. Y.-H. Ko, J.-H. Kim, S.-H. Gong, J. Kim, T. Kim, and Y.-H. Cho, *ACS Photonics*, **2**, 515 (2015).
42. S. Pereira, M. Correia, E. Pereira, K. O'donnell, C. Trager-Cowan, F. Sweeney, and E. Alves, *Phys. Rev. B*, **64**, 205311 (2001).
43. N. Gardner, G. Müller, Y. Shen, G. Chen, S. Watanabe, W. Götz, and M. Krames, *Appl. Phys. Lett.*, **91**, 243506 (2007).
44. S. Li, S. Fündling, Ü. Sökmen, R. Neumann, S. Merzsch, P. Hinze, T. Weimann, U. Jahn, A. Trampert, H. Riechert, E. Peiner, H.-H. Wehmann, and A. Waag, *Phys. Status Solidi C*, **7**, 2224 (2010).
45. H. Ono, Y. Ono, K. Kasahara, and J. Mizuno, *Jpn. J. Appl. Phys.*, **47**, 933 (2008).
46. B. J. Kim, H. Jung, H. Y. Kim, J. Bang, and J. Kim, *Thin Solid Films*, **517**, 3859 (2009).
47. J. Bai, Q. Wang, and T. Wang, *J. Appl. Phys.*, **111**, 113105 (2012).
48. C.-Y. Wang, L.-Y. Chen, C.-P. Chen, Y.-W. Cheng, M.-Y. Ke, M.-Y. Hsieh, H.-M. Wu, L.-H. Peng, and J. Huang, *Opt. Express*, **16**, 10549 (2008).
49. J. Bai, Q. Wang, and T. Wang, *Phys. Status Solidi A*, **209**, 477 (2012).
50. J. H. Yi, C. Kim, M. H. Kim, M. Yang, and J. Jeon, *J. Korean Phys. Soc.*, **39**, S364 (2001).
51. L. Wen-How, H. Kuo-Chin, H. Kai Feng, L. Jia-Ching, C. Yi-Cheng, and L. Wen-Jen, *J. Phys. Chem. Solids*, **69**, 719 (2008).
52. D. Paramanik, A. Motayed, M. King, and J. Y. Ha, *arXiv:1311.0321*, (2013).
53. D. Stocker, E. Schubert, and J. Redwing, *Appl. Phys. Lett.*, **73**, 2654 (1998).
54. J. Ma, L. Wang, Z. Liu, G. Yuan, X. Ji, P. Ma, J. Wang, X. Yi, G. Wang, and J. Li, *Opt. Express*, **21**, 3547 (2013).
55. R. Collazo, S. Mita, A. Aleksov, R. Schlesser, and Z. Sitar, *J. Cryst. Growth*, **287**, 586 (2006).
56. D. Zhuang and J. H. Edgar, *Mater. Sci. Eng., R*, **48**, 1 (2005).

57. S. K. Hong, T. Yao, B. J. Kim, S. Y. Yoon, and T. I. Kim, *Appl. Phys. Lett.*, **77**, 82 (2000).
58. B. Heying, E. J. Tarsa, C. R. Elsass, P. Fini, S. P. DenBaars, and J. S. Speck, *J. Appl. Phys.*, **85**, 6470 (1999).
59. N. Thillosen, K. Sebald, H. Hardtdegen, R. Meijers, R. Calarco, S. Montanari, N. Kaluza, J. Gutowski, and H. Lüth, *Nano Lett.*, **6**, 704 (2006).
60. P. Lefebvre, A. Morel, M. Gallart, T. Taliercio, J. Allegre, B. Gil, H. Mathieu, B. Damilano, N. Grandjean, and J. Massies, *Appl. Phys. Lett.*, **78**, 1252 (2001).
61. Y. Wu and P. Yang, *J. Am. Chem. Soc.*, **123**, 3165 (2001).
62. C. Chèze, L. Geelhaar, O. Brandt, W. Weber, H. Riechert, S. Münch, R. Rothemund, S. Reitzenstein, A. Forchel, T. Kehagias, P. Komninou, G. Dimitrakopulos, and T. Karakostas, *Nano Res.*, **3**, 528 (2010).
63. A. I. Hochbaum, R. Fan, R. He, and P. Yang, *Nano Lett.*, **5**, 457 (2005).
64. G. Kipshidze, B. Yavich, A. Chandolu, J. Yun, V. Kuryatkov, I. Ahmad, D. Aurongzeb, M. Holtz, and H. Temkin, *Appl. Phys. Lett.*, **86**, 033104 (2005).
65. S. F. Li, S. Fuendling, X. Wang, S. Merzsch, M. A. M. Al-Suleiman, J. D. Wei, H. H. Wehmann, A. Waag, W. Bergbauer, and M. Strassburg, *Cryst. Growth Des.*, **11**, 1573 (2011).
66. B. Leung, Q. Sun, C. D. Yerino, J. Han, and M. E. Coltrin, *Semicond. Sci. Technol.*, **27**, 024005 (2012).
67. V. Jindal and F. Shahedipour-Sandvik, *J. Appl. Phys.*, **106**, 083115 (2009).
68. Y.-H. Ko, J.-H. Kim, L.-H. Jin, S.-M. Ko, B.-J. Kwon, J. Kim, T. Kim, and Y.-H. Cho, *Adv. Mater.*, **23**, 5364 (2011).
69. T. Kim, J. Kim, M.-S. Yang, S. Lee, Y. Park, U.-I. Chung, and Y. Cho, *Appl. Phys. Lett.*, **97**, 241111 (2010).
70. Y.-H. Ko, J. Song, B. Leung, J. Han, and Y.-H. Cho, *Sci. Rep.*, **4**, 5514 (2014).
71. S.-H. Lim, Y.-H. Ko, C. Rodriguez, S.-H. Gong, and Y.-H. Cho, *Light Sci Appl*, **5**, e16030 (2016).
72. C. J. Barrelet, A. B. Greytak, and C. M. Lieber, *Nano Lett.*, **4**, 1981 (2004).
73. H. Baek, J. K. Hyun, K. Chung, H. Oh, and G.-C. Yi, *Appl. Phys. Lett.*, **105**, 201108 (2014).
74. D. Snoke and P. Littlewood, *Phys. Today*, **63**, 42 (2010).
75. J. Kasprzak, M. Richard, S. Kundermann, A. Baas, P. Jeambrun, J. M. J. Keeling, F. M. Marchetti, M. H. Szymanska, R. Andre, J. L. Staehli, V. Savona, P. B. Littlewood, B. Deveaud, and L. S. Dang, *Nature*, **443**, 409 (2006).
76. A. Das, P. Bhattacharya, J. Heo, A. Banerjee, and W. Guo, *Proc. Natl. Acad. Sci. U.S.A.*, **110**, 2735 (2013).
77. S.-H. Gong, S.-M. Ko, M.-H. Jang, and Y.-H. Cho, *Nano Lett.*, **15**, 4517 (2015).
78. H. Deng, G. Weihs, D. Snoke, J. Bloch, and Y. Yamamoto, *Proc. Natl. Acad. Sci. U.S.A.*, **100**, 15318 (2003).
79. J. L. O'Brien, A. Furusawa, and J. Vuckovic, *Nat. Photonics*, **3**, 687 (2009).
80. M. Pelton, C. Santori, J. Vučković, B. Zhang, G. S. Solomon, J. Plant, and Y. Yamamoto, *Phys. Rev. Lett.*, **89**, 233602 (2002).
81. T. Heindel, C. Schneider, M. Lermer, S. Kwon, T. Braun, S. Reitzenstein, S. Hofling, M. Kamp, and A. Forchel, *Appl. Phys. Lett.*, **96**, 011107 (2010).

82. A. J. Shields, *Nat. Photon.*, **1**, 215 (2007).
83. P. Ramvall, S. Tanaka, S. Nomura, P. Riblet, and Y. Aoyagi, *Appl. Phys. Lett.*, **73**, 1104 (1998).
84. S. Kako, C. Santori, K. Hoshino, S. Götzinger, Y. Yamamoto, and Y. Arakawa, *Nat. Mater.*, **5**, 887 (2006).
85. M. J. Holmes, K. Choi, S. Kako, M. Arita, and Y. Arakawa, *Nano Lett.*, **14**, 982 (2014).
86. J.-H. Kim, B.-J. Kwon, Y.-H. Cho, T. Huault, M. Leroux, and J. Brault, *Appl. Phys. Lett.*, **97**, 061905 (2010).
87. M. E. Reimer, G. Bulgarini, N. Akopian, M. Hocevar, M. B. Bavinck, M. A. Verheijen, E. P. A. M. Bakkers, L. P. Kouwenhoven, and V. Zwiller, *Nat. Commun.*, **3**, 737 (2012).
88. J. Bleuse, J. Claudon, M. Creasey, N. S. Malik, J. M. GÈrard, I. Maksymov, J. P. Hugonin, and P. Lalanne, *Phys. Rev. Lett.*, **106**, 103601 (2011).
89. J. Claudon, J. Bleuse, N. S. Malik, M. Bazin, P. Jaffrennou, N. Gregersen, C. Sauvan, P. Lalanne, and J. M. Gérard, *Nat. Photon.*, **4**, 174 (2010).
90. J.-H. Kim, Y.-H. Ko, S.-H. Gong, S.-M. Ko, and Y.-H. Cho, *Sci. Rep.*, **3**, 2150 (2013).
91. S.-H. Gong, J.-H. Kim, Y.-H. Ko, C. Rodriguez, J. Shin, Y.-H. Lee, L. S. Dang, X. Zhang, and Y.-H. Cho, *Proc. Natl. Acad. Sci. U.S.A.*, **112**, 5280 (2015).

Index

U

V

W

X

Y

Z